Ecology for the 21st century

Ecology for the 21st century

Orianna Carter
Biological Science, Ohio University
West Virginia, USA

ELSEVIER

Academic Press is an imprint of Elsevier
125 London Wall, London EC2Y 5AS, United Kingdom
525 B Street, Suite 1650, San Diego, CA 92101, United States
50 Hampshire Street, 5th Floor, Cambridge, MA 02139, United States

ISBN: 978-0-443-24031-7

For Information on all Academic Press publications
visit our website at https://www.elsevier.com/books-and-journals

Publisher: Peter Linsley
Acquisitions Editor: Rakhshan Rizwan
Editorial Project Manager: Vicky Sharma
Production Project Manager: Prasanna Kalyanaraman
Cover Designer: Greg Harris

Typeset by MPS Limited, Chennai, India

Printed in Great Britain

Last digit is the print number: 9 8 7 6 5 4 3 2 1

Contents

For additional information on the topics covered in the book, visit the companion site: https://www.elsevier.com/books-and-journals/book-companion/9780443240317

For additional materials for qualified instructors, please visit the instructor site: https://educate.elsevier.com/9780443240317

CHAPTER

1 Introduction to Ecology

Chapter Outline

1.1 Defining Ecology

Spaceship Earth is a term used to express concern over use of limited resource availability. The earliest known coining of the phrase Spaceship Earth is from a passage in Henry George's "Progress and Poverty." The metaphor of Spaceship Earth was popularized in the 1960s–1970s by environmentalists after inventor R. Buckminster Fuller revamped the term to describe the planet as a spaceship traveling in a cosmos wholly dependent on the machinations of its systems working together to survive. The concept of a contained spaceship environment is useful to understand resource limits and to encourage everyone to act as harmonious crew working toward the greater good. As one single act of resource loss may constitute dire consequences for the entire spaceship, we can extend this image to the broader concept of ecological systems and its overlapping

Ecology for the 21st Century. DOI: https://doi.org/10.1016/B978-0-443-24031-7.00001-5

Fig. 1.1

Rainy weather in the southwest United States causes an increase in pinyon nuts that, in turn, causes the deer mouse population to explode. This event coincides with outbreaks of hantavirus in humans in the region.

characteristics that are the subject of this course. In Chapter 1 we will overview key ecological areas of interest to scientists: ecosystems, energetics, and biogeochemical cycling.

Why study ecology? Perhaps you are concerned about the role humans have played in loss of biodiversity or want to better understand the factors that lead to species extinction. To do so, learning about our natural world and how species have adapted to the physical conditions of their environment would be a good place to start. Perhaps you have heard that while most medicines are derived from exotic natural products, the tropical rain forests are being destroyed at breakneck speed. A future biochemist or pharmacist might prioritize learning what steps are needed to protect the Amazonian's medicine chest until we unlock its mysteries. It is the ecologist who identifies the fascinating complexity of life in an ecosystem, which lays a framework to ensure it remains robust through the delicate balance of its many players (**Fig. 1.1**). Today more than ever humans have a major role to play in protecting those diverse ecosystems. The field of ecology is ripe with purpose and can be rewarding to the scientist as well as the public citizen. In this course you will have the opportunity to explore these different perspectives of ecology.

Humans are connected to the ecological landscape—its physical and living worlds—in ways that exemplify just how important those interactions are to human health. In **Fig. 1.2**, the vector-borne Lyme disease, officially known as *Lyme borreliosis*, emphasizes the benefit of being aware of species that coinhabit our

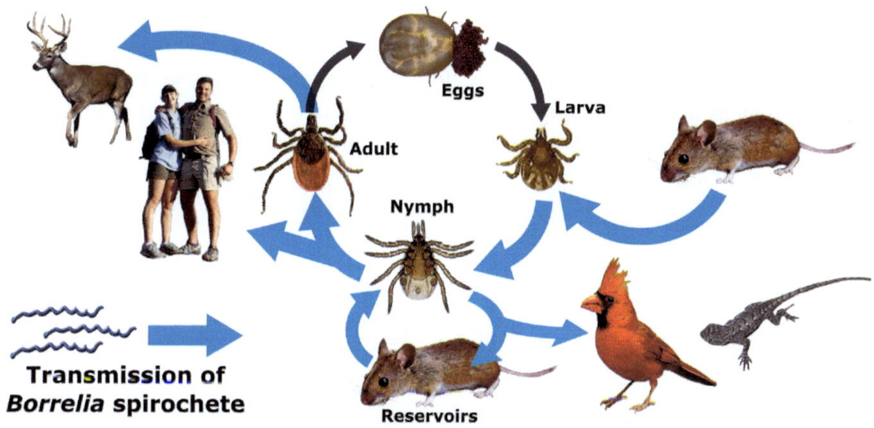

Fig. 1.2

Lyme disease is contracted by the deer tick, which is a vector for the infectious bacterium.

surroundings. Lyme disease is caused by the bacterium, *Borrelia burgdorferi*, and is transmitted to humans by an infected deer tick (*Ixodes scapularis*). The deer tick comes within proximity to humans through deer and other vertebrate animals. It has been suggested that the deer tick living on white-footed mice is more likely to be infected with *B. burgdorferi* bacterium than in the deer population. The role of an ecologist might be to help the CDC understand the environments and population densities in which infectious disease vectors, like deer ticks, are abundant. An investigation of this nature helps us better understand how Lyme disease is transmitted in the United States and how its incidence could be managed. Knowing the lifestyle of an infectious agent in its natural habitat provides valuable clues on its eradication. Understanding the ecological landscape is a critical component of the information epidemiologists need to monitor diseases transmitted by animals or water-borne diseases. The interrelationships of humans with the natural world are infinite.

Working in the field of ecology, it is possible to focus on studying just one organism in its habitat, a whole population of a species, a mixed community of species, an ecosystem, or even the entire biosphere. Ecology isn't simply about monitoring species-rich forests, pristine wilderness, or scenic vistas. Throughout nature, or in one's household for that matter, we can witness ecology in action. Think of the mold growing in your basement, the microbiota flourishing in the human gut, or biofilms forming near bathtubs. Even a mound of ants descending on a melted marshmallow are equally valid examples of ecology in action.

The remaining sections of this chapter provide an overview of ecological fundamental concepts including the three major ecosystem types, how science is done, and an introduction to food chains and energy flow. Throughout this course we will focus on energy flow as it travels through the trophic levels of a food web and as it cycles through the physical world. Energetics is an important theme and an inherent aspect of any ecological study.

1.2 The Scope of Ecology

By the end of this section students should be able to:

- Define the hierarchal levels of ecology.
- Explain how ecological research might integrate different scientific disciplines.
- Identify the abiotic and biotic components of the environment.

Ecology is the study of the interactions of living organisms with their environment, including the distribution and abundance of living things in the natural world. In Section 1.2 on Lyme disease epidemiologists relied upon ecologists to help their efforts to eradicate the disease. This is not particularly unique because the broad implications of ecological studies often involve a multidisciplinary effort. For example, an ecologist might collaborate with other scientists including a biochemist, physiologist, evolutionist, or even a molecular biologist. But it is the ecologist and conservationist that are most closely aligned in their quest to understand and combat species loss and its impact on the ecological systems.

Depending on the scope of study ecologists may consult with experts such as geologists, climatologists, chemists, and physicists regarding our physical world. One example is using chemistry to identify water pollutants detrimentally impacting birth rates of amphibians. Ecologists frequently apply mathematical models for data interpretation, which may require consultation with statistical analysts. A population ecologist interested in knowing what factors may improve survival odds of an endangered species uses mathematical models to predict the effect of an ongoing conservation management efforts. The conservation strategy would rely on accurate data collection methods for determining population size, the factors affecting survival and reproduction, habitat requirements, or potential interspecies influences (including humans) on the endangered population and its habitat.

In this course you will be encouraged to think globally across the disciplines of science, integrating life, and natural and social sciences to understand the complexity of ecological issues. Ecology in the 21st century encompasses several fundamental themes, including:

- ecosystem health
- complex hierarchical systems
- ecosystem services
- integrated modeling and assessment
- science and decision making
- quality of life and the distribution of wealth and resources

Levels of Ecological Study

Ecology is generally studied from one of four basic levels: organism, population, community, and ecosystem (**Fig. 1.3**). Yet the field of ecology is quite broad and many subdisciplines are found in each of those basic levels, including global ecology, field ecology, aquatic ecology, microecology, and even human ecology.

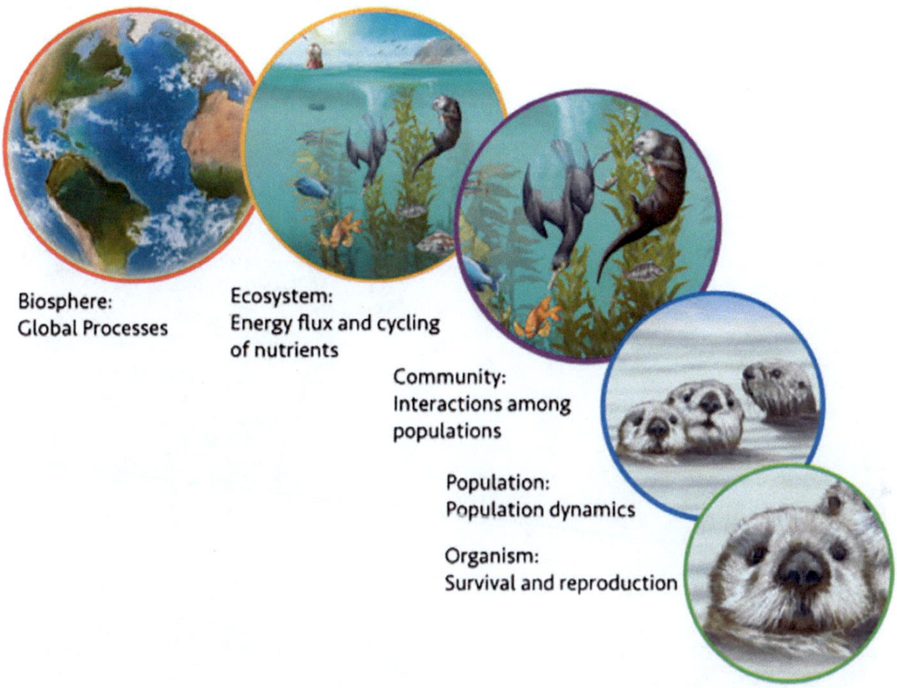

Fig. 1.3

The biological levels of organization.

Adapted from Relyea, R. and Ricklefs, R.E. (2019). Ecology: the economy of nature, eighth ed. New York: W.H. Freeman. Illustration A. James Gustafson

Organismal Ecology

Ecology studied at the organismal level would concern species adaptations that enable its survival in a unique habitat. The adaptations could be morphological, physiological, or behavioral. The small **Karner blue butterfly** (*Lycaeides melissa samuelis*) of the family Lycaenidae (**Fig. 1.4**) is known to be a specialist species, meaning it has strict requirements for its continued survival. Karner females preferentially **oviposits** on wild lupine. The Karner butterfly's habit has historically ranged in open areas in the northern United States from Minnesota to the eastern seaboard where the wild lupine is also known to grow in the wild. Preferential adaptation of larvae feeding solely on the lupine results in a highly dependent relationship between the Karner blue butterfly reproductive success and one wild plant's growth.

The Karner blue butterfly larval caterpillars will hatch on the lupine floral parts and spend 4 to 6 weeks feeding (**Fig. 1.5**) until they undergo metamorphosis to emerge as butterflies. Once released as adults, the butterflies will no longer be restricted to feeding solely on the nectar of the wild lupine flower and can visit other plant species. An organismal ecologist whose research is focused on this butterfly species likely would have several questions that need to be addressed. Those could including

Fig. 1.4

Karner blue butterfly (*Lycaeides melissa samuelis*).

Source: Modification of J & K Hollingsworth, USFWS under CC BY 4.0

Fig. 1.5

Wild lupine (*Lupinus perennis*).

Source: ZeroOne under CC BY-SA 2.0

questions about the timing of the morphological transition of the caterpillar to butterfly and optimal physiological conditions, such as temperature, at the various larval stages or the specific benefits of the rich nectar produced by the wild lupine flower.

While food sources might be a key focal point for the organismal ecologist to explore, endangered or threatened species like the Karner blue butterfly often are impacted by multiple factors. In these studies the ecologist will extend their investigations using a broader approach, perhaps even across disciplines. For example, while there are no sightings of the Karner blue butterflies existing outside the range of wild lupine growth, as we see in **Fig. 1.6** existing butterfly populations are found isolated in pockets, which are diminishing over time.

Population Ecology

A population is defined as a group of interbreeding organisms of the same species living in the same region at a specific time. The perimeter of the region where the population of interest exists may be formed by natural or artificial barriers. A natural boundary would be a geographical construct like streams, ridges, and mountains or a distinctly different terrain such as a desert. Man-made boundaries include roads, dams, agricultural fields, or manicured (and sprayed) lawns. For a population ecologist the research topic would likely involve knowing the number of individuals in

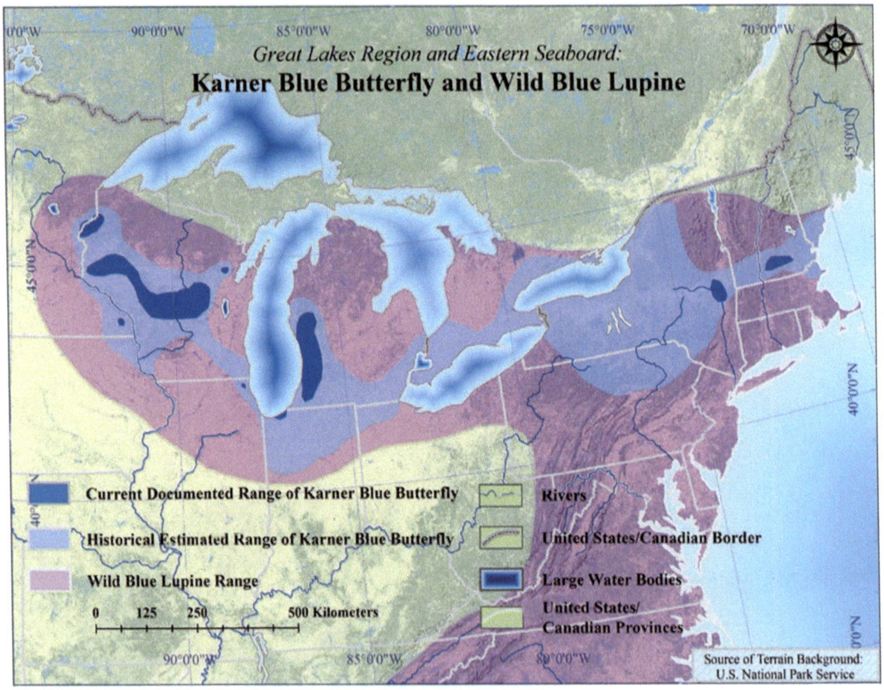

Fig. 1.6

Karner blue butterflies only live where wild lupines grow.

Source: Minnesota Dept Natural Resources; US National Park Service, Public Domain

the region of interest, understanding the ways in which that population size changes over time, and investigating what causes the changes. For example, a population ecologist might undertake data collection on the number of Karner blue butterflies, a federally endangered species, in a state where they were once known to flourish. And because it is known that the Karner blue butterfly is highly dependent on the distribution and abundance of wild lupine, any study on the distribution and density of the Karner species wouldn't be complete without mapping the flower as well. The types of questions that might be addressed in this study would include a discussion of any causes of the decline of the wild lupine. It is known that wild lupine flourish in open sunny areas of dry woodland glades, savannas, grasslands, and roadsides. Intermittent natural wildfires help to maintain open areas free of shrubs and trees. But the optimal habitat for blue lupine is shrinking due to habitat loss and increasingly harsher conditions. This includes decreasing rainfall, decreasing snowpack, and increasing temperatures. The application of mathematical models helps the population ecologist understand how any of these factors—such as wildfire suppression by humans—leads to the decline of the endangered Karner blue butterfly directly through loss of the wild lupine.

Community Ecology

A biological community is defined by the different species living and interacting together within an area (think in terms of three-dimensional space). For the community ecologists it is the processes that drive species–species interactions that are most interesting. A key focus might be competition among different species for a limited resource. Other species–species interactions include predation, parasitism, coevolution, or adaptation and competition, which all have the potential to effect population sizes and biodiversity within the community.

Using our example of the Karner blue butterfly, its larvae are involved in a **mutualistic** relationship with certain ant species. In a relationship between species involving mutualism a long-term relationship has coevolved in which both species benefit from the interactions. Research studies have demonstrated that survival of Karner blue butterfly larvae (caterpillars) increases when there are certain ant species also present on the blue lupine floral parts. It has also been shown that the Karner blue butterfly larvae have higher survival with the presence of ants, presumably due to lower predation. In return the Karner larvae release a carbohydrate and amino acid-rich liquid substance from specialized glands that will be consumed by the ants as their main energy source. A mutual, synergistic relationship is the consequence of this interaction because both the Karner larvae and ant benefit.

In another fascinating example of insect-plant mutualism, the Bathurst copper butterfly (*Paralucia spinifera*) endemic in New South Wales, Australia, produces sweet rewards for its benefactor ants. The ants in return function as goat-herders of sorts to the caterpillar. During the evenings the ants herd the caterpillar to where they can feed on Black Hawthorne plants (**Fig. 1.7**). Then in the morning the ants will herd the caterpillars back down into its ants' nest, protecting the caterpillar from predators. During ecological studies of this dynamic researchers discovered that any intruders threatening the caterpillars during feeding will fall under attack by half the ants while the other half quickly round up the caterpillars to return them to the ants' nest. The ant in this scenario is the goat-herder, or farmer, of its precious resources, the caterpillar, and its production of sweet treats.

Ecosystem Ecology

Ecosystem ecology is more inclusive than either organismal, population, or community ecology. In the ecosystem all **biotic** components (living things) in an area along with all **abiotic** components (nonliving things) within that area are considered. The ecosystem biologists will be asking all-encompassing questions about that ecosystem, such as how the species interact and how various factors impose restraints or favorable conditions on the various species, including whether nutrients and energy are stored and how they "cascade" among organismal levels and cycle through the surrounding atmosphere, soil, and water.

Consider the habitat where the Karner blue butterfly and the wild lupine are naturalized. What does the availability, or lack of, nutrients do to the distribution of plants

Fig. 1.7

Bathurst Copper Butterfly's caterpillars are herded by *Notancus* ants.

Source: David Roma, Australian Dept of Climate Change, Energy, the Environment and Water (DCCEW), public domain.

in this habitat? These types of questions would be asked by an ecosystem ecologist. Resource limitations impact the capacity and structure of most ecosystems and the movement of its resources, including nutrients and energy through the biotic and abiotic components of an ecosystem that would be of vital interest.

The Ecologist

A career in ecology provides opportunity to contribute to human society in many ways. Ecological issues are often at the very core of how humans meet their basic needs for food, shelter, and even health care. The ecologist may choose to work in natural environments or may investigate within the confines of a research laboratory. Since ecological issues are global, their endeavors may involve international travel to remote places, for example, the hydrothermal vents at the bottom of the Pacific Ocean. Ecologists work with conservationists and government officials to meet ecosystem services and resource management goals, like determining wildlife resources for hunting or timber for paper production. Ecologists might focus on informing the public of ecological issues, biodiversity loss and climate change, serving as educators, a career in health policy, nature centers, or national parks. Ecologists may serve on advisory boards at the local, state, and federal level to help develop policies that are ecologically sound. The ecologist will have completed, at the minimum, an undergraduate degree in a natural science with specialized training depending on an area of ecology. Most ecologists will be broadly trained in the physical sciences, with a solid foundation in mathematics and statistics.

Contrary to popular news, not all detrimental impacts to ecosystems are human derived. Throughout this course, we will look at multiple factors that have potential to alter the ecological landscape of a region.

Natural Historians were arguably the first ecologists, with appreciation for our relationship to the environment tracing back to the philosophers, Aristotle and Hippocrates. As the modern science emerged in the late 19th century, ecology incorporated newly established rigors of the scientific method. Indeed, the cornerstone of ecology relied upon evolutionary concepts first articulated by Charles Darwin in his treatise on adaptation and natural selection.

1.3 Ecologist Use the Scientific Method to Investigate Nature

By the end of this section, students should be able to:

- Describe the process of scientific inquiry.
- Outline the steps of hypotheses testing.
- Compare and contrast inductive and deductive reasoning.
- Describe the goals of basic and applied science in ecology.

The scientific method is the process by which questions in nature can be understood, resolved, or mitigated. Ecologists can be involved in study of the microscopic to the cellular, organismal, and ecosystems as well as the expansiveness of the planet and our biosphere (**Fig. 1.8**). In our daily lives many aspects of ecology are present in conversations and events. Annually there are reports in the news about an *Escherichia coli* (**Fig. 1.9**) outbreak in agricultural products (vegetative, dairy, or meat) due to contaminated waters. With the ever-present concerns of climate change and limitations of fossil fuels for energy, we learn of the ecological impacts of exploring biofuels while preserving the water supply and quantity and managing atmospheric carbon emissions. With alarming losses of biodiversity increasing exponentially, efforts to discover ways to protect the planet's species and ecosystems have been amplified. Inevitably these measures will involve reducing the effects of climate change. The physical causes of climate change are well established but its impact in an ecological sense is not fully understood by scientists nor yet fully shared with the general public. To do so requires thorough and irrefutable evidence and data that links causes to their effects using the scientific method.

The Process of Science

Science (Latin *scientia*, meaning "knowledge") is the acquisition of knowledge or the laws of nature by applying the scientific method to test and verify facts. Its goal is to comprehend the nature of the universe. In this process the scientific method is

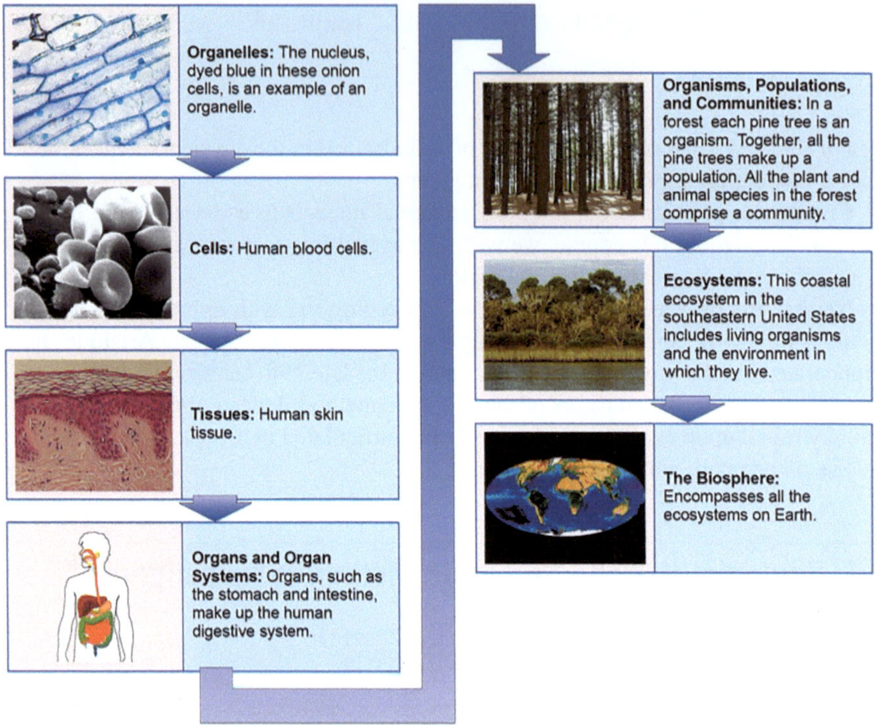

Fig. 1.8

Hierarchy in biological levels of organization.

Source: Modification of B. Wetzel & H. Schaefer, NCI; U. Salvagnin; Crystal, USFWS; NASA, Public Domain

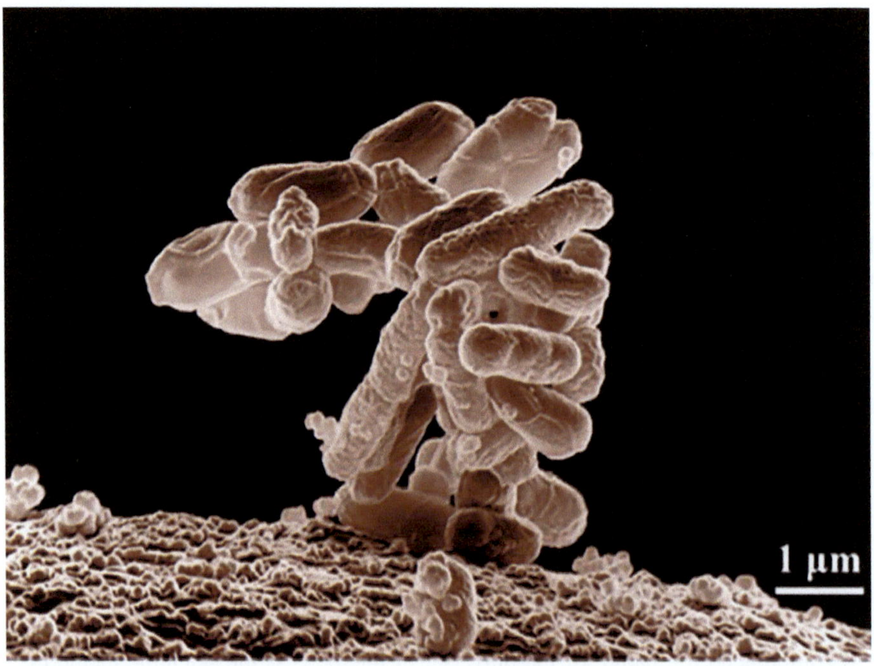

Fig. 1.9

Escherichia coli bacteria, scanning electron micrograph.

Source: Erbe, Pooley. USDA-Agricultural Research Service (USDA-ARS), Public Domain

applied, which will test a **hypothesis** by means of repeatable experiments (**Fig. 1.10**). Commonly misunderstood, a **scientific theory** is a *tested* and *confirmed* explanation for observations or phenomena. This contrasts with a **general theory**, which is defined as a hunch or guess without supporting evidence. By this definition there is no science involved.

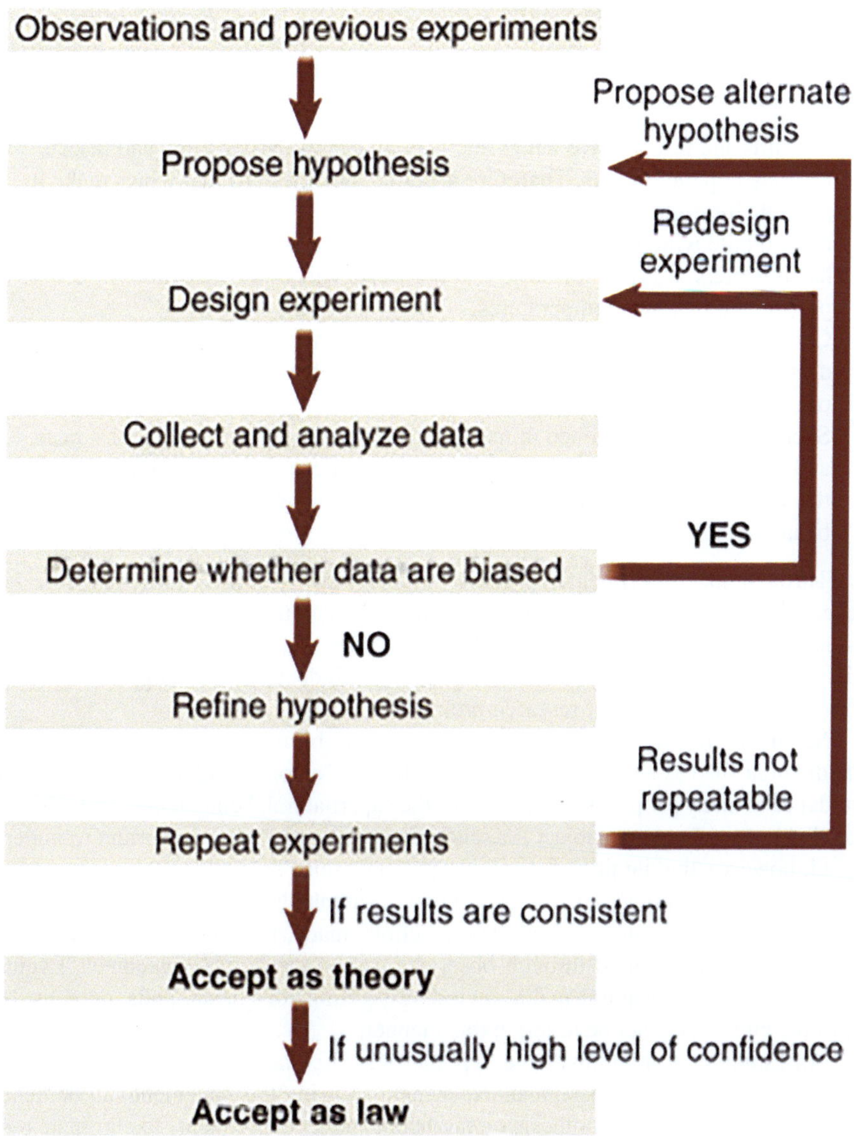

Fig. 1.10

The Scientific Method.

In any hypothesis-based science results are predicted from a generalized premise by a type of reasoning called **deductive reasoning** (proceeds from the general to the particular). In contrast scientists come to general conclusion from a number of specific observations by **inductive reasoning** (proceeds from the particular to the general). Both inductive and deductive reasoning are used in the process of advancing scientific knowledge (**Fig. 1.11**). Typically, forming the hypothesis is accomplished through inductive reasoning while the testable experimental steps follow the deductive reasoning approach.

Decide if each of the following statements is reached by a process of inductive or deductive reasoning:

- All birds and insects have wings and they all can fly. When birds and insects fly, they flap their wings. Therefore it can be hypothesized that wings make it possible for flight in any creature.
- During mild winters more insects have been observed to survive compared to in harsh winters. It is hypothesized that the world will experience more insects as pests based on rising temperatures.
- Chromosomes are the known carriers of DNA. They also separate into two cells during cell division. It can be hypothesized, therefore, that the genetic material of organisms must be DNA.
- Social behavior is exhibited in many diverse species, including humans, bees, and wild animals like the wolf or chimpanzee. It can be hypothesized from this observation that there is an evolutionary advantage to developing social behavior in a species.

Through the experimental process entirely new scientific questions may be revealed and lead to new explorations and testable hypothesis. Given the complexity of ecological systems, rather than following a linear sequence the researcher ecologist may prefer to draw inferences and generalize (inductive reasoning) but continue to seek out patterns as their research proceeds.

Scientific inquiry does not seek to displace faith, intuition, and dreams, which are distinguished as ways of knowing through use of emotion and may provide moral guidance to many people. The presence of the supernatural, being neither testable nor falsifiable, is not a good subject for scientific investigation. It is important to understand, however, that intuitive feeling, deep convictions, or traditional beliefs cannot be given scientifically validity. Recall that in the arena of scientific investigations the only questions that can be posed, or predictions made, must be limited to *testable ideas* that can be verified through observations and measurable outcomes. Events that are in the supernatural realm, including sightings of ghosts, gods, demons, or spiritual entities, cannot be tested in this manner.

An example of an unfalsifiable hypothesis is "Leonardo's Mona Lisa is beautiful." It is not possible to design an experiment showing this statement can be false. During the testing of a hypothesis researchers conduct experiments to eliminate one or even all the posed hypotheses. It is important to realize that hypothesis can be disproven and eliminated but they can *never* be proven.

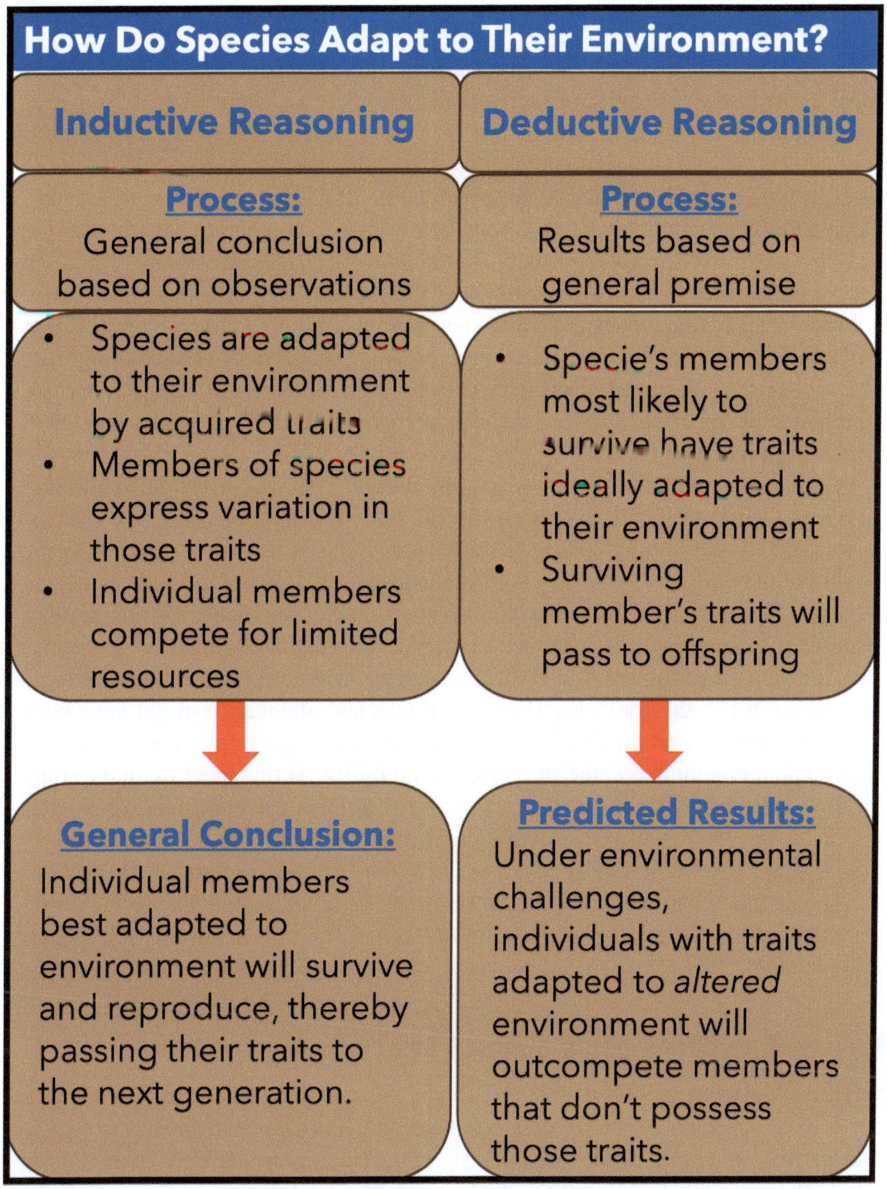

Fig. 1.11

Conclusions from inductive reasoning may become the premise for inductive reasoning.

Source: Image courtesy of author

The Ecologist Investigates Both Process and Patterns

Ecological systems are complex, covering many different hierarchical levels (individual to the biosphere) and overlapping interactions (e.g., implications of one species upon another). Unique patterns and processes emerge and lend themselves to different sets of questions and approaches to studying the patterns and processes in nature. Methods employed during an investigation might be experimental in nature or comparative or a combination of both. For example, comparative methods were used by Darwin in his landmark publication "*On the Origin of Species*" where he observed and recorded the similarities and differences in species as evidence for evolution. These initial comparative studies were essential to observe patterns that are the basis for understanding natural processes, or mechanisms of evolution. More recently analyses of ancient gene sequences by molecular experimentation became important to deciphering evolution through altered biochemical processes that led to novel phenotypes.

In other examples the dynamics of populations may be investigated through observing distribution patterns over a specific period of time in a given geographical location. A community ecologist studying all organisms that occupy a specific ecosystem may observe patterns and processes that only reveal themselves during the study. Species coexisting may impact the relative abundance of one another because of resource limitations. An ecologist studying a community of species takes into consideration the role of the physical environment and the flow of energy into and from the environment. The ratio of species' abundance might have been altered because of certain abiotic factors deeply integrated in an ecosystem, such as seasonal temperature, water, or nutrient availability. At this juncture in the study the flow of energy into the collective system becomes the emphasis. Ecologists may ask, "At what rate did available energy and nutrients incorporate into biomass and what were the limiting factors?" Patterns of species distribution across a geographic landscape may lead ecologists to question the factors that led to a species dispersal or emigration, or why some biomes are more favorable for species biodiversity.

Making Predictions Using a Model System

Some ecologists work in abstract situations, for example, analyzing past events that altered an endangered species' chances for survival in a particular region. Alternatively an ecologist may be interested in predicting future outcomes, such as canvassing events that may in time impact a region's biodiversity under unfavorable conditions, including reduced rainfall or rising temperatures. A mathematical model is useful to represent the ecological system under study. The appropriate model ranges in scale from populations, communities, or biomes. This will depend on the purpose of the study and what questions are being asked. Data is gathered from field studies to delineate the interacting relationships of species with respect to solar light availability, precipitation averages, or predator population sizes. Models would then be constructed using computer algorithms that predict real-life dynamics. Between the mathematical models and empirical observations ecologists hypothesize about the relationships that are not well understood.

Ecologists may choose to simulate larger-scale experiments that are impractical to perform in a real ecosystem. Reasons to work in a simulated system might include prohibitive costs, ethics, or challenging logistics imposed by time constraints. Using a computer simulation program ecologists can study a multitude of biotic and abiotic factors occurring in a complex ecosystem. This is often accomplished by limiting the system components to the most relevant and well understood of the process for simplification. In this manner events that occur over hundreds or thousands of years could conceivably be analyzed. Solutions may reveal themselves to complex ecosystems, often with unpredictable processes found in nature.

Model accuracy requires validation within the scientific community. This can be done by applying multiple data sets with known answers applied to the model. Another important validation method would be to compare field observation and model data outcomes. For example, the predator–prey model designed in 1926 by Alfred J. Lotka and Vito Volterra used differential equations to represent a predator and its prey in a continuous relationship. Following a relatively simple formula, the Lotka–Volterra Predator Prey mathematical model (L–V model) has been used reliably to describe population growth, dynamics, and death rates (**Fig. 1.12**).

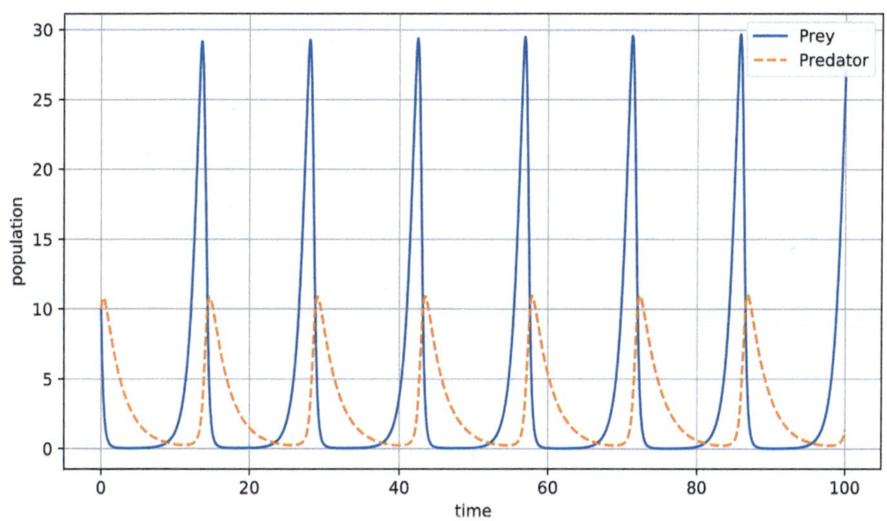

Fig. 1.12

Population dynamics for a rabbit and fox predator–prey relationship

Source: Ian Alexander under CC BY SA 4.0

Solving Problems Using Ecology

Field and simulation studies, mathematical models, and other ecological tools help us to deal with a variety of complex environmental problems such as the impact of abiotic materials and energy sources on the environment.

However, viable solutions may require a societal understanding of the human dimension, worldview decision-making processes, and the driving forces of economics and politics. During the COVID pandemic the Kayapo peoples of the Amazonian River basin struggled with habitat loss and impacts of defunding Brazil's environmental institutions and protections (**Figs. 1.13a, b**). Their home is 9 million hectares of indigenous pristine forest, savanna., and riverine ecosystems enriched with fish and

Fig. 1.13a

Kayapo people of the Amazon River basin inspired the script for Avatar.

Source: Kmusser under CC BY-SA 3.0

Fig. 1.13b

Iriri River basin.

Source: Image courtesy of Jeff Courier, Global Fly Fishing

timber resources. Unfortunately the territory is also rich in gold and soon became difficult to defend. Through strong intervention and alliances with international organizations, as of March 2021 the Kayapo Project reports they are successfully defending their land. To do so meant barricading against goldmining, commercial fishing, and logging by land and air surveillance. They guarded their lands to prevent building of roads and bridges by poachers to gain access to their resources. The Kayapo alliance safeguards the largest single tract of tropical forest and the last refuge for biodiversity in the entire southeastern Amazon. None too surprising, the Kayapo people were the inspiration for David Cameron's Avatar films.

Which of the following authentic biological problems might have called upon an ecologist to fully explore and analyze the facts of the case?

- *Declining Biodiversity*—In a field experiment the coyote density was dropped significantly to determine what impacts its diminished presence would have. The diversity of rodent species declined drastically, with most entirely disappearing (down from 10 to 2 species remaining) compared to control areas where the coyote population retained its high density. What was the likely correlation in population densities between the two species?

- *Epidemic*—A mysterious outbreak of thyrotoxicosis with 121 cases reporting in the Midwest in the 1980s was traced to consumption of meat from one slaughtering facility. How can thyroid issues in humans be related to meat processing?
- *Short-Lived Male*—The mite genus *Adactylidium* has a life strategy where its males live for only a few hours and then die while the females live their full life (4 days). What selection pressures in nature might have led this mite species to produce a male sexual type with no function to live other than to mate?
- *Unusual Pregnancy*—A 94-year-old woman was admitted for pneumonia and an enlarged abdomen. Upon examination the doctors were stunned to find she was carrying a fetus! Because the woman had dementia it was impossible to derive an explanation. What was the cause of the woman's physiological state and should this alter how we view human reproduction?
- *Falling Cats*—In this case a cat survived a fall of more than 30 stories from a skyscraper in New York City. The question then is why is it not possible for humans or other species to jump these distances?
- *Tuskless Elephants*—Elephants are poached to near extinction for their teeth-like ivory tusks. Increasing tusklessness in female elephants has been reported in African countries where rampant civil wars and fighting are present. How do human wars cause the females to lose their tusk genes? Why doesn't this occur in male elephants?

1.4 Ecosystem Types

By the end of this section, students should be able to:

- List the basic components of an ecosystem.
- Compare and contrast the basic types of ecosystems.
- Describe at least one methods ecologist use to study ecosystems.
- Identify methods to model ecosystems.

An **ecosystem** is a community of living organisms and their interactions with the surrounding abiotic (nonliving) environment. The living organisms (biotic factors) include microscopic to macroscopic organisms of the same and different species. The abiotic factors include the physical and chemical components, including wind, light intensity, moisture, temperature, concentration of gasses (O_2, CO_2), pH, and nutrients. Ecosystems do not have a standard size. They can be as small and finite as the tide pools of a rocky shoreline or as massive as the Amazonian tropical rainforest (**Fig. 1.14**).

A constant in any ecosystem is that species are in competition for limited resources. This is the defining characteristic of Darwin's theory of natural selection. The competition for finite resources is found in all communities within ecosystems and occurs between individuals of the same species as well as among different

(a) (b)

Fig. 1.14

(a) Tidal pool ecosystem. (b) Amazon rainforest ecosystem.
Source: Modification of works "takomabibelot"/Flickr; Ivan Mlinaric, Public Domain

species. There are organic and inorganic types of limited resources that all organisms must compete for to ensure their survival. This includes organic matter (from living or previously living organisms); inorganic substances like mineral nutrients from soil, rocks, and atmosphere; and energy-generating sunlight. An environment's resources not only provide food but also make up the physical structures that organisms inhabit. The physical and geographic environment (altitude, water, temperature, or oxygen availability) are critical factors that meet the variable needs and protection of living things for them to exist in a specific area.

Major Ecosystem Types

The abiotic components of the terrestrial ecosystem include climate and soil, which directly influence the growth and survival of forest trees. Among the biotic factors that interact in the forest, herbivores feed on the canopy (overhead foliage) while predators feed upon smaller prey. The forest canopy determines light availability to understory plants. A variety of organisms known as decomposers will feed on dead organic matter on the forest floor, which in turn releases nutrients to the soil that provide for the growth of plants.

The three broad categories of ecosystems based on their general environment are:

- freshwater
- oceanic
- terrestrial

Within the broader categories of land and water environments there are several distinct ecosystem types, classified by the organisms present and the type of environmental habitat, including the desert, estuary, boreal forest, or tropical rain forest. Chapter 4 will cover in more detail the terrestrial environments and Chapter 5 will delve into the various aquatic environments.

Ocean ecosystems cover 75% of Earth's surface and are generally divided into three basic ocean ecosystems: shallow oceans, deep oceans, and deep ocean surfaces. Shallow ocean ecosystems are the most biodiverse and include the coral reefs and kelp forests. Deep ocean ecosystems are also diverse with many large marine organisms. Aquatic diversity changes with ocean depth as increasing pressure, decreasing temperature, and sunlight dictate. The bottom of the ocean, impenetrable to light, still harbors living things uniquely suited for its survival. Living organisms at the surface of the ocean include plankton, small crustaceans, and other species adapted to wave action and seasonality. There are two surface ecoregions (shallow ocean and deep ocean surfaces) that contain aquatic photosynthesizers providing more than 40% of all photosynthesis on Earth. These aquatic species are interposed at the forefront of any human impacts (climate change, pollution, oil spills, plastic waste dumping).

Freshwater ecosystems cover approximately 3% of Earth's surface. Fewer than half of freshwater sources are in liquid form, which includes lakes, rivers, streams, and springs. The freshwater ecosystems are generally biodiverse (more than 140,000 species), including a variety of bacteria, phytoplankton, fungi, insects, fish, amphibians, and reptiles. The rest of Earth's freshwater is locked away as frozen ice in the glaciers and polar caps.

Terrestrial ecosystems are broadly diverse, generally grouped into categories based on animal and plant distributions called biomes. Biomes are not generally classified by their abiotic factors with distinct patterns of temperature and precipitation. Classification systems divide the terrestrial landscape into 12 or more biome groups, for example, boreal or temperate forests, savannas, deserts, chaparral, and tropical rain forests. Each of these biomes has its own ecosystem diversity. Desert landscapes diversity is shown in **Fig. 1.15**, where the Sonoran Desert of the southwest United States has abundant plant life, including saguaro cacti. In contrast the Boa Vista Island of Cape Verde, Africa, is practically a barren desert landscape.

Ecosystem health involves many interacting and moving parts. There may be variation in rainfall and temperature from season to season or from year to year. This in turn exposes plants and animals to disturbances with potential to alter their environment and geographical composition. Whenever an area experiences higher than normal levels of precipitation increased plant productivity should lead to increased resources. But sometimes the effect is the opposite. For example, robust new plant growth may shade out low-lying species and limit their survival in an altered environment. In general regional disturbances are the result of natural processes and occur within the range of variability for that ecosystem. The occurrence of a forest fire from a lightning strike may temporarily destroy a forest ecosystem and its inhabitants. But it's a common disturbance by natural causes. Regrowth of the forest occurs through the process of succession over a few generations. Beginning with emerging grasses and then shrubs,

(a) (b)

Fig. 1.15

(a) Saguaro National Park, Arizona. (b) Boa Vista Island, Cape Verde, Africa.

Source: Modification of works by Jay Galvin; Ingo Wölbern, Public Domain

the forest will ultimately be restored to its former living components based on its geographical terrain and climate. Sometimes the impact of disturbances has the capacity to wreak havoc in a manner that is not reparable through natural processes. As we move through the chapters we will discuss human activities known to be detrimental to ecosystems, including commercial agricultural practices, atmosphere/terrestrial/water pollution, deforestation, and overharvesting from the land and oceans, all of which are of major concern to conservationists monitoring biodiversity and extinction rates.

One way scientists measure ecosystem health is by investigating the transfer of energy, nutrients, and biota between and among ecosystems—all key factors that influence the health of an ecosystem. By establishing a baseline and monitoring it any perturbations to the ecosystem equilibrium can be measured. This implies an initial steady state, or **homeostasis** of the ecosystem. Ecological homeostasis describes the balance between the energetic input and output among organisms and the environment that provides its stability. Two ecological parameters to measure the impact of changes occurring to an ecosystem are resistance and resilience. In the forest that has experienced a forest fire both resistance and resilience may be operating in tandem to restore the steady state of the forest. The ability of an ecosystem to remain at equilibrium in spite of disturbances is called **resistance** (e.g., resistant species of the forest that the fire did not kill off). The speed at which an ecosystem recovers equilibrium after being disturbed is its capacity for **resilience** (the stages of vegetative succession to restore the forest). Taken together resistance and resilience determine whether an ecosystem can survive a severe or long-standing disturbance. In the case where the ecosystem is altered to a degree that it is no longer resilient to disturbances, irreversible destruction or an alteration of the ecosystem will be the outcome.

1.5 Energy Flow in an Ecosystem

By the end of this section, students should be able to:

- Describe energy acquisition in organisms.
- Give an example of a food web and associated food chains.
- Define productivity as a function of energy.
- Discuss trophic levels and how energy moves between them.

Life is an energy-driven process. All living things must obtain energy in one form or another. Most complex metabolic processes, especially those responsible for building large molecules from smaller compounds, require an input of energy. Living organisms require a continual energy input to assemble macromolecules (proteins, lipids, carbohydrates, and nucleic acids) and perform their cellular activities.

Food webs are used to illustrate the flow of energy directionally through ecosystems. Food webs illustrate how efficiently organisms acquire their energy, utilize it, and to what extent their own biomass are used by other organisms further up the food web.

"All flesh is grass." —Isaiah

What could the biblical prophet Isaiah have known about primary productivity or trophic levels in ancient times (750 BC)? Whether an inspired guess or divine inspiration, those few words encapsulate the concept of ecological food webs. It just took 2000 more years for scientists to realize how accurately those words describe the importance of primary productivity.

In the early 20th century scientist Charles Elton applied a colloquial idiom, "big fish eat little fish," as he constructed a pyramid of plants and animals organized into trophic feeding chains. In his 1927 publication "Animal Ecology" Elton introduced the principal of a pyramid, with numbers of organisms decreasing as you move up the trophic levels. The large base at the bottom of a food web represented ubiquitous producers like plants, with fewer herbivores grazing on those plants in the next level, and still fewer carnivores eating the herbivores in the upper level. Elton knew from observing the world around him (e.g., a pattern), that bigger, fierce creatures were rarer in any ecosystem. Today we understand that diminishing numbers at the top of the pyramid occur through the very nature of energy flow within ecosystems. Some biomass energy will inherently be used up by living organisms and not available to move up the food chain. Energy is continually cycled into Earth's biosphere in the form of solar radiation, which photosynthesizers make available to keep energy flowing moving through the food webs.

"Three hundred trout are needed to support one man for a year. The trout, in turn, must consume 90,000 frogs, that must consume 27 million grasshoppers that live off 1000 tons of grass."

—G. Tyler Miller, Jr., American Chemist (1971)

Food Chains and Food Webs

The term food chain is sometimes used as a social construct to describe competition in the human world for survival (Who eats who?) as individuals progress and prosper through life. Individuals who are the most successful are viewed as being at the top of the food chain. From this vantage point, they consume all others for their benefit, whereas the less successful are seen as being at the bottom of the food chain and will consume much less of the available resources. Not all societies operate under this mindset but most capitalist societies would have no problem accepting that analogy as realistic.

The ecological **food chain** is a linear sequence of increasingly larger organisms through which nutrients and energy pass. It starts at the bottom with primary producers, then the primary consumers (of the producers), and on toward the higher-level consumers at the top of the food chain. Food chains help ecologists to describe the ecosystem structure and its dynamics. Because it is linear there is a straightforward path through the chain. The species or groups of species are assigned to a **trophic level**, based on their role as producers or consumers. What do you suppose would be the trophic level for humans?

The bottom of the food chain will always be the photosynthetic organisms (plants and phytoplankton) because they are **primary producers**. All primary producers can synthesize organic molecules from simple inorganic precursors (water, CO_2, nutrients). The biochemical energy is stored in their tissues (biomass). The process by which they convert simple inorganic precursors to stored energy is usually photosynthesis, which, as its name implies, requires light. The sheer number of primary producers is usually very high, forming a wide base to support the increasingly fewer numbers of consumers above it, which in turn has captured less and less of the original energy source, as illustrated in the food pyramid (**Fig. 1.16**).

Heterotrophs are organisms that can only obtain their energy by feeding on other organisms. Heterotrophs are also called consumers, and they are consumers of any organism, including plants, animals, microbes, or even dead tissue. Because these organisms consume the primary producers they are herbivores and are called **primary consumers**.

Secondary consumers are carnivores that eat primary consumers but can also consume primary producers.

Tertiary consumers are always carnivores that eat other carnivores. They are higher-level consumers and feed on the next lower tropic levels. There can be several trophic levels of carnivores consuming carnivores but the top of the food chain will always be the **apex consumers** (top of the food chain). Apex consumers are unique to the food chain and will be discussed in several scenarios throughout this textbook. In the Lake Ontario food chain, shown in **Fig. 1.17**, Chinook salmon is an apex consumer.

The limiting factor on food chains is the amount of energy captured. At every level some energy is utilized, some is stored as biomass, and some is lost as heat as it flows between the trophic levels. Due to the laws of thermodynamics no

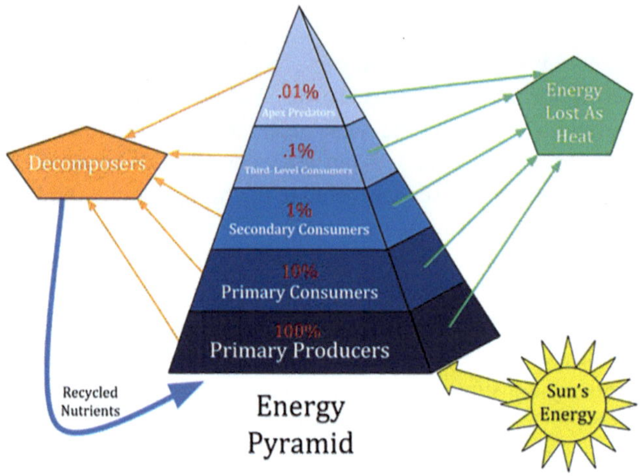

Fig. 1.16

The sun's energy passing through various trophic levels.

Source: Swiggity.Swag.YOLO.Bro under CC-BY-SA-4.0

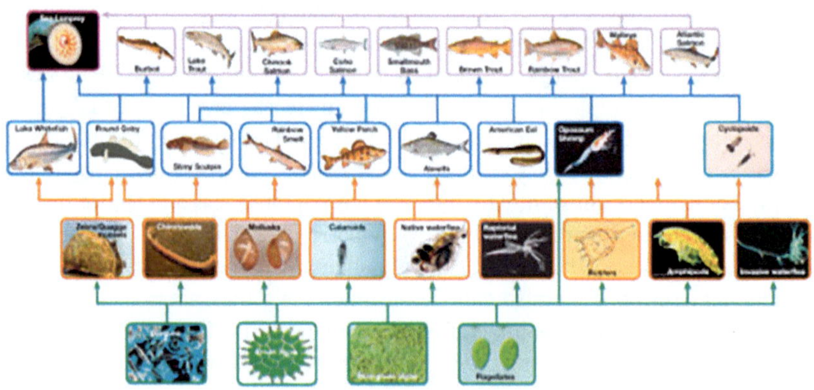

Fig. 1.17

Trophic levels of food chain NOAA. National Marine Fisheries Service. Fisheries of the United States, 2014. NOAA Current Fishery Statistics No. 2014. https://www.st.nmfs.noaa.gov/commercial-fisheries/fus/fus14/index/

Accessed on 7-May-2025

energy can be created and eventually the amount of energy still residing in the food chain, after losses at each transfer, will not be sufficient to support a higher trophic level.

Each of the trophic categories reflect levels at which energy and nutrients transfer up the chain. However, assigning organisms to trophic levels isn't always clear cut. Organisms that are **omnivores** eat both plants and animals, and an organism may consume another organism on its same trophic level. While food chains are straightforward, following the path of energy, a **food web** consists of many intersecting food chains and is more representative of how organisms consume prey and primary producers.

There are several distinct differences between a food chain and a food web:

- Food chains are linear pathways in which food energy and nutrients travel within the ecosystem.
  - Food webs represent several interconnected food chains.
- Food chains depict higher-trophic-level organisms feeding on one lower trophic level.
  - Food webs depict higher-trophic-level organisms feeding on multiple lower trophic levels.
- Food chains represent isolated or separate activities, which leads to instability within ecosystems.
  - Food webs increase stability within ecosystems because of their interacting, overlapping complexities in moving energy and nutrients.
- Food chains have no effect on improving the adaptability and competitiveness of the organisms.
  - Food webs are complex and multifaceted depictions of organismal interactions leading to enhanced adaptability and competitiveness of the organisms.

How Organisms Acquire Energy in a Food Web

The energy needed to support life at any trophic level is acquired by primary production and by consumption of the biomass of other living or previously living organisms. It is accurate to state that ecosystems are linked by the flow of energy and matter (organic and inorganic).

Photosynthetic and chemosynthetic organisms are grouped together as **autotrophs** because they can synthesize their own food. As we will see in the chapter on photosynthesis, autotrophs acquire inorganic carbon to utilize as their carbon source. Autotrophs are critical for all ecosystems because once the carbon has been converted into chemical energy it becomes the source of carbon for all consumer organisms. Consumers are known as the **heterotrophs** and cannot acquire carbon directly from the soil or atmosphere. Without autotrophs capturing solar energy and carbon life itself would not be possible on Earth.

There are two groups of autotrophs:

Photoautotrophs include plants, algae, and photosynthetic bacteria and are the major autotroph type for most of the world's ecosystems. These ecosystems are sometimes described as grazing food webs. Photoautotrophs harness the solar energy of the sun and convert it to chemical energy in the form of adenosine triphosphate (ATP) and nicotinamide adenine dinucleotide phosphate (NADP). The energy that is temporarily stored in ATP will be used by the organisms during their daily activities to synthesize complex organic molecules, such as glucose, throughout their lifespan. Energy flow also moves through dead biomass in what is called the detritivore food web.

Fig. 1.18

The Eifuku volcano hydrothermal vent, at 1535 m ocean depth, is where chemoautotrophic bacteria and particulate organic matter support aquatic life.

Source: National Oceanic and Atmospheric Administration (NOAA), Public Domain

Chemoautotrophs consist mostly of prokaryotes found in extreme environments where sunlight is not available, for example, dark caves or hydrothermal vents at the bottom of the ocean (**Fig. 1.18**). These producer organisms, mostly specialized autotrophic bacteria, can convert inorganic nutrients to organic compounds in marine or freshwater environments, using the rich supply of inorganic sulfur or hydrogen sulfide gas in those habitats. Regardless of whether solar radiation is available to convert to chemical energy, autotrophs must be able to synthesize complex organic molecules, such as glucose, to energize their own metabolic activities and to support the trophic levels above it.

Productivity Within Trophic Levels

Productivity within an ecosystem is measured as the percentage of energy (in Joules or Kilocalories) that enters the ecosystem and is converted into biomass. Productivity is based on the average rate of organic matter produced per unit area at a specific period of time, for example, kcal/m^2/year. **Biomass** is always related to productivity and can be determined for each trophic level in an ecosystem by totaling all the living and nonliving members of that level.

The rate at which photosynthetic organisms convert energy from the sun, including what is used by the plants own metabolism, is known as the **gross primary productivity (GPP)**. GPP is in the form of chemical energy. The highest GPP rates are possible under conditions of unlimited sunlight, water, and nutrients and with a climate that is conducive for long growing seasons and warm temperatures. Because all organisms need to use some of this energy for their own functions (like respiration and resulting metabolic heat loss) the amount of energy that passes up the food chain is calculated as the **net primary productivity (NPP)** of an ecosystem. NPP is the amount of assimilated and available energy after subtracting the amount of energy used by the producer to perform its own metabolic respiration (R). The value of R accounts for energy lost as heat during respiration.

The formula for average NPP for any geographic region is calculated in calories or joules:

$$GPP - R = NPP$$

Biomes with the highest NPP values are estuaries and swamps (12,000 kcal) or tropical rainforests (15,000 kcal). The lowest NPP values are found in tundra biomes (600 kcal), deserts (500 kcal), and the open ocean (800 kcal). Agricultural crops have a wide range in NPP values. There are highly productive sugar cane crops in Hawaii (25,000 kcal) and alfalfa fields (15,000 kcal) compared to low-producing crops of corn (4500 kcal). The impact of the human footprint is noteworthy when discussing the global NPP, given that humans have consumed, coopted for land development, or otherwise destroyed an estimated 40% of potential terrestrial NPP.

NPP accounts for the biomass available to primary consumers in the next trophic level. Thereafter **secondary productivity** is the term that describes the rate of energy stored at decomposer and consumer trophic levels. Productivity is not associated at any trophic levels above the primary level. Food assimilation is measured at each subsequent trophic level.

Productivity is directly related to energy flow rates and measures the richness of an ecosystem. The ratio of productivity to biomass (P/B) decreases as one moves up the food chain to larger organisms and higher trophic levels. In Chapter 8 the bioenergetics of productivity, food biomass, and its flows through trophic levels will be discussed in greater detail.

Glossary

Karner blue butterfly
null & alternative hypothesis
experimental groups
control groups
variable groups
dependent variables
independent variables
deductive reasoning
inductive reasoning
food chain
resistance
resilience
homeostasis
organismal ecology
community ecology
ecosystem ecology
population ecology
omnivores
carnivores
herbivores

Citations

1. B. Fuller, Operating manual for Spaceship Earth, Lars Müller Publishers, Zurich, Switzerland, 2020.
2. Foste, H. Mice aren't nice, they help transmit Lyme disease. *Entomology Today: Amazing Insects*. March 30, 2015. https://entomologytoday.org/2015/03/30/understanding-the-complex-life-cycle-of-the-blacklegged-tick-to-combat-lyme-disease/. Accessed Feb 15, 2025.
3. "U.S. Fish, Wildlife Service, threatened and endangered species Karner blue butterfly *Lycaeides melissa samuelis.*" US Fish and Wildlife Service. Retrieved (Accessed 20 June 2018).
4. K. Fiedler, Ants that associate with Lycaeninae butterfly larvae: diversity, ecology and biogeography, Divers Distrib 7 (1/2) (2001) 45–60. https://doi.org/10.1046/j.1472-4642.2001.00096.
5. Haack, R. (1993). The endangered Karner blue butterfly. (Lepidoptera: Lycaenidae): biology, management considerations, and data gaps. In: Gillespie, A.R., Parker, G.R., Pope, P.E., Rink, G., eds. Proceedings of the 9th Central Hardwood Forest Conference; Gen. Tech. Rep. NC-161. St. Paul, MN: U.S. Department of Agriculture, Forest Service, North Central Forest Experiment Station: 83-100. https://research.fs.usda.gov/treesearch/15307 Retrieved Feb 14, 2025.
6. Bathurst Copper Butterfly (Paralucia spinifera) Recovery Plan, 1 June 2001 by NSW National Parks and Wildlife Service, 78 pages. https://nswdpe.intersearch.com.au/nswdpejspui/bitstream/1/12993/1/bathurst-copper-butterfly-paralucia-spinifera-recovery-plan.pdf. Retrieved Feb 14, 2025.
7. Carter, O. (2017). "When Harry met Gabby – hypoxia and the scientific method." Flipped Classroom. National Center for Case Study Teaching. 3–4.
8. E.O. Voit, Perspective: dimensions of the scientific method, PLoS Comput Biol 15 (9) (2019) e1007279. https://doi.org/10.1371/journal.pcbi.1007279.
9. A. Dean, J. Thornton, Mechanistic approaches to the study of evolution: the functional synthesis, Nat Rev Genet 8 (2007) 675–688. https://doi.org/10.1038/nrg2160. Retrieved Feb 14, 2025.
10. Abakuks, A. The mathematical theory of the dynamics of biological populations II. J R Stat Soc Ser A. 145(4) (1982) 512. https://rss.onlinelibrary.wiley.com/doi/abs/10.2307/2982110. Retrieved Feb 14, 2025.
11. Ommundsen, P. Problem-based Learning. Taylor & Francis Group, 1st Edition eBook.
12. S.E. Henke, F.C. Bryant, Effects of coyote removal on the faunal community in Western Texas, J Wildl Manag 63 (4) (1999) 1066–1081.
13. C.W. Hedberg, et al., An outbreak of thyrotoxicosis caused by the consumption of bovine thyroid gland in ground beef, N Engl J Med 316 (1987) 993–998. https://doi.org/10.1056/NEJM198704163161605.
14. S.J. Gould, The Panda's Thumb: More Reflections in Natural History, W.W. Norton & Co, 1980, pp. 73–75.
15. D.I. Miller, J. Dillon, An unusual abdominal mass in an elderly woman, N Engl J Med 321 (23) (1989) 1613–1614.
16. *Natural History Magazine*. August 1989, 20–26.
17. EcoWatch: Olivia Rosane. Oct. 22, 2021, 01:16 pm EST.
18. R.R. Helm, The mysterious ecosystem at the ocean's surface, PLoS Biol 19 (4) (2021) e3001046. https://doi.org/10.1371/journal.pbio.3001046.

19. Oliver Wurl, Werner Ekau, William M. Landing, Christopher J. Zappa, "Sea surface microlayer in a changing ocean – a perspective.", in: Jody W. Deming, Jeff Bowman, (Eds.), Elementa: Science of the Anthropocene, University of California Press (2017), http://doi.org/10.1525/elementa.228.
20. J.S. Albert, et al., Scientists' warning to humanity on the freshwater biodiversity crisis, Ambio 50 (1) (2021) 85–94. https://doi.org/10.1007/s13280-020-01318-8
21. C.S. Elton, Animal Ecology, The Macmillan Company, New York, 1927.
22. O. Venter, et al., Sixteen years of change in the global terrestrial human footprint and implications for biodiversity conservation, Nat Commun 7 (2016) 12558. https://doi.org/10.1038/ncomms12558.
23. https://kayapo.org/report-2020/.

CHAPTER

Biogeography and Climate 2

Chapter Outline

2.1 Biogeography

By the end of this section, students should be able to:

- Define biogeography and discuss its variation in terms of species habitation.
- Use ecosystems and biomes as distinct ways to view a biogeographic region.
- List and describe the primary abiotic factors that affect global distribution patterns.
- Summarize the effect of abiotic factors on net primary productivity.

Biogeography is the study of how living things are distributed across landscapes, with all the abiotic factors that play a role of their geographic distribution. A biogeographic region's **climate** is generally classified according to its abiotic factors, as they will have a major impact on living things that can inhabit a specific region. The presence and quantity of biotic factors (light intensity, wind, temperature, moisture,

Ecology for the 21st Century. DOI: https://doi.org/10.1016/B978-0-443-24031-7.00002-7

etc.) can vary widely depending on latitude and elevation. Ecologists who study biogeography will be examining patterns of species distribution. Large biogeographically defined regions are known as **biomes**. Biomes are often classified according to their dominant climate and major vegetation type. The global map in **Fig. 2.1** illustrates a distinct biome gradient across the latitudes.

In a journey that begins at the equator you would be immersed in a terrain of tropical wet forests and broad-leaved evergreen trees that capture intense sunlight. But as your journey moves northward a gradual change becomes noticeable in **flora** and, on closer inspection, the **fauna** cohabitating those spaces. The rich greenery eventually gives rise to seasonally drier forests, more scattering of trees and shrubs, and a drop in precipitation and humidity. Upon reaching the latitude of 30°N more than likely the terrain has converted to deserts, with very low precipitation. Continuing in a northerly direction, if the area is still untouched by human development, it will likely be grasslands and prairies with mild temperatures and seasonal precipitation. Pursuing your hike farther north would gradually require a jacket as there will be cooler temperatures, in a seasonal pattern, with deciduous, temperate forests. Even colder temperatures, requiring warmer coats and protective gear, are a certainty as your journey reaches the boreal forests of the subarctic. This is just south of the Arctic Circle. Should you persist to reach a destination furthest from the equator you would encounter a frozen Arctic tundra at the most northern latitudes on planet Earth.

What this journey makes clear is that there is a gradual change in climate and the types of organisms that have adapted to the environmental factors associated with its latitude. This can be explained by the fact that the greater the distance from the equator, the less sunlight reaches Earth's surface. Whether your trek began at the Amazonian rainforests of South America or the Borneo lowland forests of Indonesia, you would experience a similar equatorial biome type. However, the precise latitude where a biome begins or ends will vary. This is because biomes are also influenced by geographic features like altitude and by the ocean currents and jet streams.

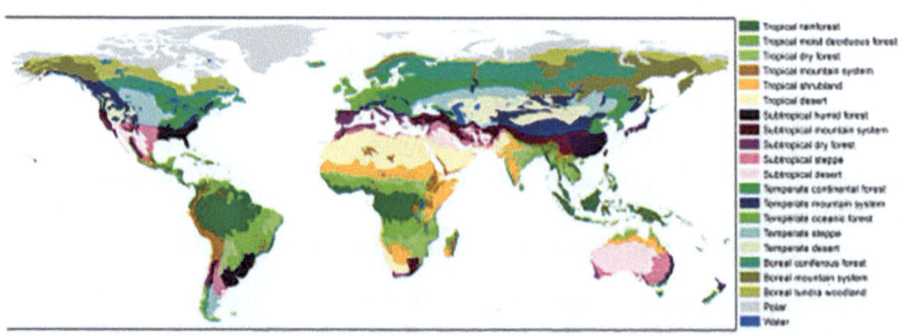

Fig. 2.1

Biomes are a major type of climate. They are often distinguishable according to latitudinal gradient.

Ruesch and Gibbs (2008). Courtesy: Carbon Dioxide Information and Analysis Center (CDIAC).

A pattern of species diversity—increasing toward tropical, equatorial regions and decreasing toward temperate regions—is known as the **latitudinal diversity gradient** (**LDG**). This observation is thought to be first described in 1807 by Alexander von Humboldt, a world-renown naturalist who headed many expeditions to the tropics of South America and to the northern latitudes of Russia.

Ecosystems and Biomes Describe a Biogeographic Range

Any given area can be classified as an **ecosystem** according to its environment, specific species of animals that are known to exist under its climate, the vegetation present, and any other life forms. The ecosystem is also defined by its weather, terrain, and other abiotic factors such as precipitation and temperature. Biotic and abiotic factors are linked in an ecosystem by their nutrient composition and energy flow. Is ecosystem just another name for a biome? No, the biome is a much larger region and can enclose more than one ecosystem. For example, an ecosystem under study by an ecologist might be a tiny pond in a much larger prairie grass biome. Then again, ecosystems can be on a macro scale, for example, the entire Amazonian rainforest. In comparison, the ecosystems are defined more by species interactions and habitat while a biome is defined more by its climate and vegetation type. The biome's climate is generally broader in scope and will have an *influence* on its ecosystems' climate patterns and local biodiversity.

The Fossil Record

The fossil record is a physical imprint of the past and dates to a time when all of Earth's continents were combined as one large land mass, known as **Pangaea** (335 to 175 Mya [million years ago]). Using modern tools to date prehistoric evidence, like radioactive decay, fossils can be dated and compared by a biogeographer to similar fossils found on different continents. Fossils are created when something dies while embedded in dirt and rock in a manner that will preserve it over time. While the record is not complete for several reasons, there is an extensive collection of fossils providing a chronological record of events of the past. The fossil record can be applied to recreating the process by which different groups of animals are related through evolution. For example, if the record places similar fossils from the time of Pangaea in different locations around the world, it is reasonable to assume those organisms lived across Pangaea without barriers, at least until the continents split into different biogeographic regions. Thereafter the organisms would have begun to evolve and adapt to their specific climates, diverging into different species by means of different adaptive strategies. The timing of fossilization is therefore very informative on the main drivers of evolution and adaptation.

Biogeography and the Evolutionary Map

Biogeography specializes in distribution patterns over time and space and can provide insight on how animals and plants have changed over time. Indeed, some of the earliest biogeographic studies were instrumental to emerging evolutionary thought during the 19th century. The study of biogeography remains vital to our understanding of adaptation and evolution as the influence of abiotic and biotic factors on species evolution requires long periods of time.

An example of a biogeographic study by an ecologist might be to compare primates that live in the eastern hemisphere with primates of the same species that migrated to the western hemisphere. Of particular interest would be any significant differences in growth rates, food sources, and behaviors, such as whether they are active during the day or night (nocturnal). The observations and other forms of data collected will reflect favorable or unfavorable conditions within their respective biogeography.

Viewing the natural world through a biogeographical perspective is not a new approach. Some of the world's most famous naturalists—like **Alexander von Humboldt**, an explorer that could arguably be called the first ecologist, and his contemporary, **Charles Darwin**, the father of the theory of evolution and natural selection—used a biogeographical context for their groundbreaking discoveries (**Fig. 2.2**). Both men were fortunate to be wealthy enough for travels to new worlds, where they meticulously documented and published their observations.

von Humboldt began his long and illustrious career when the King of Spain hired him to catalog and chart locations of gold, silver, cocoa, and other riches in Spanish-owned territories of the New World. By the 1860s von Humboldt had written more than 30 books with detailed maps. Everywhere he traveled von Humboldt took meticulous measurements of altitude, air temperature, and barometric pressure. Through his first-hand experience, he wrote on topics ranging from latitudinal species distribution, the cosmos, and biodiversity, describing Earth as "a living organism where everything was connected, from the tiniest insect to the tallest tree." He was an early proponent of LDG. With remarkable foresight for the 1830s, von Humboldt began to describe the phenomenon of human-induced climate change!

In the 1830s Darwin examined the patterns of species distribution to understand variation in finch beak size on the Galapagos Islands, located off the coast of South America. He determined the finches had originated from one location on the mainland before migrating to the Galapagos Islands, where food limitation became a key factor in their survival. Only the finches that successfully adapted to eat the available food types were able to survive in their new location. Across the islands from the Galapagos, there were distinct microenvironments with diverse climate, vegetation types, and predator–prey relationships defining the bird's food sources. Some of the finch's heads grew larger over time, some of their beaks got thicker (depending on the size of the food), and the feather coloring even changed to suit the island's fauna. Darwin's finches are a classic example of how animals adapt to new environments (Chapter 5 for Darwin's evidence for natural selection).

Today biogeography is an integrative field of inquiry and relies upon the expertise of many disciplines: ecologists, taxonomists, geologists, paleontologists, climate scientists, or other disciplines. Biogeographic studies may focus on historical,

Fig. 2.2

Alexander von Humboldt wrote Cosmos as a lecture series in Berlin, Germany, between 1845 and 1862.

Source: Alexander Humboldt, Joseph Karl Stieler under Public Domain

ecological, or regional biogeography with an emerging subdiscipline focused on the environmental impacts due to forest die-back, air and water pollution, desertification, and the overall health of rainforests.

Distribution Patterns of Plants and Animals

No species can exist in every climate. For example, the Venus flytrap is **endemic** to a subtropical wetland in North and South Carolina. Endemic species include animals, plants, fungi, or even microorganisms that exist only in a particular region. The definition differs slightly from the term **indigenous**, or native species, which occurs naturally in an area but can be found native in other areas as well. Endemic is rarer, requiring a more fastidious set of environmental conditions for its survival than normal. The indigenous plant might have a larger distribution pattern. The raccoon, for example, is a mammal native to temperate and tropical habitats throughout most of North and Central America and has been introduced to Europe. Other species that can live in a wide variety of geographic areas are called **generalist**.

Isolated land masses, like Australia, Hawaii, or even Madagascar, are large enough to have a variety of endemic species that would not likely thrive under different climatic conditions. Because endemics are highly specialized they are often found on lists of endangered species. The evergreen gardenia (*Gardenia brighamii*) endemic to Hawaii is extremely rare and endangered, with only an estimated 15 to 20 trees remaining (**Fig. 2.3**).

Because plants are immobile they are even more dependent upon the local climate and selective pressures to adapt. For this reason plants are capable of enhanced flexibility in growth habit and limited resource needs. Indeed, plant scientists study hybridization of plants to identify those better suited to a changing climate, for example, species that might be better adapted to survive in drought, high heat, colder temperatures, and harsh soil conditions. One example is the staple US crop known as winter wheat, bred through hybridization to ensure the crop will be able to survive frigid temperatures under snow in dry regions.

Species distribution patterns conform to the biotic and abiotic factors and their influence on species evolution. In Hawaii, for example, there are few native terrestrial species including one Hawaiian native mammal, the hoary bat, but no native terrestrial reptiles or amphibians. Some of the most distinctive variation in plants and animals occurs across regions that have been physically separated by geographic barriers for millions of years by geographic barriers. For example, the breaking up of the large landmass of **Gondwana** (25–66 Mya) led to Australia's approximately

Fig. 2.3

Federally endangered forest gardenia is endemic to the Hawaiian Islands.

Source: David Eickhoff, Gardenia brighamii under CC by 2.0

(a) (b)

Fig. 2.4

Endemic species of Australia. (a) Wallaby (*Wallabia bicolor*), a marsupial and member of the kangaroo family. (b) Echidna (*Tachyglossus aculeatus*), an egg-laying mammal.

Source: jalexartix under CC BY 2.0

700,000 species of living plants and mammal being 75% endemic and found only in Australia (**Fig. 2.4**).

In New Guinea marsupial mammals were the first to cross over from Australia (25 Mya) and evolve independently after the Gondwana land mass separated. Australia and New Guinea were once connected by a land bridge, which later flooded with rising sea levels to form the Torres Strait. Marsupials are the dominant mammal type but ecologists now know that nearly 10 million years later placental mammals arrived in New Guinea. Native bats and rodents are thought to have crossed in New Guinea from the Asian continent by way of the Indonesian archipelago.

Solar Radiation and Nutrient Uptake for Primary Production

In Chapter 1 we learned that all living things require a source of energy and that the initial source of all energy on Earth originates from solar radiation. Energy from the sun is captured by the primary producers, plants, and certain aquatic microorganisms. The photosynthetic microorganisms include cyanobacteria, algae, and protists. All photosynthesizers retain the capacity to convert solar energy into a form of chemical energy. Because light availability varies, primary producers have developed extensive strategies to be able to access that energy source for its own growth habits. In a shaded forest ground plants beneath the tree canopy must find ways to capture limited solar radiation. Both plants and algae have the capacity to modify their light-capturing apparatus, the chlorophyll molecules that capture solar radiation energy.

Under low light conditions it is possible for the chlorophyll to utilize low-energy red or infrared solar radiation instead (**Fig. 2.5**). One strategy involves seasonal life cycle timing.

Spring **ephemeral** plants are rapid growers that bloom in early spring before the bare deciduous trees surrounding them leaf out and block their sunlight. Spring ephemeral plants will disappear by late spring (**Fig. 2.6**). In its short life the ephemeral's welcoming blooms of spring complete its reproductive cycle needed for the following spring.

In aquatic ecosystems light energy is plentiful close to the water's surface but at the bottom of the lake or ocean light waves have already been absorbed by the water, organisms, and any floating particulates. Photosynthesis cannot take place without light energy. This explains why submerged aquatic plants with photosynthetic tissue are positioned close to the water's surface (e.g., sea grasses, flowering water lily) and only grow in shallow waters.

For primary producers to continue growing and be the energy source for other trophic levels, they must acquire nutrients. As aquatic organisms die, they sink to the bottom of the ocean. By the process of decomposition they begin to release nutrients that had been stored as their biomass. Decomposed nutrients would stay on the ocean floor but for a process called ocean upwelling. **Ocean upwelling** occurs as prevailing winds blow toward the coastline, pushing the surface waters out into the ocean, which is then replaced by deeper waters moving upward and toward the coast (**Fig. 2.7**). In the process the deposited nutrients are swept up and made available as

Fig. 2.5

Plants and algae that grow in deep shade compete for light energy by changing their chlorophyll's capacity for solar radiation.

Source: Energy.gov under Public Domain

Fig. 2.6

Spring beauty ephemeral flowering plant completes its life cycle in early spring to avoid competition for sunlight under a leaf canopy.

Source: USFS, Eastern Region, Public Domain

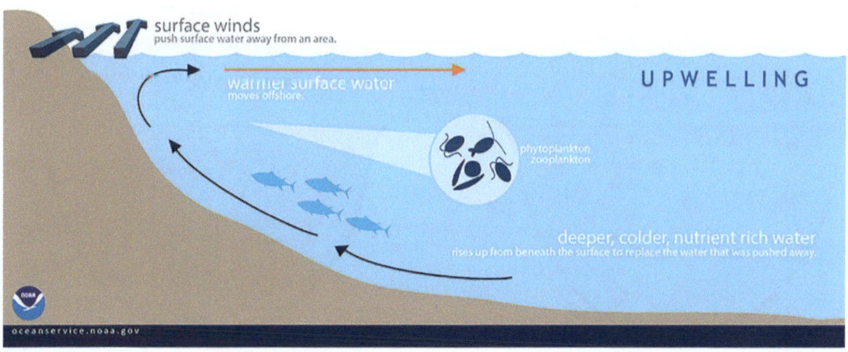

Fig. 2.7

Ocean upwelling recycles nutrients and energy in the ocean.

Source: https://oceanservice.noaa.gov/facts/upwelling.html

nutrients for other living organisms. The exception is within deep ocean hydrothermal vents, where organisms at the bottom of the food chain extract energy directly from inorganic chemicals (rock minerals) in a process called chemosynthesis. This is because it is pitch dark without access to either light energy or decaying organic matter. Nutrient acquisition is an important component of primary productivity and will be discussed further as part of the constraints species encounter in their terrestrial and aquatic environments.

In freshwater systems recycling of nutrients occurs in the spring and fall as a response to atmospheric temperature changes. During spring and fall turnover the nutrients that have sedimented on the bottom of lakes are recycled as water is mixed. In this physical process nutrients and oxygen are brought back to the surface (**Fig. 2.8a**) with seasonal changes. Together the spring and fall events are known as **lake turnover**. Lake turnover is due to the formation of a **thermocline**—a layer of water beneath the surface with a gradually declining temperature as the depth increases. In the warm summer months it will be increasingly cooler further from the surface. In the freezing temperatures of winter the water below the ice is slightly warmer and gradually more toward the bottom of the lake until the temperature is stabilized around 4°C (40 °F). Water at the lowest level will be at its most dense mass. The deepest water is also oxygen poor because decomposition of dead organisms at the bottom of the lake occurs by a process that consumes oxygen, aerobic respiration,

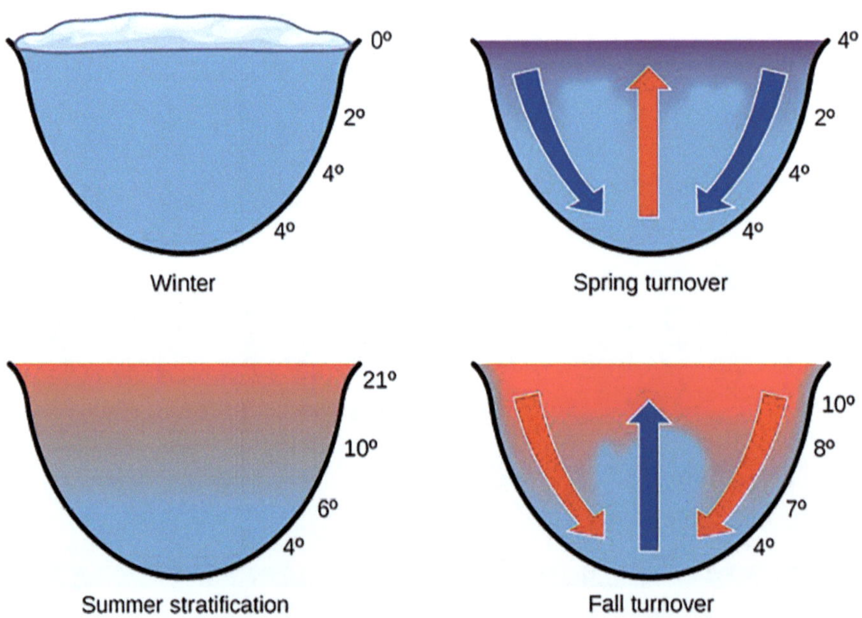

Fig. 2.8a

Spring and fall turnovers are important processes in freshwater lakes.

with corresponding oxygen production (photosynthesis). The limited oxygen that diffuses from the atmosphere down to the bottom of the lake is inadequate, especially when solid ice forms a barrier to gas exchange at the surface.

Spring turnover: Surface ice on seasonal lakes begins to melt with rising air temperatures. Once the surface water warms to 4°C it will be at its heaviest density (**Fig. 2.8b**) and will sink, displacing any water beneath it. As newly displaced water begin to rise from the bottom of the lake it carries with it lake sediments and nutrients. This change in water temperature and corresponding water density occurs once again during the fall season, as a second turnover event that remixes nutrients from the bottom of lakes, making them accessible to aquatic animals and, importantly, the primary producers.

Fall turnover: Seasonal lake turnover in the fall occurs as water begins to cool below 4°C due to temperature exchanges with a cooling atmosphere. As shown in the water density graph, water has less density at temperatures both higher and lower than 4°C. The cooler water will begin to drop once again, displacing water beneath it and as it reaches the surface will be cooled by the atmosphere. Under freezing temperatures the uppermost water becomes solid ice that insulates the lake and halts the turnover process. Solid ice acquires a new property that is lighter and more buoyant than liquid water at any temperature. The turnover cycle will pick up again with melting temperatures of the next spring season. Lake turnover occurs only in temperate

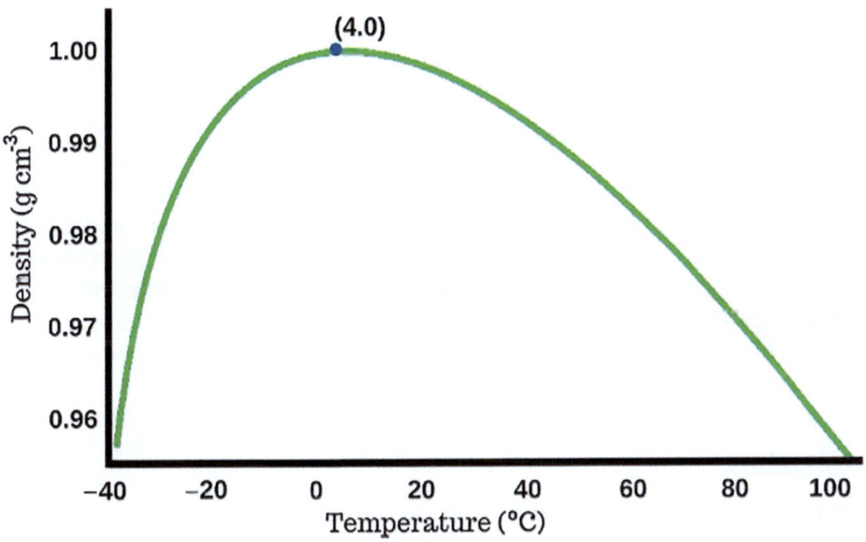

Fig. 2.8b

Water density is temperature dependent and is the cause of fall and spring turnover events.

Source: Lumen Biology under CC BY 4.0

Fig. 2.9

Mature cones of the jack pine (*Pinus banksiana*) open only when exposed to high temperatures.

Source: Doug McGrady under CC BY 2.0

climates. Tropical climates do not experience sharp changes in atmospheric temperature that lead to lake turnover events.

Soil composition affects the rate of inorganic nutrient uptake in terrestrial plants. The biogeographical distribution of plants is dependent upon soil structure (particle size of soil components), pH, and mineral content (see Chapter 4, Terrestrial Environments, for detailed information on soil quality). The biogeographic distribution of animals is more related to what they eat in predatory–prey relationships and may need to stay mobile in pursuit of their diet. Oxygen availability becomes restricted at very high elevations. In aquatic systems oxygen is dissolved into water at rates dependent on water temperature and the speed at which the water moves. Wind becomes a terrestrial abiotic factor because it influences the rate of evaporation and transpiration and its ability to move soil. Fire is not often thought of as an abiotic factor but it remains an important agent of disturbance in terrestrial ecosystems, thus impacts biogeographic distribution. Some organisms are resilient to fire and may even need the occasional fire to complete their life cycle. The jack pine tree requires high heat for its seed cones to open (**Fig. 2.9**).

Temperature

Temperature is an abiotic factor that clearly imparts a constraint on how species are distributed geographically. Temperature exerts a strong influence on places where living things can exist, with only a few living things, called **extremophiles**, that are able to withstand extreme environments.

Most organisms have metabolic constraints at temperatures below 0°C (32°F) due to the molecular nature of working cells. Proteins called enzymes do most of the work in the cells of organisms and their biochemical structures are maintained and active within a narrow temperature range. Outside that range most enzymes are denatured and no longer function as required. This is one reason why organisms develop strategies to maintain internal temperature within specific ranges or make modifications in their molecular machinery. But across Earth's biogeography we find ecosystems inhabited with animals have adapted to survive under significant temperature fluctuations.

Sometimes this is accomplished by adapting behaviors that help species survive in hostile climates. Strategies include the capacity to **hibernate** in cold climates or **estivate** in hot, dry climates. Animals that hibernate or estivate enter a state known as **torpor**—a sort of suspended animation as their metabolic rate mostly shuts down and their body temperature drops to just above freezing. Either hostile environment (temporarily too cold or too hot) is tolerated using these strategies until the environmental conditions return to conditions favorable that support their lifestyle.

Another example of temperature-driven behavioral modification is the annual migrations of birds to warmer climates using environmental cues and wind currents. Not all birds in temperate climates need to migrate to warmer climates during the winter months. For those that do migration solves many winter-related problems, like sourcing food and reproduction. One of the most impressive migratory paths is taken by the Arctic Tern (*Sterna paradisaea*), a mid-sized bird that makes its annual 55,000 mile round trip, almost pole to pole, from the feeding grounds of Weddell Sea in the Antarctic to its breeding grounds in Greenland. The endangered Monarch butterfly (*Danaus plexippus*) migrates from eastern Canada to the forests of Sierra Madre Mountains of Mexico (3000 miles). Other butterflies are capable of overwintering in their larval life stage during the cold winter months. Given the risks of migration, perhaps it is not surprising that Monarch butterfly populations have diminished by 53% since 2019. According to the World Wildlife Fund (WWF), which monitors preservation of endangered species, the Monarch butterfly migratory patterns are being impeded by abnormal patterns of drought and rainfall that disrupt normal food supply and breeding sites.

A nonbehavioral adaptive strategy to severe temperature is the production of an antifreeze-like chemical by cells that serves to prevent their bursting or the misfiring of cellular signals. Natural antifreeze products are seen in broadly different species and geographic regions, including the wood frog (*Rana sylvatica*), cold-tolerant plants, and arctic ice fish (Antarctic *Notothenioids*), which thrive at 4°C. Extremophile bacteria have adapted to extremely hot temperatures found in geysers, hydrothermal vents, and volcanoes (temperatures exceeding 45°C [113 °F]). Other extremophiles inhabit the extremely cold climates of the arctic permafrost.

Water

Water is the solvent of life and is a basic requirement for all living things. The human body, for example, is made up of 60% water and some plants have more than 80% water content. All terrestrial primary productivity requires an adequate supply of precipitation. Plants uptake water through their roots with limited ability to move toward sources of water. Organisms have evolved many adaptations to retain their required moisture level without becoming dehydrated from water loss or bloated from too much water intake. In Chapter 7 we will review in more detail other types of adaptations that are used by plants and animals to survive in their specific environments.

Examples of adaptations to maintain water balance include:

- Plants produce a waterproof barrier, the waxy cuticle to prevent loss of water to the environment.
- Freshwater organisms are surrounded by water and need to avoid excess water intake from osmotic forces. A water balance system in freshwater organisms involves excretion of dilute urine with minimal mineral loss.
- Marine organisms risk dehydration by osmotic forces to the high solute seawater. A water balance system in marine animals involves excretion of very concentration urine while retaining a stable internal moisture level.

2.2 Global Climate and Weather Patterns

By the end of this section, students should be able to:

- Define climate and weather and give examples of each.
- Explain the role of solar radiation in climate, both globally and regionally.
- Describe three natural factors affecting climate and name the predominant causes of long-term global climate change.

Climate can be thought of as a climate *zone* because it will be influenced by not just geographic features and latitude but also the surrounding atmosphere's ability to process and distribute heat and moisture across the landscape. The global climate determines what can grow, the timing of its life span, and where it can grow. It also plays a role in the quality of soil and the type of vegetation it can support. The **Köppen–Geiger climate classification** system starts with five basic types and is divided into climate subtypes (**Fig. 2.10**). The five categories, based on precipitation and temperature (the two most important determinants of climate), are designated by letters as follows:

- *A*—tropical (mild and rainy year round)
- *B*—dry (arid, semiarid)
- *C*—subtropical; this group was reclassified in the 1960s as it was too broad
- *D*—temperate or continental
- *E*—polar or boreal

In recent years climate scientists have begun monitoring the gradually changing average climate known as **global climate change**. This phenomenon describes altered global weather patterns, increasing average temperatures, and rising levels of atmospheric carbon dioxide.

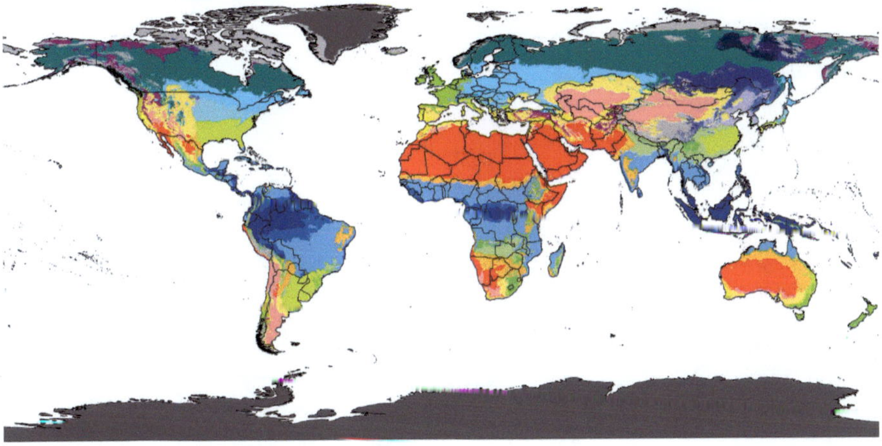

Fig. 2.10

Updated Köppen–Geiger climate classification system.

Source: DOI:10.1038/sdata.2018.214 under CC BY 4.0

Climate and Weather

Global climate change is not about short-term changes in weather that occur in a localized area, such as experiencing the first winter storm in your country one week earlier than the previous year. Evidence for global climate change has never been based on weather. The misconceptions on what is evidence for a changing climate are due to the confusion over those two terms: climate and weather.

Climate is the long-term prevailing patterns of weather and gives predictable ranges of atmospheric conditions in a specific area or climate zone. **Weather** refers to the conditions during a short period of time. Weather forecasts are usually made in 48-hour cycles. If you live in the Northern Hemisphere, you could plan an outdoor wedding for the month of July and predict it will be mild enough not to wear outer garment. But when planning the wedding it would be impossible to predict the weather, such as chances of windiness, high humidity, or precipitation. Weather is based on atmospheric conditions that are in flux. Climate is based on Earth's solar radiation and its rotational axis.

Solar Radiation Determines Earth's Surface Temperature

Solar radiation is a stream of photons, **electromagnetic energy**, that emanates from the sun and reaches Earth through a vacuum of space (**Fig. 2.11**). Once it enters Earth's atmosphere the waves of solar radiation will continue to emit some of its energy in the form of radiant or thermal energy, and it will reflect some radiation back into space. The sun is approximately 5800°C and emits high-energy **shortwave radiation**.

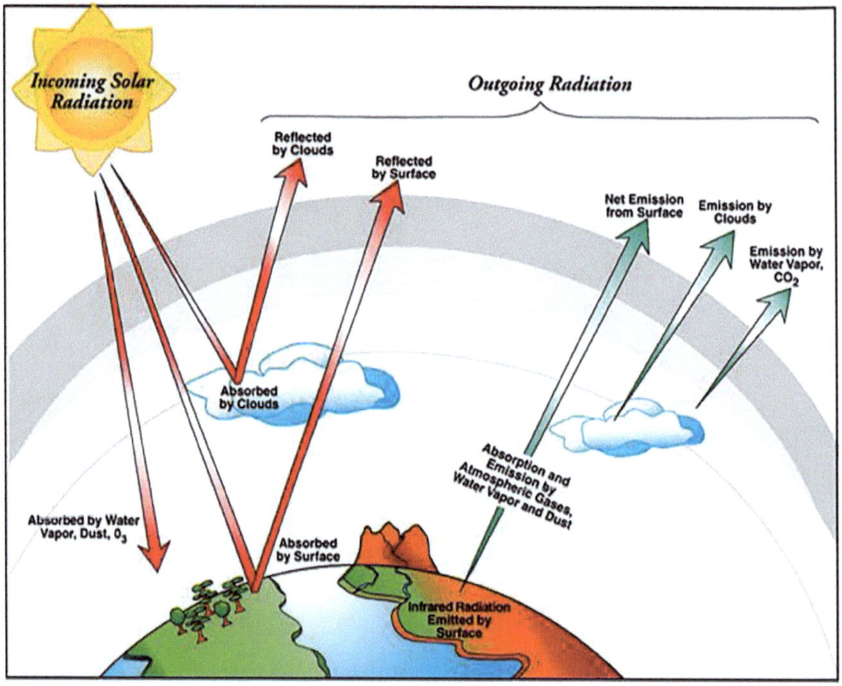

Fig. 2.11

Earth's radiation balance.

Source: Radiation Balance, US Department of Energy (ARM), Public Domain

The degree of shortwave solar radiation reflected back into space rather than absorbed is a function of a surface's reflectivity known as **albedo**. For example, solar radiation landing on the snows of the tundra have a high albedo rating (approximately 0.8–0.9, or 80%–90% reflection of incoming solar radiation). In contrast solar radiation that lands on broad green leaves of the tropics would have a low albedo (approximately .05, or only 5% reflection of incoming solar radiation). The broad green leaves were able to absorb much more of the solar radiation. Globally, Earth's overall albedo rating average is 0.30 (30% reflection).

Earth's surface has an average temperature of 15°C and emits lower energy longwave radiation, known as **terrestrial radiation**. The temperature of Earth's surface affects the amount of terrestrial radiation, with hotter surfaces emitting more radiant energy than cooler surfaces. The difference between emitted terrestrial longwave radiation and incoming shortwave solar radiation is known as the **net solar radiation** and determines temperature at any locale on Earth's surface. As terrestrial longwave radiation is emitted it can be trapped in the atmosphere by atmospheric gases and water vapors and emitted back toward Earth's surface as longwave **atmospheric radiation**. This emission of radiant energy results in the

greenhouse effect if there is overabundance of absorbent gases and water vapors in the atmosphere.

Typically, the incoming shortwave radiation vs outgoing longwave radiation balance out, leaving Earth's surface temperate at an average of 15°C (slightly higher near the equator and lower near the poles). This is due to the angle and distance the shortwave radiation travels to each as well as any albedo affect. But when levels of atmospheric gases rise due to fossil fuel burning and other industrial processes, more radiant energy is trapped. The radiant energy can be emitted back toward Earth again, which in turn raises the surface temperature above the net radiation average.

Biomes Experience Seasonal Variation in Temperature

The axis of Earth's rotation as it travels around the sun is the main cause of the planet's seasons and temperature fluctuation. At each of Earth's poles (north and south), one half of the year the pole is closer to the sun with greater solar radiation. During the other half of the year, the pole is farther away from the sun and experiences less solar radiation. Earth is split into two hemispheres, one above and one below the equator, called the Northern Hemisphere and Southern Hemisphere, respectively. Seasonal patterns occur at the opposite times of the year in each of the hemispheres (shorter days with colder weather vs longer days of warmer weather). Every year on two separate occasions Earth's poles will be positioned at equal distances from the sun. During these events the length of day and night (light cycle) will be roughly equal (12 hours) in both hemispheres. These singular events are reference points called the **equinox**. The moment of alignment is recorded in standard time, called Greenwich Mean Time (GMT). The GMT will be converted to relevant local time zones so that people around the globe will experience the equinox at different times of the day.

Earth's axis of a 23.5° angle (from north pole to south pole) gives rise to variation in the length of the days (diurnal: day/night light cycle). The tilt of the axis and the gravitational pull of other planets result in an elliptical pattern (rather than circular) as Earth travels around the sun once over the course of the year (**Figs. 2.12a, b**).

The sun is directly overhead the equator during the spring (vernal) and autumnal equinoxes (March 20–21 and September 22–23, respectively). During the equinox the equator receives the greatest shortwave radiation and other surface locations experience an equal 12-hour day/light.

There are two times during the year when the opposite occurs, and the 24-hour diurnal cycle has the most *variation* in length of the day and night. These events are called **solstices** and occur when Earth's axis is at its most oblique angle to the sun. During the Northern Hemisphere's summer solstice, solar radiation is directly overhead in a location known as the **Tropic of Cancer** (23.5° N), which will have the longest days and shortest nights of the year. Because it receives extended solar exposure this is also the warmest time of the year (June 20–21). During the same time period the Southern Hemisphere will be in a reverse situation, experiencing its shortest days and longest nights of the year (i.e., winter). By December 21–22,

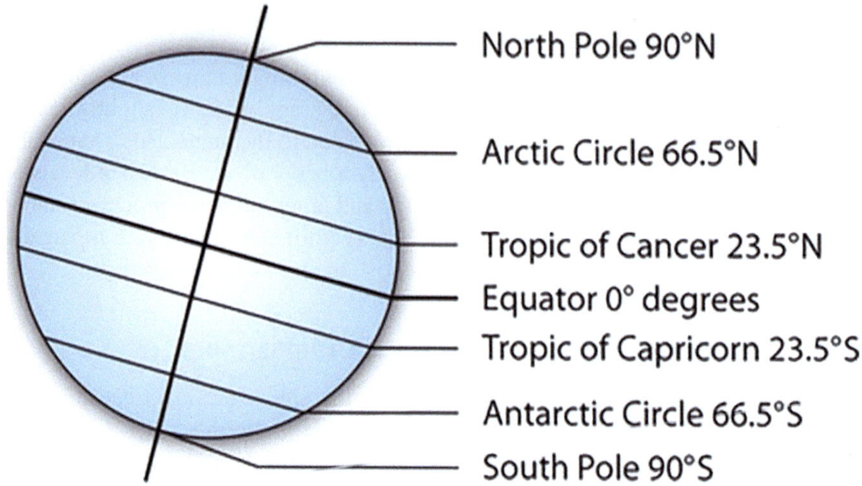

Fig. 2.12a

The angle of Earth's axis is 23.5° from north pole to south pole.

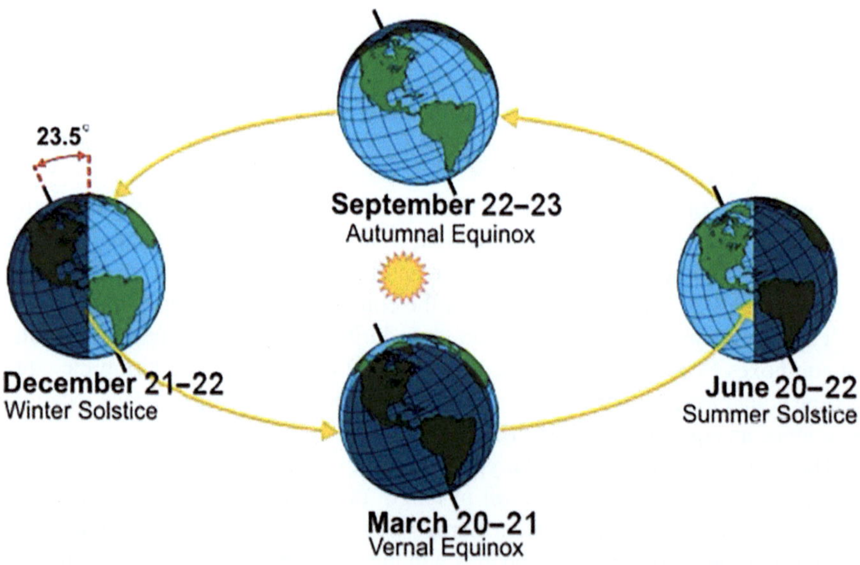

Fig. 2.12b

Earth's spinning rotation around its axis results in the diurnal (day/night) cycle, whereas its orbit around the sun with its axis tilted at 23.5° results in seasonality due to variation in exposure to solar radiation.

Source: US Weather.gov, Public Domain

however, solar radiation in the Southern Hemisphere will be directly overhead midday at the location known as **Tropic of Capricorn** (23.5° S) with the longest day and shortest night of the year (summer). Solstices occur in June and December and depending on whether in the Northern or Southern Hemisphere it will be summer or winter weather. If not for the tilted axis the phenomena of solstices would not even occur. Earth would have an equal 12-hour day/night cycle similar to the equinox phenomena.

In summary, the effect of the axis tilt and diurnal cycle has a major impact on length of days, temperature, and seasons. In turn the degree of solar radiation felt will impact an ecosystem's temperature and productivity. The axis is predominantly fixed at a 23.5° tilt. But the shape of the elliptical orbit slightly diverges over the course of millennia and contributes to historic climatic events.

A compilation of 48 photos of the sun taken during the year, once per week, in the same place and at the same time of day is shown in **Fig. 2.13**. The highest point is the summer solstice and the lowest is the winter solstice. The composite image forms the lemniscate, a symbol of infinity.

In **Fig. 2.14** there is a pale blue circle surrounding the planet Earth that represents an expanse of atmosphere that hugs Earth's surface. The image also

Fig. 2.13

An intriguing view of the sun's movement in the sky.

Source: Image courtesy of Gyorgy Soponyai.

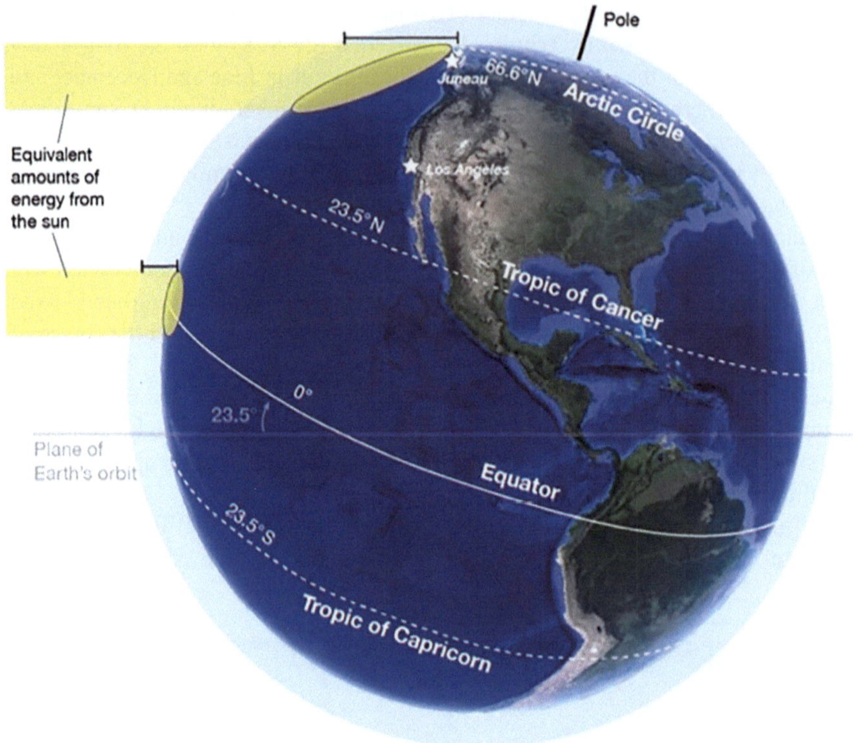

Fig. 2.14

Diagram showing sunlight hitting different latitudes of Earth's surface.
Map data: NOAA, US Navy, NGA, GEBCO, US Dept State Geographer; Image: Landsat, Google

highlights two yellow bands of solar radiation reaching the planet at two different locations, the equator and near the north pole. The sun's energy must pass through the ring of atmosphere before arriving on land surfaces. The illustration demonstrates that solar radiation passes through a much greater expanse of atmosphere at the north pole (shown as distance markers) due to the angel of Earth's circumference. Furthermore, because incoming rays of light are spread across a wider surface area in the north pole the solar radiation is more diffuse (shown as ovals on Earth's surface). In the lower, equatorial latitudes solar radiation travels along a more direct path and will be concentrated within a smaller surface area.

To recap, net solar radiation is not equally distributed on Earth, which leads to a wide variance in biome temperatures. In the previous section we noted that there is a warming effect associated with longer days at different times of the year due to Earth's rotational axis. Further, temperature varies latitudinally on any given day of the year due to the effect Earth's circumference has on light intensity and its capacity for productivity (energy flow). Both phenomena are additive to the differences in net solar radiation and surface temperature caused by differences in its reflectivity (albedo).

The Effect of Surface Temperature on Precipitation and Wind Patterns

Solar radiation directly affects temperature and indirectly affects precipitation and prevailing winds. This relationship is most obvious at the equatorial zone, where the Tropical biomes are positioned as a band between 23.5° N latitude (Tropic of Cancer) and 23.5° S latitude (Tropic of Capricorn). A significant portion of Earth's weather patterns can be explained by tropical air flow, as atmospheric conveyor belts called Hadley circulation or Hadley cells, which straddles the equator. **Fig. 2.15** illustrates how the **Hadley cell** drives continental climate. A pattern of global tropical air circulation is created due to uneven heating with rising latitudes (less intense solar

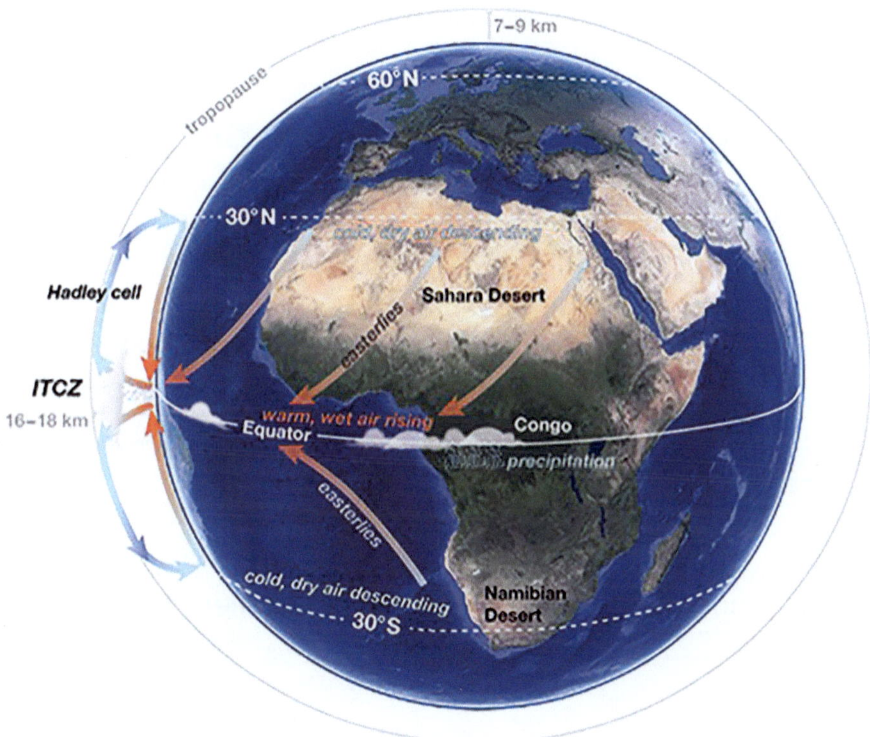

Fig. 2.15

Coriolis effect as Hadley cells drive continental climate.
Source: Satellite image of Earth from space courtesy NASA-Goddard Space Flight Center

radiation moving further away from the equator). Starting at the equator where solar radiation is at its greatest intensity, the air is warmed and rises up into the atmosphere (10–15 kilometers). This warmed air then flows poleward for a distance (approximately 2100 miles) before it descends back down. Once the air reaches the surface again it moves back toward the equator. This circular air flow pattern is caused by the Hadley cell effect, which also controls precipitation in the subtropics and trade winds in the **Intertropical Convergence Zone** (ITCZ). It is also responsible for highly precipitative storms like tropical rains and hurricanes, the jet streams, and even the placement and types of biomes. Climate scientists monitor Hadley cell circulation, which is believed to be weakening and threatening a rise in subtropical desertification. How does the equatorial band of circular air movement accomplish all of this?

The positive net solar radiation in the equatorial region is the cause for air to rise and creates the Hadley cell. As air warms, it becomes less dense than cooler air it displaces above it. When the air rises, a zone of low pressure is created below at the surface. To compensate air will move in from the north and south directions toward the equator. The point where incoming winds converge is the ITCZ. A column of warm air forms at the ITCZ that is continuously rising due to the positive net radiation (**Fig. 2.16**). The exact center of the ITCZ varies seasonally because it follows the sun but is

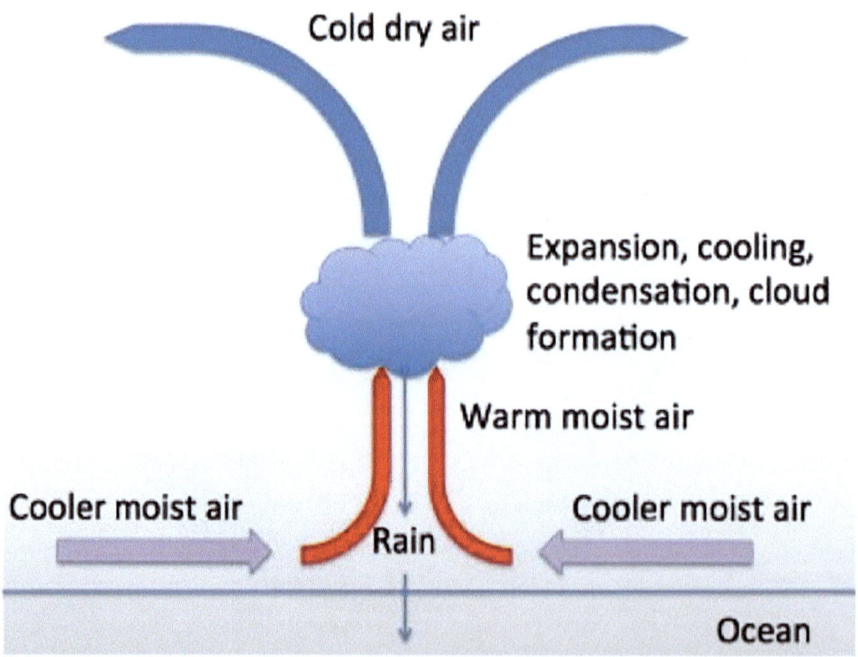

Fig. 2.16

ITCZ is the region of ascending air, high cloud cover, and heavy rains.

Source: CC BY-SA 2.5

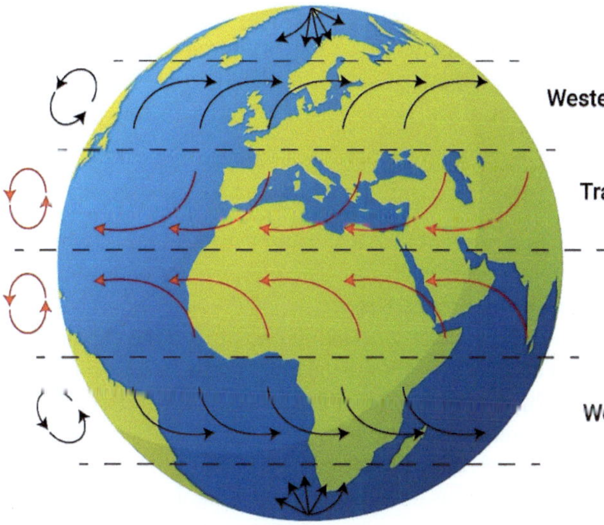

Fig. 2.17

Effect of the Coriolis force is an apparent deflection of the path of an object. The object does not actually deviate from its path but appears to do so by the motion of Earth's rotation.

Source: Kaidor based on File:Earth Global Circulation under CC BY-SA 3.0

essentially the thermal equator. As a column of air rises in the ITCZ it is dispersed to the north and south where it begins to cool; by the time it reaches 30° N and S latitude it has become dense again. Solar radiation is less intense and a drop in temperature causes the air to sink back to Earth's surface. This will lead to a rise in air pressure, which will cause air to flow back toward the low-pressure column of rising air, completing the Hadley cell cycle.

Along its circulation path air flow is deflected slightly by Earth's rotational axis. The pathway of mass air movement in the Hadley cell is not due north or south but will curve clockwise toward the Northern Hemisphere and counterclockwise toward the Southern Hemisphere (**Fig. 2.17**). Air flow patterns created by the Hadley cell and Earth's rotation are described as the **Coriolis effect**. The Coriolis effect includes the three tandem circulation patterns (Hadley, Polar, and midlatitude cells) and any air flow fluctuations due to changing temperatures (red lines), with resulting winds forming the subtropical jet streams. The Coriolis effect results from a combination of factors but is ultimately determined by net solar radiation and surface temperature.

Ocean Currents Are Determined by Wind Patterns

Prevailing winds lead to systematic water movement, or ocean currents, heading in the same direction. There are two major oceanic circulation patterns, or **gyres**. This pattern originates at the equator, where trade winds interact with air pressure to move warm waters in a westerly direction until they reach land. Wherever the incoming currents hit land masses the water course is altered to flow northerly and southerly, forming the north and south gyres. At middle latitudes currents change course and again at the polar latitudes, consistent with three airflow patterns.

Precipitation Is Influenced by Surface Temperatures

The warm, moist air converging at the ITCZ also causes tropical rains, which can be torrential, and the occasional oceanic cyclones. These events are understood best by first examining the physical properties of water.

On land and in the atmosphere water is found in one of its three chemical phases (liquid, water vapor, or ice) according to the surrounding temperature. Liquid water is transformed to water vapor by **evaporation** and ascends into the atmosphere. The air can become saturated with water molecules, at which point no further transformation will occur. Whenever water vapor exceeds the capacity of air to hold onto water molecules, **condensation** occurs (visible as dew), as some water molecules transform to its liquid state. The point of saturation can vary, dependent on temperature. Higher atmospheric temperatures have a higher point of saturation (can hold more water molecules). The tropics are both moist and warm because the air supports a greater water vapor pressure. **Humidity** is the amount of water contained in a given volume of air. When relative humidity is at 100% for a specific temperature the vapor pressure is saturated to its fullest extent possible. As air cools water vapor will transform to its liquid form. This is because the water vapor has reached its saturation point sooner. Transformation back to liquid water explains the high precipitation surrounding the Hadley air circulation. As a warm column of humid air rises to higher altitudes it begins to cool, causing condensation, cloud formation, and its release back to Earth as precipitation.

Water exchanges between its liquid and gas phase occur on land as well. For example, in the early morning hours when dew and mist formation dissipates (**Fig. 2.18**). Condensation of moist, warm air into dew or mist occurs whenever it encounters cooling surfaces and temperatures. The **dew point** (relative humidity at 100%) reflects the evening's lower water vapor saturation point, causing the transformation of water vapor into dew or mist. As the new day's solar radiation begins to warm the air, its water vapor saturation point increases and any dew or mist will quickly dissipate into the atmosphere.

When either the solid or liquid form of water falls to Earth from clouds it is called **precipitation**. Across the major biome types average rainfall is either high or low and reflects the fauna and flora of that region. Incoming warm trade winds gain moisture as they converge at the ITCZ and rise. The result is high precipitation in South and Central Americas, Africa, and regions of Asia, as depicted in the red areas of the map in **Fig. 2.19**. Around 30° to the north and south of the equator the landscape is arid. This is the territory where Hadley cell air descends and warms, which increases its capacity to hold onto water molecules. Observe the biomes located around 30° global latitude, where you will find the world's most famous deserts located (Sahara, Arabian, Mojave, etc.).

Fig. 2.18

Tree ferns in a cloud forest on Mount Kinabalu, Borneo (latitude 3°).

Source: NepGrower~commonswiki under CC BY-SA 2.5

Topography Affects Local Climate Patterns

Global patterns of climate are relatively consistent from year to year and form the basis of regional conditions and the species that are well adapted to inhabit and thrive. There is clearly a correlation between net solar radiation, surface temperature, wind patterns, ocean currents, and precipitation that define a climate. Yet this combination of factors doesn't entirely explain why areas in close proximity might be so different that they support entirely different species. It turns out that topographical features like large bodies of water or mountains can have a drastic effect by forming barriers to wind, impacting how thermal energy is exchanged, and influencing average rainfall.

Along the northwest coast of the United States residents might seek to warm up during a cold winter day with a trip to the Pacific Ocean. And in the sweltering

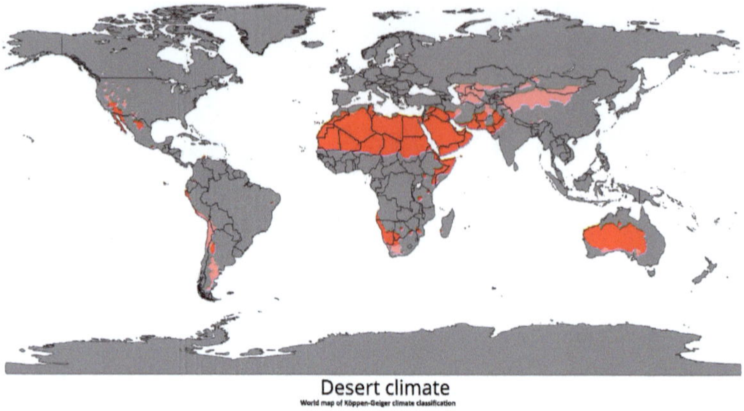

Fig. 2.19

The world's deserts are mostly situated at 30 N and 30 S latitude.

Source: Worldwind, ARC, NASA, Public Domain

summer heat outdoorsy residents might cool off by hiking the windward high-altitude coastal Pacific Northwest Cascade Mountains. The northwest outdoorsmen's lifestyle reflects the common, empirical knowledge known as the **maritime influence**.

Another unique chemical property of water is its ability to hold heat better than most liquids. Large bodies of water contain a high specific heat capacity, or thermal energy, and can act as a buffer for temperature alterations. It is this ability to absorb and retain heat from the atmosphere that creates the maritime influence. In contrast large land masses experience greater fluctuations in local temperature, known as **continentality**.

The topography of mountains doesn't only affect temperature according to altitude. Mountains have a specific pattern of precipitation, with one side supporting vegetation from ample rainfall and the other side being mostly arid and possibly desert conditions, known as the **rain shadow effect**.

Rain shadows are strongest near large bodies of water with their prevailing winds of warmed air reaching land (**Fig. 2.20**). Similar to Hadley cell air circulation, as the warmed air begins to rise up the mountain air pressure and temperature will drop. The corresponding drop in vapor saturation point causes condensation, cloud formation, and precipitation going up the terrain altitude. The prevailing winds carry the now-cooled air down the other side of the mountain, where it is warmed and retains moisture as its saturation point has increased. The rain shadow occurs on the **leeward side** of the mountain. The **windward side** of the mountain faces the prevailing winds and any maritime influence. Note the topography of the Cascade Mountain range of the northwest territory in the United States with its lush vegetation but drastically different semiarid terrain on the wind protected side (**Fig. 2.21**). One of the hottest and driest places on Earth is Death Valley, Nevada, situated on the leeward side of the Sierra Mountains in the United States. The windward side is dominated by giant

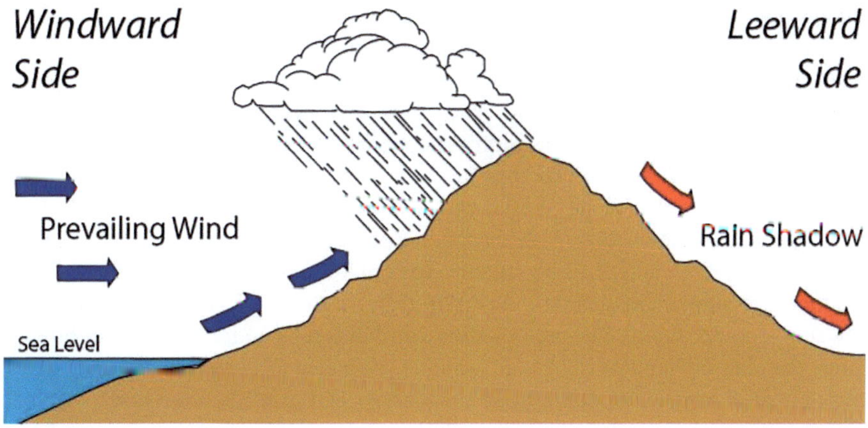

Fig. 2.20

Illustration of a rain shadow forming on the wind-protected side of a mountain.

Source: Meg Stewart under CC BY-SA 2.0

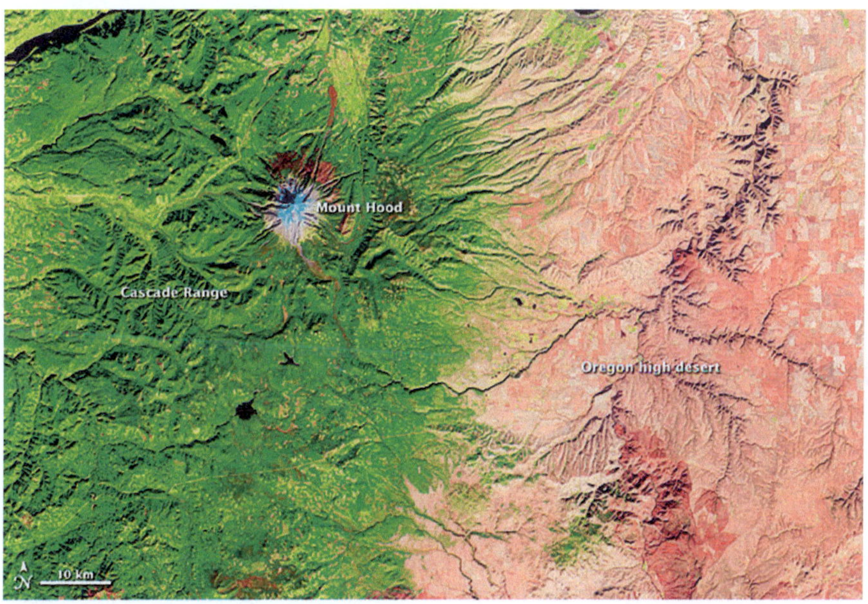

Fig. 2.21

High desert and sparse vegetation of the leeward side of the Cascade Mountain are shown in brown shades. The windward side is shown in green. The blue spot identifies the glacial cap of Mount Hood in Oregon.

Source: Landsat 5 satellite acquired October 27, 2011, Public Domain

sequoias that thrive in dense forests. Most deserts are on the leeward side of mountains, including Argentina's Patagonia Desert and the Gobi Desert of Mongolia. The Appalachian Mountains has a rain shadow but lacks semiarid conditions because winds from the Atlantic Ocean do not reach it. Which oceans and mountains created the conditions of the Gobi and Patagonia Deserts? The maritime influence and rain shadows are examples of a **microclimate**.

A smaller ecological niche is classified as a **microenvironment** when it is distinctly specialized and isolated. Atmospheric conditions can create a microenvironment significant enough to be suitable for only certain species of wildlife and unsuitable for similar species that generally thrive in the region. Slight changes in light, temperature, wind, and precipitation can have substantial impacts on biodiversity that can be supported. As an example, consider the south-facing side of a building that is exposed to sunlight and is warmer and drier than the north face of the building, where less exposure to sunlight means it is cooler and moister. The growth of moss or beetles in the expanse of just a few feet along the building creates an entirely different microclimate for other species to inhabit. A forest canopy is another example of a microenvironment. The presence of a microenvironment would be important to recognize when defining an individual organism's lifestyle.

Glossary

Flora
Fauna
Ephemeral
Albedo
Equinox
Solstice
torpor
estivate
gyres
evaporation
condensation
rain shadow
leeward
windward
microenvironment
indigenous
endemic
generalist

Citations

1. The Ecosphere and Environmental Issues, 4: Biogeography and Biomes. https://viva.pressbooks.pub/theecosphereandenvironmentalissues/chapter/chapter-3-biogeography/
2. Ecology and Distribution of the Endangered Hawaiian Hoary Bat. Pacific Island Ecosystems Research Center. https://www.usgs.gov/pacific-island-ecosystems-research center/science/ecology-and-distribution-endangered-hawaiian
3. Kerr, Isaac Alan Robert. (2022). This giant kangaroo once roamed New Guinea: descended from an Australian ancestor that migrated millions of years ago. *Paleontology*. https://www.australiangeographic.com.au/news/2022/07/this-giant-kangaroo-once-roamed-new-guinea-descended-from-an-australian-ancestor-that-migrated-millions-of-years-ago/.
4. Bista, I., Wood, J. M. D., Desvignes, T., et al. (2023). Genomics of cold adaptations in the Antarctic notothenioid fish radiation. *Nat Commun*, *14*, 3412. https://doi.org/10.1038/s41467-023-38567-6.

CHAPTER

Climate Change

3

Chapter Outline

3.1 Evidence for Climate Change

By the end of this section students should be able to:

- Provide one or more examples of historic climate change events.
- Summarize how the Industrial Revolution affected global atmospheric CO_2 levels.
- List two or more greenhouse gases that are on the rise and explain why.

The Intergovernmental Panel on Climate Change (IPCC) prepares comprehensive assessments that cover the state of scientific, technical, and socio economic knowledge on climate change. The summary reports detail its impacts and future risks and make recommendations on how the global community can reduce the rate of climate change and its impact on local ecology (**Fig. 3.1**). Thousands of scientists contribute to the panel's findings, which is the internationally accepted authority on global climate change. Between 2021 and 2022 the IPCC released the Sixth Assessment Reports. The Assessment Reports are released in stages by three working groups that are tasked with specific aspects of climate change. The 2021–2022 special reports focused on global warming of 1.5°C rise; climate change and terrestrial impacts (desertification, food security, sustainable land management); and climate change in the seas, oceans, and cryosphere (ice caps). To understand the scope of the IPCC's contribution more than 14,000 scientific publications were assessed before releasing

Ecology for the 21st Century. DOI: https://doi.org/10.1016/B978-0-443-24031-7.00003-9

the 2021–2022 report. The conclusion reached by the working group is that it isn't possible to mitigate an irreversible 1.5°C to 2.0°C average global rise in temperature without achieving massive changes in greenhouse gas emission. This is an endeavor that can only be accomplished by rapid, unprecedented transitions in energy and land use, leading to a 45% decline in atmospheric CO_2 before year 2030. As a result of the latest report release the United Nations has issued a **Code Red for Humanity**.

Code Red for Humanity is a dire warning and prediction from the conservative IPCC and the scientific community concerning continued existence of life on Earth, quality of life, and loss of biodiversity in general. In Chapter 14 we'll review the basis of biodiversity loss and sixth mass extinction driven by human-caused events. The goal of this chapter is to present evidence for climate change in the past, present, and future as it pertains to biogeography and ecosystems.

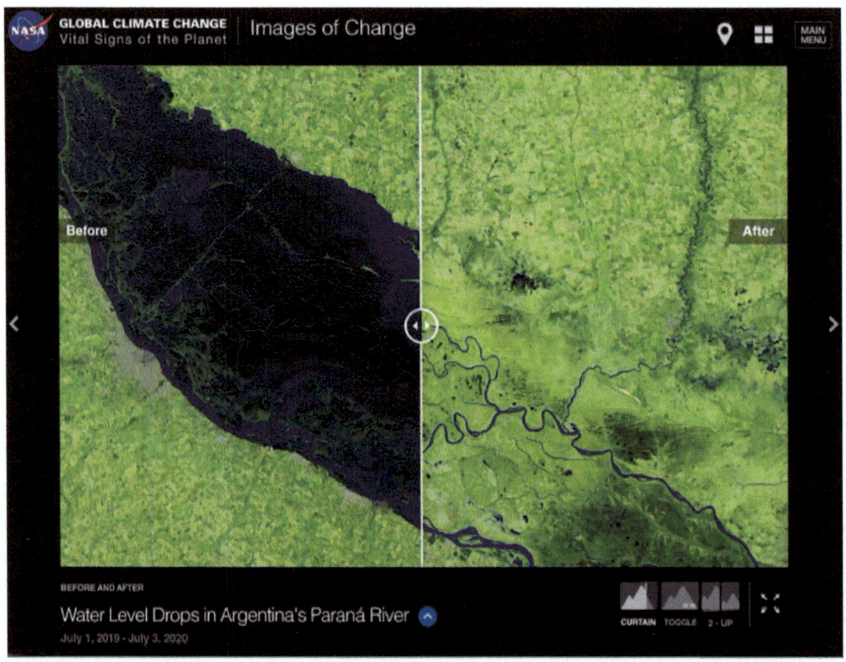

Fig. 3.1

Climate change is monitored in the United States by the National Aeronautics and Space Administration (NASA). Vital Signs of the Planet is a public web-based catalog of before and after climate events around the globe. In South America, Argentine's Paraná River is experiencing its lowest water level in 77 years due to a drought.

Source: Global Climate Change: Vital Signs of the Planet, NASA, Public Domain

Evidence for Global Climate Change

Since scientists cannot go back in time to collect direct measures of climatic variables, it is necessary to establish valid indirect measures of definitive changes on Earth that occurred over millions of years ago. A number of methods have been established that climatologists and other scientists use to gather reliable geological and historical evidence of Earth's climate cycles. One important measure from prehistoric times is collected from the **cryosphere**. Glaciologist **Claude Lorius** first applied a technique to study climate change called **ice core drilling** in the mid-20th century (1950s). From this early research Dr. Lorius connected the dots between global warming and human-induced pumping of CO_2 into the atmosphere at rates that did not exist before the industrial age. Ice core analysis indirectly estimates the atmospheric CO_2 concentrations and global temperature over the past 400,000 years. To obtain ice core samples of polar ice, it is necessary to drill beneath thousands of meters, into either ice sheets or mountain glaciers (**Fig. 3.2**). In Lorius' documentary, "Sky and Ice," he says, "One evening, while drinking a whiskey on rocks chipped off an ice core sample, I watched bubbles get liberated from the ice, and realized, these bubbles must contain air from the past. Like tiny fossils, they contained a sample of atmosphere from the time the snow fell. And the oldest ice would reveal Earth's climate history

Fig. 3.2

Ice core, NASA Goddard Space Flight Center.

Source: Climate NASA.gov, Public Domain

from its very beginning." The deeper the sample, the earlier the time record of temperature and CO_2 data. The actual oldest ice core records to date (800,000 years ago) were collected from East Antarctica.

During the Cold War between the United States and the Soviet Union Dr. Lorius heard that the Russians had an ambitious drilling program established at the Antarctic **Vostok Station**. After successfully making contact to participate Lorius joined their research project in 1974. He became instrumental in helping to set up an accurate weather monitoring station and ice core drilling operation. Today the Vostok ice core drilling is recognized as the most important record of the relationship between atmospheric CO_2 and temperature (**Fig. 3.3**). Vostok ice core data

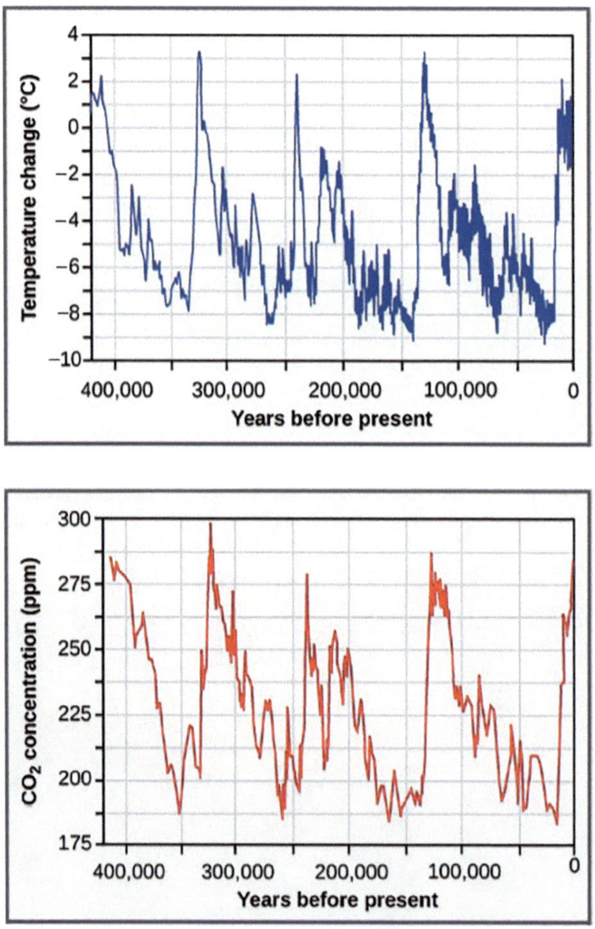

Fig. 3.3

Vostok station is more than 3000 meters deep and shows CO_2 historically cycles between 180 ppm and 300 ppm by volume.

Source: Modification of Vostok, Public Domain

demonstrate that cyclic rising and falling CO_2 levels are proportional to temperature fluctuations during the same period. Since 1960 atmospheric CO_2 data has also been collected by the Scripps Institute of Oceanography monitoring sites in Mauna Loa Observatory. In 2019 Scripps' data revealed atmospheric CO_2 peaking at 419 ppm, well above the average cyclic pattern (180 ppm to 300 ppm) that ice cores have established (**Figs. 3.4a, b**). The last time Earth has seen CO_2 levels this high was 4.5 Mya

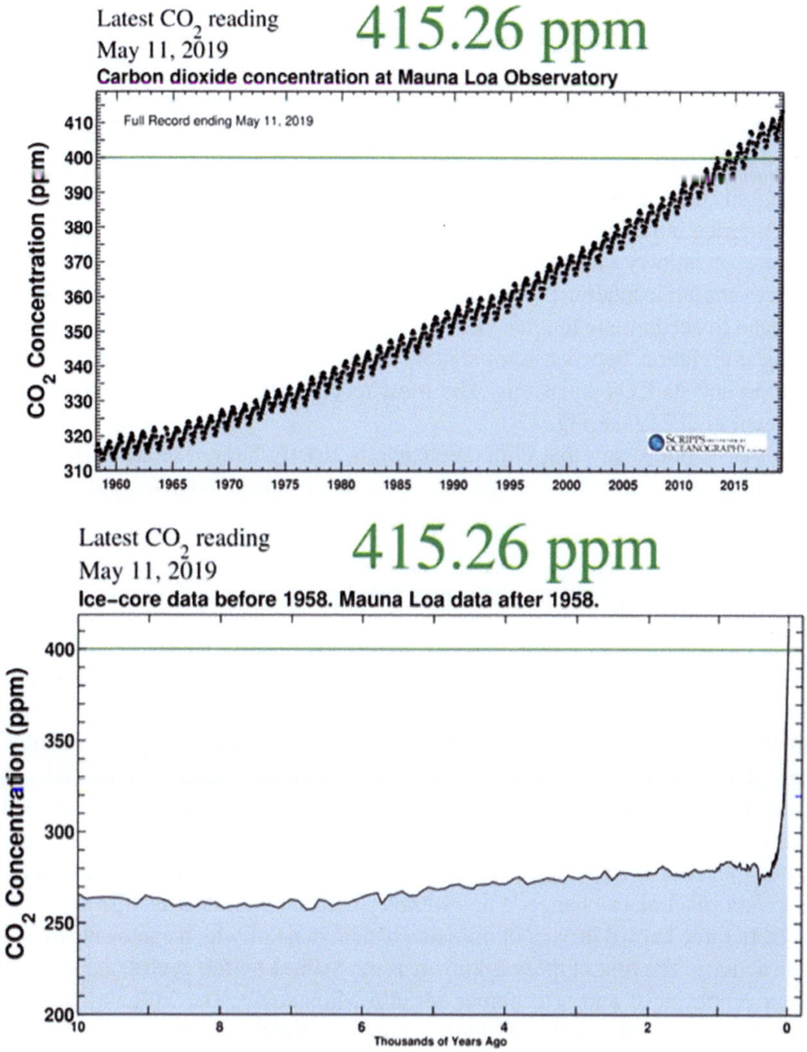

Figs. 3.4 a, b

Atmospheric CO_2 greater than 415 ppm captured at the Scripps Institution of Oceanography, Mauna Loa Observatory, Hawaii.

Source: Research News, NOAA, accessed Aug 2021, Public Domain

during the Pliocene Epoch. The geologic record reveals that the sea levels rose 80 ft during the Pliocene, a foreboding of potentially dire consequences from excessive atmospheric CO_2 levels. This is the Code Red for Humanity, which began with the industrial era and our reliance on fossil fuel consumption.

A 2000-Year Record of Climate Change Events

The **Industrial Era** of the 19th century continues in the 21st century as a period of landmark advances for human society. Industrialization ushered in societal advancements with an improved standard of living for many people throughout Europe, the Americas, and Asian countries. Progressive changes included innovative production to provide cheaper goods and labor markets with the means to acquire them. These new technologies were powered by machines using a new energy source, the burning of fossil fuels in the form of oil, coal, and gas. Along most rivers in Europe and North America industrial cities were centers of emerging communities surrounding the fumes of billowy smokestacks from factories incessantly pumping CO_2 emissions into the atmosphere. During the Industrial Era record levels of atmospheric CO_2 began to accumulate that has never slowed down. Climate monitoring indicates a strong correlation between temperature anomalies that align with the industrial revolution and its CO_2 emissions. The most recent data was released in the IPCC assessment of 2021 (see **Fig. 3.9**).

But that is not to say that climate change is strictly human related. There are a number of natural causes for climate anomalies that are revealed in the **paleoclimatic** record.

Current and Past Drivers of Global Climate Change

Without the handiness of a time machine to directly observe causes of climate fluctuations, scientists are often forced to rely upon data gathered from other sources to determine the major drivers, or factors, that may be responsible for climate change. Indirect *geological evidence* is collected from ice cores and terrestrial boreholes (a narrow shaft bored into the ground) and by measuring glacier lengths and ocean sediments. Indirect *biological evidence* is collected from the analysis of pollen remains, tree rings, and fossilized specimens. Altogether data presents a compelling story describing a strong correlation between the timing of temperature changes and drivers of climate change. For example, before the Industrial Era (pre-1780) there were three known drivers of climate change, none of which can be attributed to human activity. The first of these is known as the **Milankovitch cycles**, named for the Serbian geophysicist and astronomer Milutin Milankovic. The Milankovitch cycle describes climatic effects due to Earth's tilt and slightly changing elliptical orbit due to gravitational forces (mostly from Jupiter and Saturn). Milankovitch orbital cycles range in length from 19,000 years to 100,000 years and result in cyclic alterations in solar intensity and its effect on climatic patterns (**Fig. 3.5**). According to the

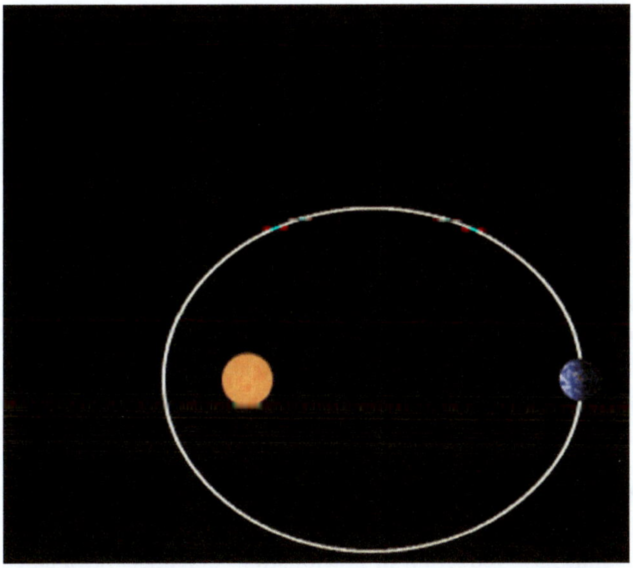

Fig. 3.5

Planets orbiting the sun follow elliptical orbits that rotate gradually over time.
Source: WillowW under CC by 3.0

Milankovitch theory, it is possible to predict changes in Earth's climate at a minimum of every 19,000 years.

A comparison of data sets collected from Vostok Station and the Milankovitch cycles juxtapositioned to major and minor extinction events gives some credence to this ideology.

The second natural driver of some of the aspects of climate change is the variation in the sun's intensity. **Solar intensity** or the sun's brightness is the amount of solar power or energy the sun emits, which is not a perpetually exact amount. As solar intensity increases (or decreases) Earth's temperature correspondingly increases (or decreases). Solar intensity changes are generally thought to explain climate anomalies.

The third natural factor driving climate change is **volcanism**. And while volcanic eruptions may last only a few days, the released emission of solids and gases can have a huge impact on the climate and sun-blocking ash over a much longer period of years. Atmospheric gases and solids are released by volcanic eruptions in massive quantities. Volcanic emissions can include CO_2, sulfur dioxide, hydrogen sulfide, and carbon monoxide. Generally, the effect of a volcanic eruption is cooling. This was the case in 1783, when a string of volcanoes in Iceland erupted and caused the release of large volumes of sulfuric oxide. During an 8-month period an estimated 14 km^3 (3.7 *quadrillion* gallons) of lava poured from 135 fissures and volcanic craters, covering an estimated 2500 km² (965 mi²) of land (**Fig. 3.6**). The Icelandic volcanic eruptions, known as Laki, led to **haze-effect cooling**. During the aftermath of the Laki eruptions sulfur aerosols reflected solar radiation back into space and throughout Europe and North America and haze-effect cooling persisted for several years. The result was some of the lowest average winter temperatures on record occurring between 1783 and 1784. It has even been suggested that the 1783 Laki eruption, subsequent cooling, and crop failures led to the socioeconomic conditions in Europe that triggered the French Revolution.

Volcanic eruptions have been the underlying cause of several earlier ice ages, including a near extinction of early humans in many regions of Earth 74,000 years ago. This climatic event was triggered by Mount Toba in Indonesia (Sumatra) and is considered the largest volcanic eruption in the last 28 million years. A **supervolcano** event is rare, as it requires an explosive magnitude of 8 M (1000 kg of volcanic material) (**Fig. 3.7**). The effects of the Mount Toba supervolcano spanned at least a decade, bringing a "long winter" across much of Earth. The tons of ash and volcanic gases that covered Indonesia and the Indian Ocean with 16 cm (6 in) of debris also blocked solar radiation with a likely global cooling range between 3.5°C and 9°C and an overall 25% reduction in precipitation. Scientists have since identified regions unaffected by the mass extinction events caused by the Toba supervolcano. This includes the South African coast; however, significant genetic biodiversity was lost worldwide. Volcanism continues to create havoc with climate patterns, bringing unseasonably cold weather. Hopefully, Earth won't experience another supervolcano in the next millennia.

There were two significant climate anomalies that occurred within the last 2000 years (preindustrial era). These are known as the **Medieval Climate Anomaly** (or

Fig. 3.6

Crater fissures along Laki.

Source: Mirra fotograf og kunst under CC BY-SA 2.0

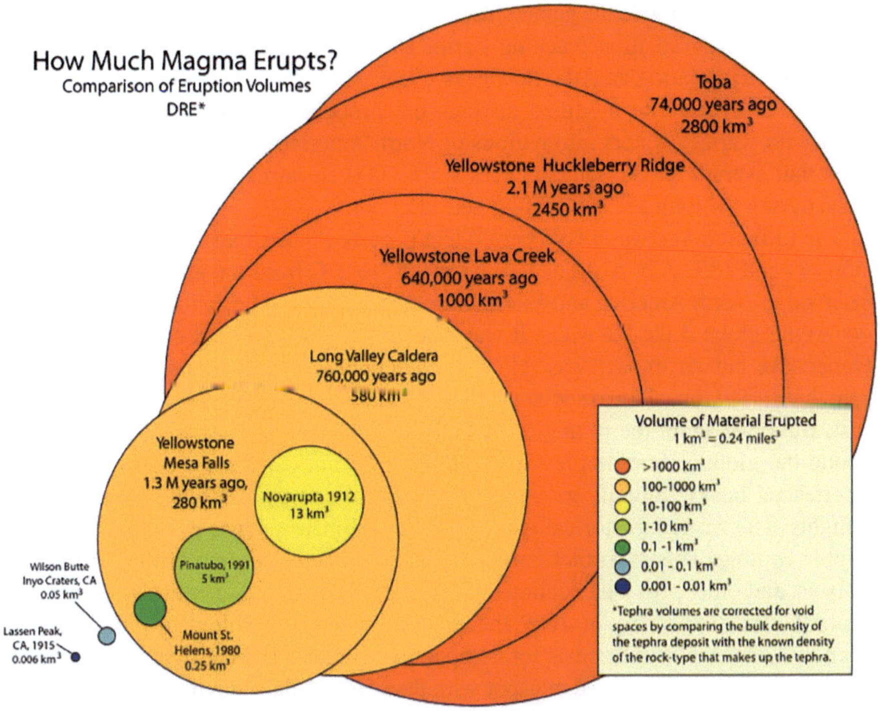

Fig. 3.7

Volcanic eruption by magma-ejected material. Mount Toba is registered as 8 M, and Mount St. Helens is registered as 5 M.

Source: Public Domain by US Geologic Survey

the Medieval Warm Period) and the subsequent **Little Ice Age**. Each altered climate trends for a significant length of time, 400 years and 300 years, respectively.

The Medieval Climate Anomaly (MCA) occurred across a period of 400 years, between 900 AD and 1300 AD, mostly affecting the climate of the North Atlantic. Fluctuations occurred in other regions of the world during the MCA, including a proposed cooling in the South Pacific. Today's climate scientists have estimated that the MCA period had slightly warmer weather conditions with higher-than-average temperatures between 0.10°C and 0.20°C above the norms of the time. It was never a global phenomenon. And while 0.10°C does not appear to be significant enough to create biogeographical changes, it was still enough to break up Arctic ice that allowed the Vikings to colonize Greenland. The slightly warmer climates in Western Europe and the Arctic lands were aligned with increases in exploration, food production, and prosperity. Yet there was a much darker side to the MCA. Warming temperatures led to a rise in infectious diseases and the black plague brought on trade ships, which was responsible for wiping out half of the known world's human population.

Using climate knowledge gained from Chapter 2, we can relate the cause of MCA to levels of solar radiation reaching Earth. During this period solar radiation had increased, precipitating an Atlantic Ocean cycle with unusual weather patterns that resemble La Niña. The modified air pressure brought stronger-than-normal trade winds with warmer waters into regions of North America. Even so, it's sobering to know that average global temperatures since 2000 are already higher than experienced during the long MCA warming period.

The Little Ice Age was a cold period that occurred more than 300 years, between 1550 AD and 1850 AD. A slight cooling of 1°C to 2°C from the global average was observed in North America and Europe during this time and possibly in other areas around the globe. Like the MCA it was a seemingly insignificant change in global temperature but its impact was felt in socioeconomic arenas. Historical accounts from this period reveal a time of exceptionally harsh winters with heavy snowfall and frost. But once again the timing and severity of climate changes were not uniform around the globe. For example, Eastern China did not experience cooling, Europe experienced heavy rainfall, and Africa and central and southern Asia experienced droughts. The written, historical record provides climate science evidence of a geographic cooling. This is further corroborated by modern ice core drilling, tree ring analysis, and other proxy paleoclimatic measures that climate scientists have at their disposal. And, beginning in 1659 there were direct meteorological measurement taken by European astronomists for comparative studies. The cause of the Little Ice Age is proposed as a reduced sunspot with decreased solar output, compounded by *increased* volcanism, which emitted sun-blocking volcanic ash into the atmosphere.

3.2 Tracking Climate Change

After reading this section students should be able to:

- Discuss current methods in monitoring climate.
- Cite major contributors of climate change in the 21st century.
- Discuss current methods in monitoring climate.
- Understand the significance and history of the biome.

Greenhouse gases are significant drivers of climate change today. When heat energy from the sun strikes Earth atmospheric gases absorb and emit radiant energy as heat in the atmosphere and are known as **greenhouse gases**. This name was coined because gases that trap heat are analogous to the phenomena of how panes of glass in a greenhouse prevent heat from escaping. The greenhouse gases that affect Earth include a mixture of varying concentrations of ozone and CO_2, methane, and nitrous oxide, which are largely released as combustion products from burning fossil fuels. Approximately half of the radiation emitted from the sun passes through these gases and arrives at Earth's surface. There the solar radiation is converted into thermal radiation, with a portion of that energy radiated back into the atmosphere. If there are higher-than-normal levels of greenhouse gases the reflected radiation now becomes trapped and directs its thermal energy back toward Earth's surface for a second time. It goes without saying that the more greenhouse gases present in the atmosphere the more thermal energy will be trapped and reflected back, heating up Earth's surface.

Evidence collected over the past century supports a tightly correlated relationship between atmospheric concentrations of CO_2 and a rise in net radiation, or global temperature. Between 1950 and 2006 the concentration of atmospheric CO_2 increased from 280 ppm to 382 ppm. In 2011 the atmospheric CO_2 concentration was 392 ppm and in 2019 it was 415 ppm. Note that this is significantly higher than the last recorded cyclic CO_2 high of approximately 300 ppm, which also occurred over a far shorter period of time.

Extensive atmospheric CO_2 data was used to graph a norm (**Figs. 3.3 and 3.4**) with a recurring cycle of approximately 50,000 years between the lows of 180 ppm to the highs of approximately 300 ppm, a cyclic pattern that has continued for the past 400,000 years. Today this pattern appears to be broken. The atmospheric CO_2 concentrations have breached their historical maximum of 300 ppm and the rate of increase has sped up, occurring now over hundreds of years rather than thousands of years. For climate scientists a new hypothesis must be formulated to describe this phenomenon. One key factor is prominent when comparing historical data to current data: the role of modern human society. This conclusion considers the extent and magnitude of climate change taking place without the presence of other known drivers capable of the same degree of impact.

Human activity has been responsible for the release of CO_2 and methane—two of the most important greenhouse gases—into the atmosphere in unnatural ways. The primary cause for abnormally high CO_2 gas released is the burning of naturally deposited fossil fuels as gasoline, coal, and natural gas (**Fig. 3.8**). Fossil fuels are nonrenewable as they are formed over thousands of years from accumulated organic biomass from past eras. Other human activities, like the burning of forests during land clearances and industrial manufacturing, also contribute to CO_2 emissions. Agriculture, animal herding, and decomposition in landfills and wastewater treatment facilities all contribute to methane gas emissions. Methane gas (CH_4) occurs chemically when carbon is hydrated and is naturally produced whenever microorganisms break down organic matter under anaerobic conditions, such as in swampy waters, underground, and even in the intestines of mammals. Because of the size of the human footprint, large-scale decomposition of our wastes leads to excessive methane emissions. The melting of **clathrates** (structures in which methane gas is trapped in cavities of crystallized ice) dispersed throughout the bottom of the ocean floor occurs with increasing ocean temperatures. Normally CH_4 release is appropriate for use by methanotroph organisms that require it as a carbon source. As excess methane gases are released from clathrates and contribute to greenhouse gases, they will in turn lead to further melting of the clathrates and more greenhouse gases. Climate change through melting clathrates is an example of a positive feedback loop, leading to an even more rapid rate of rising global temperatures.

The IPCC's August 2021 report argues for the urgent need to immediately alter the course of human activities that have caused climate change to ensure the future quality

Fig. 3.8

Smoke plume from a power plant chimney, Helsinki, Finland.

Source: Pöllö under CC BY 2.0

of life on Earth. In **Fig. 3.9** the data illustrates most of the observed warming is human based, with greenhouse gases and deforestation as the main drivers of global warming. According to IPCC, without immediate coordinated action across governments Earth will reach a tipping point on its ability to reverse the human life-threatening direction by the year 2030 (20 years sooner than last reported). Even after achieving net zero carbon emissions there would be an estimated 20 to 30-year interval before the worst consequences of global warming impacts would be fully mitigated.

Documented Results of Climate Change: Past and Present

Climate scientists collect data from many sources (past, present, and future projections) to help understand the impact of the changing climate. The consequences of

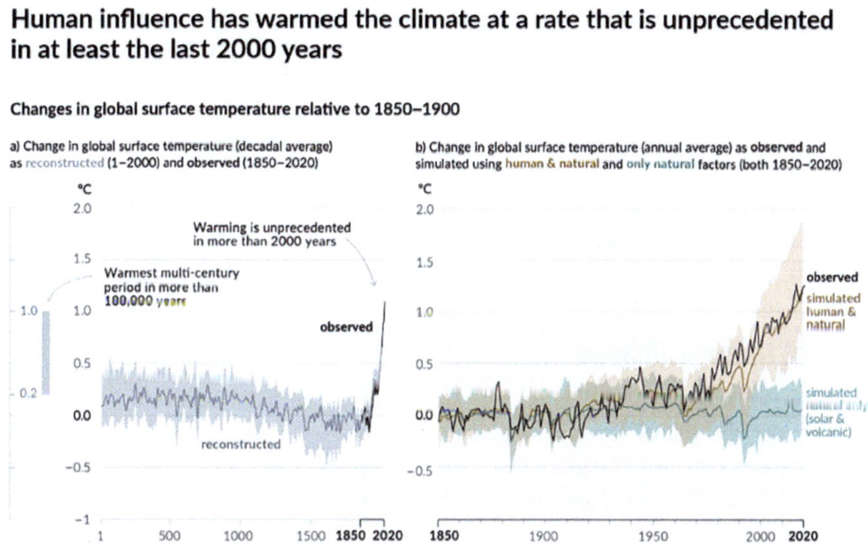

Fig. 3.9

IPPC 2021 Report, Climate Change 2021.

Source: IPCC 2021, Public Domain

long-ago climate events, namely extinction of species and significant rises in sea level, are evidenced in the geologic record. Meanwhile, modern-day phenomena like retreating glaciers and endemic species extinction at a rapid pace present another type of evidence. The current climate pattern is presenting a cause for the negative impacts to living organisms and habitats in the 21st century.

Geological Climate Change

Climate scientists have associated *global warming* with at least one planet-wide extinction event during a geological time period, though it has likely contributed to many mini-extinction events. The **End-Permian** extinction event happened about 251Mya, ending a roughly 50-million-year-long geological time period that was relatively warm and moist. The Permian Period was known as one of the third warmest periods in geologic history, abundant with tropical wet forests covering much of Earth. During the End-Permian mass extinction, which lasted several million years, Earth gradually became drier. Deserts began appearing across landscapes that were once tropical or subtropical. Scientists estimate that approximately 70% of all terrestrial plant and animal species and 90% of all marine species went extinct in the End-Permian mass extinction. Organisms that were adapted to 20°C to 30°C (68°F–86°F) moderate climates and plentiful annual rainfall of 300 to 400cm (118–157 in) were less likely to survive the Permian climate change. It is estimated that land temperatures rose 10°C to 30°C higher than today and seawater temperatures reached 30°C to 32°C (75°F–82°F). These hot tub conditions were unsuitable for most aquatic invertebrates, especially those at the bottom of the food chain. Scientists believe there were a number of factors leading to the magnitude of global warming during End-Permian, with an early trigger being a prolonged eruption of the Siberian **flood basalts** (extended low-intensity volcanism of basalt material). The result of the Siberian flood basalts was the continuous release of CO_2 and ash into the atmosphere while rising ocean temperatures released high levels of methane gases. The prolonged deluge of global CO_2 emissions led to a rise in atmospheric CO_2 levels estimated to have gone from 452 ppm to as high as 2507 ppm.

Present Climate Change

There are global events occurring presently that fall under the definition of climate change. For example, the glacial retreats occurring at the Glacier National Park in Montana. In 1850 the park contained 150 glaciers. By 2010 the glaciers had shrunk to such a degree that only 24 glaciers existed greater than 25 acres in size. This is a phenomenon known as **glacier recession**. The reduction in glacial ice has ominous consequences for local ecosystems through diminished summer melt, which leads to shortages of seasonal freshwater supplies. The glacier recession at Glacier National Park correlates with a mean annual temperature rise of 1.33°C over the past 100 years. Within the last half century the Grinnell Glacier of Mount Gould, in Glacier National Park, shrank 40% (between 1966 and 2005) (**Fig. 3.10**). Glacial recession has been observed for a long time in Greenland and the Antarctic with ice sheet shedding in the news regularly.

Throughout the ages mini-ice ages and warming periods have occurred and the ocean levels have risen and fallen in response. For example, during one ice age an exposed Beringia Land Bridge permitted large mammals like the Wholly Mammoth and early humans to cross between Northern Asia and the territory now known as Alaska. As climates warmed and ice was released from the Arctic regions, the Beringia Land Bridge became submerged, cutting off the passage. Today the sea is rising at a rate of 3.6 mm per year, with a total rise of 20.3 to 22.9 cm (8–9 in) since 1880 (Climate.gov).

Fig. 3.10

Effect of global warming on Grinnell Glacier retreat.

Source: Modification of USGS by Lumen Biology under CC BY 4.0

An unprecedented loss of biodiversity has marked the beginning of the Sixth Extinction and the **Anthropocene Epoch**, a topic we will discuss in more depth in Chapters 14 and 15. There are many causes of biodiversity loss in the oceans and on land, with climate change being one of the major factors. Currently one million species are at threat for extinction as they are forced to move out of habitats in which they have evolved to coexist. The rapid rise of temperature resulting from human-caused CO_2 emissions has an effect on biogeography by not allowing sufficient time to develop new adaptations in new locations.

A Review of How Changing Climate Affects Ecosystems

Karl August Möbius (1825–1908) was a German scientist who pioneered the field of **community ecology**. Möbius studied oyster culture and was the first to detail how a marine animal community maintains its state of equilibrium due to the limitations of resources found within its oyster bank ecosystem. The ecological terms "ecosystem" and "biome" are sometimes used interchangeably, and indeed their meanings overlap significantly. But keep in mind that biomes are extensive, finite, and often include many different ecosystems. The ecosystem is focused more on species interacting with other species in its surrounding abiotic factors. The biome concerns distinct biogeographical parameters like vegetation in response to shared physical geographical and climatic characteristics.

It wasn't until 1916 when **Frederic Edward Clements**, a 20th-century American plant ecologist first suggested **biome** as a synonym for "biotic community of Möbius" (**Fig. 3.11**). Though Clements used biome with respect to plant communities the term was soon broadened to the animal element. In 1935 the term biome was expanded even further to the climatic and soil elements. In 1964–1974 the IBP popularized the concept of the biome. The IBP effort initiated five biome projects, the largest being the Grassland Biome Project and the Eastern Deciduous Forest Biome Project, marking some of the earliest national funding for ecosystem ecology.

Biome types are matched to the survival of plants and animals distributed within them and are susceptible to climate change. When human activities are responsible for altering the ecological balance of any ecosystem, for example its biodiversity, this is an **Anthropogenic effect**. Conservationists are concerned with restoring that balance by protecting the health of ecosystems and mitigating the effects of climate change. The goal is to restore the populations of endangered plants and animals.

3.3 Living in a Bubble

Since 2007 the University of Arizona has housed the B2 research facility, initially constructed as a simulated biosphere with seven ecosystems in a closed system. The B2 was the brainchild of philanthropist Ed Bass and systems ecologist John P. Allen in 1984, and the ambitious project was constructed between 1987 and 1991 (**Figs. 3.12 and 3.13**). Currently the University of Arizona uses the 3.14-acre facilities for educational programs and smaller-scale ecological experiments.

Karl August Möbius, German zoologist was a pioneer in ecology.

Frederic E. Clements was a pioneer in ecological succession.

Fig. 3.11

Pioneers in community ecology and ecological succession.

Source: Portraits in Public Domain

Fig. 3.12

Biosphere 2 (B2) containment of different ecosystems sealed in a steel and glass structure.

Source: LukeGordon1 under CC BY 2.0

The original vision for B2 was to further our understanding of natural and man-made environments through experimentation under the guidance of interdisciplinary scientists and to address current or future challenges that affect the quality of life. B2's immense enclosed facility provided a bridge between what was currently

Fig. 3.13

B2 glass-encased artificial tropical ecosystem in the Arizona Desert.

Source: DrStarbuck under CC BY 2.0

possible through small-scale, controlled, laboratory-based simulations of Earth processes and field studies in the natural world. The goal was to overcome the limitations of working in the field, where the scientist cannot control the infinite environmental conditions. To this end they encapsulated a facsimile of a real-world, self-sustained colony in which conditions could be manipulated and analyzed and performed two missions with mixed results.

In its 2-year maiden mission crops were highly productive (83% self-sustaining) yet participants ("biospherians") reported continual hunger. There was unanticipated condensation that made the "desert" too wet, greenhouse ant and cockroach population explosions, and rainforest overgrowth that blocked 40% to 50% of the sunlight entering. The persistent engineering issues included difficulty manipulating bodies of water to create the planned waves and tidal changes and maintenance of heating and cooling systems. The biospherians reported symptoms of sleep apnea and fatigue. Tense relations occurred among the participants by month 16 as oxygen levels had steadily dropped from 20.9% to 14.5% (equivalent oxygen at elevation 13,390 ft). Between the daytime photosynthesis (generates oxygen) and nighttime high metabolic consumption (generates carbon dioxide), atmospheric carbon dioxide levels were fluctuating on average 600 ppm daily. During a short winter day, atmospheric CO_2 ranged from 2100 ppm to 3700 ppm. And during a longer summer day, CO_2 concentrations ranged from 800 ppm to 1700 ppm. The B2 project implemented a CO_2 scrubber as a means of containing potential for rising CO_2 levels. However, absorption of oxygen by the concrete architectural structures and soil hadn't been accounted for.

On B2's second, shorter-duration mission the crew reached 100% food production efficiency but continued experiencing unprecedented challenges navigating a long-term closed ecological system. Project management friction also developed, and the over-budget mission closed before the 10-month duration ended. Though neither B2 mission was deemed a success, each led to various breakthrough developments in unexpected areas, with deeper understanding of the opportunities and difficulties to maintain closed ecosystems. For example, its coral reef ecosystem (the largest artificial reef ever built) helped to develop some of the methods being applied to restoration of endangered coral reef systems. And B2 surprisingly maintained its biodiversity despite significant species losses and predation from stowaway ant species. The wetland wastewater treatment facility also successfully delivered recycled water for farm irrigation system.

B2 is currently utilized for smaller-scale studies, no longer operating as a closed system. For example, research is ongoing to evaluate water dynamics in complex landscapes that are deemed crucial to developing sustainable societal practices.

According to Nate Allen, B2 staff scientist and sustainability coordinator, "Water is the hammer that climate change will hit the Earth with." In the face of climate change it's critical to study new ways to allocate water, restoration of watersheds, enhanced recycling of water, and to develop new technologies. B2 runs a testing ground to help understand the complicated relationships of water and energy usage in Arizona, which receives its water supply from the Colorado River System. According to the USGS, Lake Mead on the lower Colorado River in recent years has witnessed severe drops in water levels. It is the largest man-made reservoir, and it supplies water for seven states of the southwest United States and Mexico, including Arizona. The impact of decades of drought, decreasing snowpack, and rising water demand has led to discussions of mandatory water cutbacks. The Colorado River system, which includes Lake Mead and Lake Powell reservoirs, serves 40 million people and 6300 square miles of farmland. There are also international treaties, interstate agreements, and court rulings to navigate the delicate balance of water demand surpassing the needs of communities involved.

The takeaways from the B2 experiments are highly relevant as we attempt to restore ecological balance in a changing climate. There appears to be a resurgence of interest in the value of B2 in that regard as it has been greenlit to mimic another plant-filled habitat resembling a new colony constructed on an alien planet. This time the focus will be to monitor effects of rising CO_2 concentrations and temperature. According to its current director, Joaquin Ruis, "By changing the concentrations of gases in the atmosphere to those that we think are going to exist in the future we will see how the planet could fare." It will be interesting to see whether the next B2 mission will help unravel the threads of climate change effect to future conditions of life on Earth.

Albert Einstein was known for his genius as a leading physicist but many people do not know that he was very ecologically minded. Excerpt from "Einstein and the Poet: In Search of the Cosmic Man" (1983) by William Hermanns:

> *The basic laws of the universe are simple, but because our senses are limited, we can't grasp them. There is a pattern in creation. If we look at this tree outside whose roots search beneath the pavement for water, or a flower which sends its sweet smell to the pollinating bees, or even our own selves and the inner forces that drive us to act, we can see that we all dance to a mysterious tune, and the piper who plays this melody from an inscrutable distance—whatever name we give him—Creative Force, or God—escapes all book knowledge.*

Glossary

Flood basalts **clathrates** **solar intensity**

Citations

1. IPCC: Updated IPCC Sixth Assessment Reports, 2023. https://www.ipcc.ch/2023/03/20/press-release-ar6-synthesis-report/
2. A. Bojanowski, Iceland's impact, Nat Geosci 4 (2011) 732.
3. S. Osipov K., et al., The Toba supervolcano eruption caused severe tropical stratospheric ozone depletion. Commun Earth Environ 2 (2021) 71. https://doi.org/10.1038/s43247-021-00141-7
4. J. Daley, Ancient humans weathered the Toba supervolcano just fine, Smithsonian Magazine (2018). https://www.smithsonianmag.com/smart-news/ancient-humans-weathered-toba-supervolcano-just-fine-180968479/
5. Lasher, G. E., & Axford, Y. Medieval warmth confirmed at the Norse Eastern Settlement in Greenland, Geology 47 (3) (2019) 267–270. https://doi.org/10.1130/G45833.1.
6. Luning, S., et al. (2019). The medieval climate anomaly in the mediterranean region. Paleoceanogr Paleoclimatology 34 (10) (2019) 1569–1670. https://doi.org/10.1029/2019PA003734.

7. Osipov, S., Stenchikov, G., Tsigaridis, K., LeGrande, A. N., & Bauer, S. E. The role of the SO_2 radiative effect in sustaining the volcanic winter and soothing the Toba impact on climate. J Geophys Res Atmos (2020) 125. https://doi.org/10.1029/2019JD031726.
8. L. Elkins-Tanton. From Russia with magma: what a volcanic explosion 250 million years ago tells us about climate change. Future Tense (2015). https://slate.com/technology/2015/07/siberian-flood-basalts-the-earths-largest-extinction-and-climate-change.html.
9. Y. Wu, et al., Six-fold increase of atmospheric pCO_2 during the Permian-Triassic mass extinction. Nature Comm, 12 (2137) (2021). https://www.nature.com/articles/s41467-021-22298-7.
10. K. Cornelius, Biosphere 2: the once infamous live-in terrarium is transforming climate research, Sci Am (2021). https://www.scientificamerican.com/article/biosphere-2-the-once-infamous-live-in-terrarium-is-transforming-climate-research/.
11. M. Nelson. Pushing our Limits: Insights from Biosphere 2, University of Arizona Press, Tucson, AZ, 2018.

CHAPTER

The Terrestrial Environment

4

Chapter Outline

4.1 Physical Constraints for Land Dwellers

By the end of the section, students should be able to:

- Identify the major adaptations required for life on land.
- Discuss the various challenges faced by land plants vs land animals.

Adaptations are physiological, morphological, or behavioral features of organisms that enhance their fitness to particular environments (**Fig. 4.1**). The development of adaptations occurs through a gradual evolutionary process. To maintain a water balance, animals and plants evolved an outer covering of specialized cells coated with watertight molecules: keratinized proteins in mammals and a lipid-based waxy

Ecology for the 21st Century. DOI: https://doi.org/10.1016/B978-0-443-24031-7.00004-0

Fig. 4.1

Leaves are complex, energy-harvesting machines.

Source: Jon Sullivan, PdPhoto under Public Domain

cuticle layer in plants. Physiological processes require a means of transport of atmospheric oxygen and carbon dioxide. The adaptations optimize controlled intake using regulatory processes while preventing significant loss of water molecules by evaporation or **transpiration** (**Fig. 4.2**). Early land plants like ferns, mosses, and liverworts lacked elaborate structures, which limited their lifestyles. And the earliest land animal, the Ichtyostega, existed hundreds of millions of years ago by slithering onto land using its two front limbs as it slowly adapted to an environment that cycled through wet and dry periods. Eventually more sophisticated processes evolved, expanding the range of climatic conditions and diverse lifestyles.

Mobile animals sometimes travel vast distances to seek out sources of water for their hydration needs and possess a wide array of adaptations for water. Plants, on the other hand, are stationary and rely on different adaptive strategies for internal water balance, especially in the driest biomes. Water uptake occurs in plants through their roots from whatever is available in the surrounding soil. Water travels throughout the plant through its vascular tissue, which functions analogous to circulatory transport in animals.

Upright mobility requires another major adaptation for transition to a lifestyle on land. This is because of the impact of gravitational forces. In water animals remain buoyant and their mobility through water is relatively easy, mostly without need of

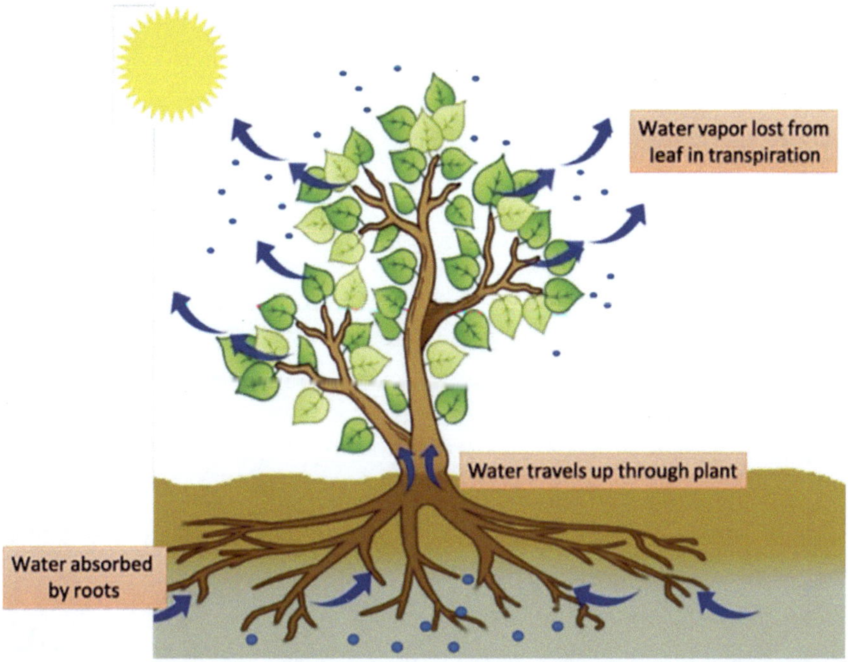

Fig. 4.2

Transpiration process in plants.

Source: Image courtesy of author.

walking across a surface. On land specialized structural supports like the skeleto-muscular system of animals or the woody tissue of plants are adaptations for upright growth. Competition for access to sunlight is a driving force in plant height. For example, the old growth forests of Northern California overcame gravitational forces to reach heights of 100m (~300ft). Mosses and liverworts lack evolved structures and remain land-hugging plants confined to moister, low light environments. Aquatic plants that grow in deeper waters are able to meet their daily light requirements while remaining rooted in the murky ocean floor. These giant sea kelps stretch upright to heights of 30m (~100ft) or more through a property of water buoyancy using tiny gas-filled bladders (**Fig. 4.3**). Gas bladders are an adaptation to deep water but would be of no use on land. The giant sea kelp would collapse immediately without the buoyancy of water to keep the bladders afloat. These giant forests of the sea never transitioned to land, which would require a gravity-defying adaptation.

Land plants are rooted and must rely on nutrients leaching from soil and a flexible growth habit to achieve sufficient exposure to solar radiation. Access to nutrients and solar radiation is less restrictive for most animals due to their mobility and diverse array of adaptive strategies to survive in all types of environments. It is the environment that ultimately drives the evolution of appropriate adaptations to meet a particular **constraint** imposed on the organism. These processes occur over many

Fig. 4.3

Channel Islands, Ventura, CA, USA kelp forest.

Source: NPS, Outreach & Education under Public Domain

generations. But when disruptions to the environment occur too quickly species cannot adapt quickly enough and may need to relocate to avoid extinction.

In Chapter 2 we identified the key role of climate, with its variations in temperature, moisture, and light availability on ecosystems. Climate and geographic locations place **constraints**, or **stabilizing parameters**, on how plants and animals evolve through adaptations to thrive. Species must cope with fluctuating temperatures and seasonality. Some animals like Arctic foxes will adapt to cold winters by growing a thick, white coat that insulates and camouflages them against the snowy background. This is an example of a physiological adaptation. Seasonal winter adaptations include behavioral adaptations, like mammals and birds seeking out temporary enclosures, dark burrows, and caves where they've cached a food reserve and huddled for warmth.

Some species tackle surviving a long winter by **hibernation**, a cold-weather strategy used by many animal groups, including vertebrates and invertebrates. Hibernation is a voluntary action designed to conserve energy in the dead of winter when food is scarce and the days are shorter. Before going into a deep sleep the animal will store food reserves or increase its body fat under hormonal changes. Then it goes into a state of near-suspended animation in a transformation that includes lowered body temperature as well as decreased breathing and heart and metabolic rates. For example, the ground squirrel's normal heart rate of 200 to 400 beats per minute

Fig. 4.4a

Virginia big-eared bats in hibernation.

Source: USFW, NE Region under CC PDM 1.0

drops to a rate of 7 to 10 beats per minute. Hibernation is an amazing physiological feat that developed as a significant adaptative strategy for seasonally cold climates. Hibernation typically lasts for months.

Torpor is a shorter-term survival strategy to combat the challenges of colder climates and food shortages and lasts for several hours. Many animals are capable of going into torpor, including mammals, amphibians, some marsupials, and birds. Torpor could occur in response to daily temperature variations, particularly in locations where there is a severe drop in temperature after the sun goes down. Like hibernation, torpor involves the lowering of body temperature, breathing rates, heart rates, and metabolic rates. Unlike hibernation, torpor is an involuntary action, in a similar fashion to muscle shivering in cool weather. In the state of torpor the animal becomes lethargic as its physiological condition adjusts to the surrounding conditions for a transient period. Also, unlike hibernation, animals are still foraging for food when going in and out of torpor. Some species, like bears, are known as "light hibernators" that fall into a state of torpor midwinter. Bats have acquired the ability to go between torpor and hibernation through the winter, depending on temperature.

The endangered Virginia big-eared bat spends its winter in a state of hibernation, with a body temperature close to freezing. It will cycle through intervals of temporary torpor to save energy on a particularly cool day with interrupted brief periods of arousal to return their bodies to normal temperatures (**Figs. 4.4a, b**). Bats hibernate in **hibernacula**, caves, mines, and other structures with ideal temperature and humidity for hibernation.

Fig. 4.4b

Torpid Virginia big-eared bats (*Corynorhinus townsendii virginianus*).
Source: Larisa Bishop-Boros under CC BY-SA 3.0 US

Bats Under Threat

Unfortunately many bat populations like the Virginia big-eared bats have become susceptible to the debilitating, infectious white nose fungal disease (WNF) during hibernation. WNF is a metabolic and respiratory disease that is currently decimating the cave-dwelling social bat populations. The US Wildlife Services approximates the loss of 5.5 million bats since the nonintentional introduction of the cold-loving fungus to the northeastern United States. WNF is threatening many bat species throughout North America with broader ecological implications. An ecologist can almost always think of a reason why the loss of one single species, no matter how lowly or pesty, disrupts the delicate balance of an ecosystem's food web. But in the case of the bat there is a long list of valuable services that would be lost. Healthy bat populations save farmers an estimated 3 billion dollars per year in pest control services and many bat species provide **ecological services** in pollination and seed dispersal. In

2008 the United States Forest Service estimated bat die-off from WNF equating to 2.4 million pounds of insects escaping predation, which led to crop damage in states hit the hardest by WNF.

Because of the importance of bat species to the world economy, conservationists are working tirelessly to protect bat caves. This includes employing novel techniques like irradiation with UV light to kill the invasive fungus and efforts to bring public awareness to hikers who inadvertently enter and contaminate the caves. There is reassuring evidence to suggest some bat species will survive this environmental challenge and develop resistance to the fungus. Other bat species may evolve to thrive in warmer climates and migrate for the winter season. In any event changes occurring in the bat species will have ecological consequences. For example, plants that rely on bat pollinators may be placed under a new selective pressure.

While the climate dictates the biogeographic placement of plants and animals across environments, there are some basic, universal adaptations that all terrestrial plants and animals must overcome. These are maintenance of a water balance, structures to bear gravitational forces, and access to light and nutrients. Developing adaptations or movement potentially creates new ecological dynamics or challenges to other species.

4.2 Soil Composition, Formation, and Physical Properties

By the end of this section students should be able to:

- Describe the steps of soil formation from starting material.
- Identify the three major components in soil.
- Relate soil composition to biogeography.
- Describe a soil profile using five horizon layers.

Plants obtain carbon directly from the atmosphere and inorganic elements from the soil by their roots. Soil is defined as the outermost loose layer, or shell, covering Earth's rocky crust, varying in depth from 9 km to 20 km thick. It is the abiotic foundation upon which all terrestrial life depends but can be teeming with organic life supporting a diverse microbial soil ecosystem. Soil is also the principal factor controlling the fate of water in land environments and productivity of plants. The soil microbial communities participate in Earth's nutrient recycling system, breaking down waste products of plants and animals. This is essential to transform organic molecules into their basic elements for re-assimilation by plant roots.

An area of soil is characterized by its three-dimensional length, width, and depth. But the degree of granulation, rock, plants, and microbial life shows great variation, making it difficult to precisely define soil. Just above unweathered bedrock, there exists a layer made up of a heterogenous unconsolidated mixture of dust and particulates (**Fig. 4.5**). Combined with the soil layers above, the region above solid bedrock is known as the **regolith** and represents an interface between air, water, and living organisms.

Soil quality is an important determinant of biogeographical distribution and growth of plants. Soil quality depends on its chemical composition, **topography**

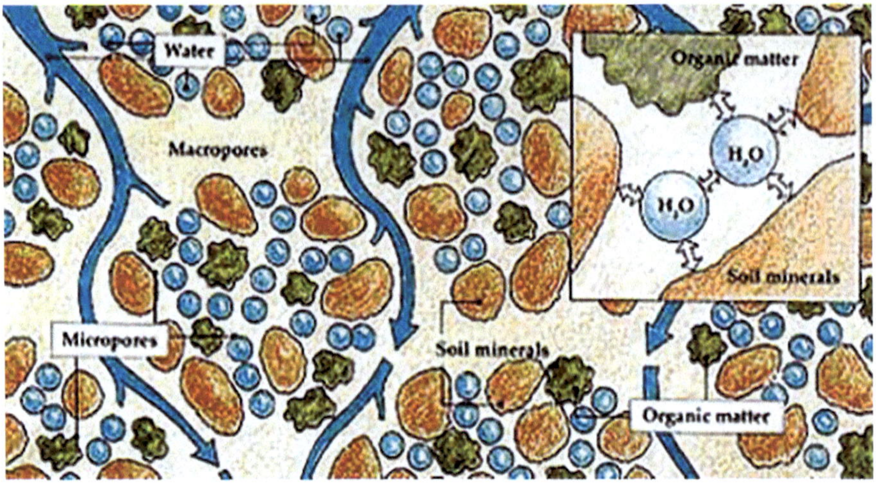

Fig. 4.5

Soil components include both biotic and abiotic factors.

Source: Adapted Brady & Weil 2002

(regional surface features), and presence of biotic factors, which aid nutrient cycling and soil aeration. Agricultural practices consider the history of the soil, including any local cultivating practices and previous plantings, when modifying soil characteristics and fertility. This is because productive soils can be ruined by improper soil management. The use of heavy mechanical equipment, heavy foot traffic, and extensive grazing all cause **soil compaction**. Compacted soil is less productive with stunted roots and aboveground growth.

Soil forms very slowly from natural and environmental forces that act upon mineral and organic compounds in the regolith. The two basic soil types are: **organic soils** formed from sedimentation of organic matter and **mineral soils**, or inorganic soils, formed from mechanical weathering processes that act upon rocks. They are considered nonrenewable because of how long they take to form—hundreds of millions of years. In terrestrial ecosystems if the soil is covered by water for at least part of the year it will be mineral soil. Together organic and inorganic content should constitute 50% of the soil volume. In good healthy soil water and air make up the other half. Each component of soil helps to support plant life and microbial communities and has its own distinct chemical composition.

Soil Composition

The major components of soil are:

- organic matter, 5%
- inorganic mineral matter, 40% to 45%
- water and air, 50%

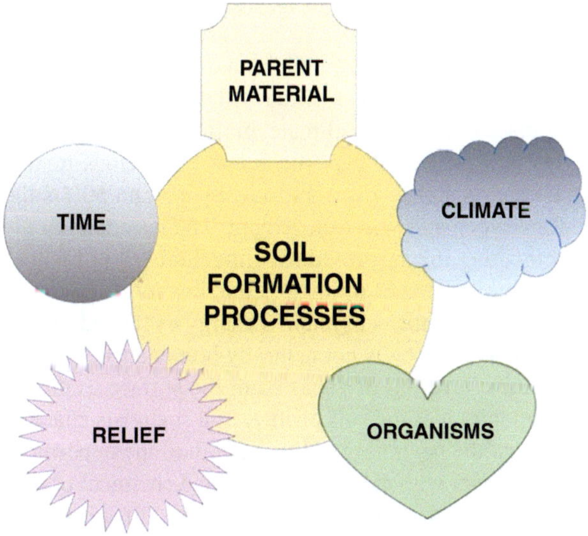

Fig. 4.6

Five primary factors account for soil formation.

Source: USDA, Public Domain

Soil is a natural resource made up of four basic components: air, water, minerals, and organic matter. The composition and presence of each component depend upon how much vegetation is growing, how compact the soil is, and whether there is any moisture in the soil (**Fig. 4.6**).

The organic material found in soil is **humus**, rich in biomatter, populated with millions of microorganisms (dead and alive), and the decomposing carcasses of dead insects, animals, and plants. Humus improves the architecture of soil and provides plants with water and mineral nutrients from the detritus. The inorganic material comes from rock that was slowly broken down into smaller and smaller particles. Soil particles between 0.1 mm and 2 mm are categorized as **sand**. Soil particles between 0.002 mm and 0.1 mm are categorized as **silt**, and particles less than 0.002 mm in diameter are categorized as **clay**. The best soil composition for plant productivity is a medium of mixed particle size, called **loam**, which contains an optimal balance of sand, silt, humus, and aeration. Intensive farming has an effect on soil quality by damaging or removing its O Horizon and altering the soil's structure.

The moisture-holding capacity of soil greatly influences the geographic distribution of land plants and animals. Soil pore and particle size together determine its ability to hold onto water molecules. When soil contains more water than it can hold in its pores it reaches **soil saturation** and excess water will drain off. Sand contains 30% to 40% of its volume as pore space while clay and loam range from 40% to 60% of their volume as pore space. This means both clay and loam are able to retain water to a higher degree than sand, which drains water the quickest.

The Process of Soil Formation

Soil formation is a consequence of a combination of biological, physical, and chemical processes (**Fig. 4.7**). Its exact composition will depend on the parent material and interactions taking place over time by climate, biological organisms, and topography. The process begins with weathering of rocks and the various mixture of minerals contained in the rocks. As soil forms it will ideally contain 50% solid material and 50% pore space for oxygen and water movement. The introduction of organic material serves as a cementing agent, providing a medium for nutrients to reach plant roots, the storing of moisture, and ultimately making soil tillable for farming. The organic component is part of the ecosystem's energy cycle and provides fuel to soil microorganisms. Soil distribution is not naturally homogeneous.

Soil has a distinctive layering of zones from the surface to the substratum. The vertical sectioning of soil gives the **soil profile**. Soil scientists characterize soil horizons, which range from the beginning of the regolith to the uppermost surface layer. Each **horizon** is a soil layer with distinct physical and chemical properties that differ from the other layers (**Fig. 4.7**). Letters designate the different soil horizons. The O horizon consists of organic material in various stages of decay. The A horizon consists primarily of mineral particles but includes some organic material moved by percolating water and soil-dwelling animals from the O horizon. The B horizon is characterized by less organic material than the upper horizons and by its accumulation of inorganic material such as clays. The C horizon is the least weathered horizon and is located just above the underlying bedrock.

Staring materials. All of the organic and inorganic material in which organic soils will be formed is called the **parent material**. Mineral soils form directly out of weathering bedrock. Soils can form in materials originating elsewhere, including sands and glacial drift. Sediments in rivers result in variable soil characteristics. For example, a fast-moving river has larger sediments of rocks and sand; a slow-moving river is more likely filtered with fine-textured clay.

Topography includes regional surface features that influence the characteristics and fertility of soil. Topography affects where water runoff occurs. Severe water runoff is known to strip away parent material, lessening the quality of the soil. Soils on steep slopes are more prone to erosion from water runoff and are likely thinner than level soils.

Biological factors include any living organisms that have the potential to affect soil formation and structure. Animals and microorganisms can bore or channel through soil, which results in pores and crevices. The roots of plants penetrate into crevices, which can fragment the rock and speed up soil formation. The decomposing matter includes leaves and other material that fall and the dead animals and plants in the process of decomposition.

Time is factored into soil formation because soils develop over long periods in a dynamic process. Materials are deposited, decompose, and transformed into usable material for living organisms or deposited onto the surface of the soil.

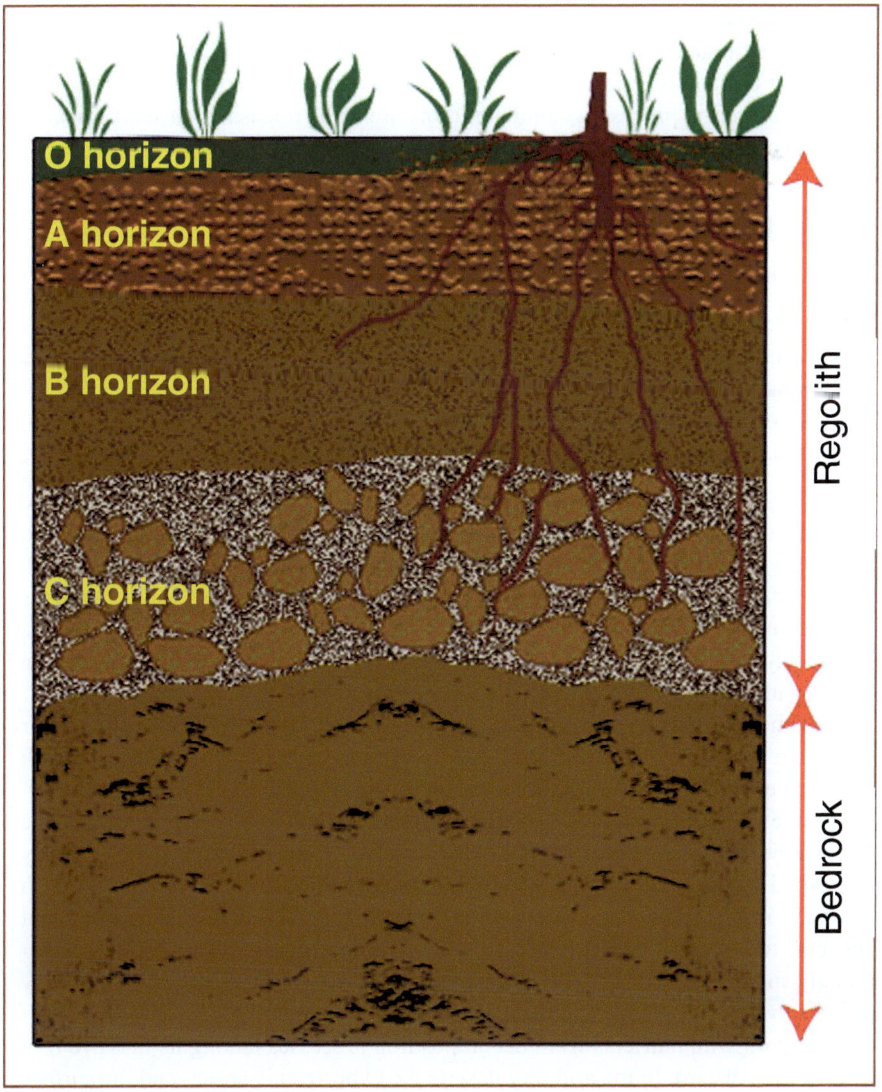

Fig. 4.7

Soil horizons. Soil scientists often identify the regolith as Horizon C only.

Source: vtorov under CC BY-SA 4.0

Climate includes abiotic forces that cause different patterns and rates of ***mechanical weathering***, the breaking down of rock into smaller particles. Water, wind, and temperature inflect freeze–thaw cycles as wind-born particles wear away at the rock surface, cracks, and crevices.

Chemical weathering results from the combined presence of water, oxygen, acids from soil microorganisms, and dead organic matter (DOM) that all interact with rainfall filtering through a series of chemical reactions that transform the particles into soil. The weathering process accumulates moisture while releasing nutrients, making it more favorable to biological activity in the newly forming soils.

Physical Properties of the Soil

Soils are named and classified according to their soil profile horizons in four distinct zones:

- horizon, loamy organic soil
- A horizon, mineral soil
- B horizon, subsoil
- C horizon, soil base (substratum)

The **O horizon** (organic) is generally a very thin layer of **topsoil** (average 10–15 cm) that indicates the soil is good quality. It has a high degree (greater than 20%–30%) of freshly decomposing organic matter (leaves and other plant material), called **humus**, which is broken down by soil microorganisms. The role of humus is to provide nutrients and moisture retention for the roots of plants and soil microbes. This top layer of soil is typically dark-colored (reds and blacks) and easily broken through with a trowel.

The O horizon depth varies considerably within an ecosystem. For instance, river deltas have deep layers of topsoil but less O horizon depth going up their riverbank where soil erosion takes place, and there is less sediment from decomposing matter. This is one of the reasons why there are different plants and species found in the delta riverbed then on its nearby banks. In the forests the O horizon is continuously replenished by partially decomposed leaves, moss, or lichen, making a topsoil rich in organic material. Forests are excellent habitats for microbial communities that contribute to nutrient cycling. However, the tropical rainforest soil O horizon is surprisingly thin at <2 cm. This occurs because of torrential rainfalls washing many nutrients too quickly to form soil. In contrast the temperate grasslands O horizon is deep and dark with nutrients from decomposition processes among its massive root system (up to 30 cm). In the arid, dry deserts the O horizon is absent and the surface soil is the A Horizon.

The **A horizon** is the true soil (average 8–20 cm) and is a mixture of different soil types, including some organic content from the topsoil above and the weathered products of inorganic particulates below. The organic content is carried into the A horizon primarily by rainwater percolating through it and any disturbances or plowing of the topsoil. The boundary of the A horizon in a soil profile is identified from the lower horizons as being darker-colored because of the presence of some organic matter. In most cases the A horizon contains a greater proportion of inorganic content, and this layer will be more packed and harder to break through than the topsoil above. In drier, arid climates the A horizon is known as **topsoil**. However, soil classifications just as often combine O and A horizons together as topsoil (approximately 13–25 cm depth).

The **B horizon** is the subsoil (average 20–30 cm), somewhat brittle and much more difficult to break through. It is found approximately 61 to 91 cm below the surface depending of course on its location, moisture, and organic content above. If the B horizon has good drainage it will be yellowish or brown. The mineral content includes irons, clay, and carbonates. There will still be a small fraction of organic material that percolated with finer materials through rainwater from the horizons, resulting in a dense layer. Distinct bands of different minerals may show movement (illuviation) through the horizon that can be geologically identified by their specific density, color, and other methods.

The **C horizon** is the soil base of the regolith, or substratum, with mostly unweathered parent material. This horizon shows little evidence of soil formation but there may be a small fraction of organic content mixed with some weathered inorganics. This layer contains pockets of silica, carbonates, gypsum, and soluble salt. While there is little coloration, geologic layering or strata is often present. Beneath the C horizon lies the completely unweathered bedrock.

Soil classification is highly technical and the definition of what constitutes the regolith is still a matter of debate. Soil scientists' opinions vary between restricting the regolith to just the C horizon and including all soil horizons (O, A, B, and C). Furthermore, not all soils are created equally, with some having additional layers or lacking one of the main four basic layers. For example, the **E Horizon** (eluviated) is an additional layer of sand and silt identified for its water-leaching capacity. If present the E horizon lies beneath the A horizon. The thickness of each layer varies, dependent upon the presence of factors influencing soil formation (starting material, topography, biological matter, climate, and time).

Soil Management

Soil scientists often seek to understand a soil's history and current health status to develop viable soil management strategies for future practices that improve crop production. The soil scientist's job is to study the biological and chemical properties of soil, its distribution, genesis, and fertility. They may be hired by federal or state agencies, academia, or the private sector seeking to increase their crop yields. Their jobs may involve inspecting soils, conducting soil surveys, and carrying out research in the field, laboratory, or office. These activities are needed to accurately interpret data to recommend sound soil management strategies. In **Fig. 4.8** a soil scientist is analyzing the soil horizons and soil quality after digging 1 to 3 m from the surface.

The training of a soil scientist according to the United States Department of Agriculture (USDA) should include "good observation and mathematical skills to analyze and determine the characteristics of different types of soils. Use of geographic information systems (GIS) help the scientist to analyze the multiple facets of geomorphology, topography, vegetation, and climate to discover the patterns left on the landscape."

Fig. 4.8

A soil scientist examines horizons and composition of soil.

Source: USDA, Public Domain

For soil to be productive it must be able to participate in nutrient exchange with plants. The plant roots can only extract mineral elements dissolved as ions. Mobility of nutrients in a soil type is closely related to its chemical properties and moisture content. Essential ions needed for plant growth are in the form of cations (positively charged) and anions (negatively charged). These include calcium (Ca^{+}), ammonium

(NH_4^+), magnesium (Mg^+), nitrate (NO_3^-), and sulfate ($SO^{2-}4$). Ions are retained by soil particles based on the principle of attraction of opposite chemical charges. Clay and organic soil particles have a net negative charge. They are attracted to water and favor dissolution of salts, specifically attracting positively charged protons (H^+) and other cations. The presence of water helps to displace ions bound to soil particles, making them available for uptake by the tiny root hairs on plant roots. The plant roots will have transporters for uptake of all needed nutrients. Water needs are also supplied through the plants' roots as the plants above ground structures are waterproof. Excessive moisture, however, is detrimental if the soil becomes saturated. In tropical rainforests frequent heavy rains result in a propensity to leach nutrients as percolating water carries dissolved nutrients out of the soil profile. A drenching rain might apply enough force to rapidly displace nutrient ions in soil before nutrients can participate in ion exchange with roots. With a working knowledge of the properties of soil and nutrient uptake, good soil management optimizes soil quality for optimal long-term crop productivity.

The Dust Bowl

Soil erosion is an ecological issue that threatens agricultural sustainability. The Great Plains of colonial America was a massive, half-million-acre grassland prairie ranging to the north into Canada and to the south through the state of Texas. It was bounded west to east by the Rocky Mountains and Mississippi River. Agriculture to feed the growing nation was possible in the Great Plains because the rich topsoil could be irrigated directly from the **Ogallala Aquifer**, a massive underground reservoir spreading across eight states of the Great Plains (high plains). These two features of the Midwest were vital to the prosperity of the new country, producing wheat, barley, and corn in sufficient quantities to feed the nation and export surplus around the world. But then disaster struck in the 1930s. The agricultural practices leading up to the Dust Bowl should have sounded the alarm bells.

Prior to the devastation of the Dust Bowl American farmers in the Great Plains had unknowingly overcultivated crops, tilling millions of hectares. In the process they destroyed surface prairie grasses that had previously formed a dense mat of root fibers, supplying a massive carbon sink of biodiverse biological activity and deep O horizon organic base. During the 1930s, a period of prolonged drought compounded by high winds throughout the Great Plains led to erosion of the overcultivated, dry topsoil that quickly depleted its fertility. This was possible because years of poor land management had already altered its topsoil. The surface layers—no longer anchored by a dense, water-retaining root structure—were carried off by strong winds. The rampant soil erosion led to massive hunger and suffering as the Dust Bowl spread across most of middle America on the heels of the Great Depression. The plight of the small farmer and the forced exodus of 3.5 million people during the Dust Bowl has been memorialized by John Steinbeck in his classic book, "The Grapes of Wrath."

Fig. 4.9

American Dust Bowl.

Source: US Forest Service, Wikimedia Commons CC

By 1934 approximately 300 million tons of soil had been redistributed by erosion. By 1938 approximately 12.5 inches of topsoil were no longer present on 4.4 million hectares of middle America (**Fig. 4.9**). One-fifth of all remaining rural families in the Great Plains were on federal emergency relief, which was as high as 90% of the population in some counties. As a result of this American tragedy Franklin D. Roosevelt established the Soil Erosion Service (now known as the **Natural Resources Conservation Services**, NRCS). This new federally funded project focused on the education of good soil management practices and planted more than 220 million trees to counteract the effect of wind damage and to maximize irrigation. Once fully grown the trees formed a "shelter belt," which extended from Texas to the Canadian border. The US food supply has remained self-sustaining ever since.

The history of the Dust Bowl serves as a lesson on the importance of land preservation. Yet only a century later there is rising opposition to the mission of NRCS. Commercial farming practices seem to have forgotten past lessons. And over the past 20 to 30 years some states have permitted bulldozing their tree line shelter belts (**Fig. 4.10**) to plant more corn, wheat, and soybean crops (particularly in Nebraska). There appear to be reasonably safe agribusiness practices, given the bulk of the Great Plains has converted to raising cattle. Given that most climate models predict extreme weather conditions with a high propensity of megadroughts in the near future, clearing the trees for greater profits appears risky. Add that conditions of a severe drought would make raising cattle too costly because of their high demands for land, animal feed, and water. Responsible land stewards understand the complexities of our ecosystems and resource limitations and the importance of planning ahead to mitigate future economic and social disaster.

It's worthwhile to note that farming communities draw water heavily from the Ogallala Aquifer. The Ogallala Aquifer alone provides water to one-quarter of all US agribusiness. The USGS estimates the Ogallala has dropped 15 m over a 5-year period of monitoring (2013–2107). The Chapter 4 case study is an in-depth look at how water resources from the Ogallala Aquifer are being drained at an alarming rate and discusses federal policies that might serve the combined interests of agricultural, farming, and residential drinking water well into the future.

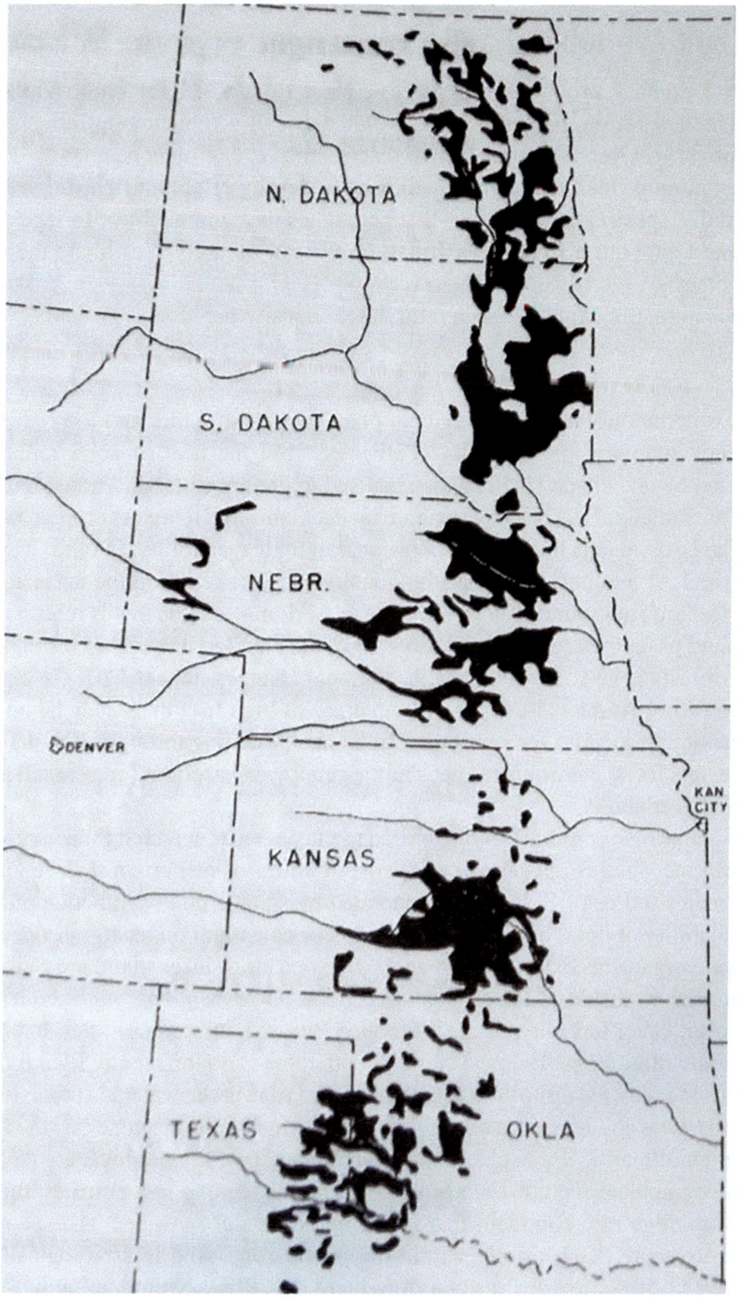

Fig. 4.10

The major planting areas of the Shelterbelt Project from 1933 to 1942.

Source: NOAA George E. Marsh Album, Public Domain

4.3 Light Availability and Distribution

Terrestrial primary productivity requires access to sunlight for photosynthesis to occur. The absorption and reflection of light through a tree canopy influences the degree of vertical light distribution that reaches the vegetation beneath. The quantity and orientation of leafy vegetation will have a varying effect on light intensity and wavelengths (spectral composition) that penetrates the canopy. Since foliage density affects the light path a **Leaf Area Index** (**LAI**) can be used to estimate the rate of photosynthesis.

LAI is normally expressed as a ratio of leaf density per area of ground:

$$\text{LAI} = \text{leaf area m}^2/\text{ground area m}^2$$

LAI is commonly used to describe a plant canopy structure and to predict **photosynthetic primary productivity (PPP)**. Because the heat of solar radiation is intercepted by leaf shade LAI can also be used to predict water and transpiration. In horticultural studies LAI is a key variable that is monitored to assess crop growth and yield. The LAI value is inversely proportional to light availability. In other words the greater the LAI number, the lower the quantity of light reaching the surface. A 4:1 ratio of leaf area to ground area would have an LAI of 4, a value that is reasonable for a shrubland or temperate forest. Desert ecosystems tend to have LAIs <1 and dense tropical forests have LAIs closer to 9. Seasonal changes impact LAI, for example, after leaf drop LAI decreases. But even in areas of high LAI there may be openings in the canopy to account for significant solar energy getting through (**Fig. 4.11**). For example, in a forest environment gaps in the canopy can account for as much as 80% of its light availability.

Aquatic primary productivity is limited to the degree to which light penetrates the water. Marine water (high salt content) cover two-thirds of Earth's surface and provides a major source of PPP energy generated by aquatic photosynthetic organisms. The availability of light, however, places a major constraint on locations where photosynthetic organisms exist. In open areas, like oceans or lakes, there are few limitations to solar exposure at the surface level. The few exceptions might be transient cloud cover, water turbidity, or algal blooms. But with increasing water depth, light accessibility diminishes. Below a certain depth there is no longer any light transmission. This is called the **aphotic zone**. The layer or zone at the water's surface is called the **photic zone** and reaches a depth of 80 to 100 m. The photic zone has the highest distribution of marinc life and is home to all aquatic primary producers.

A topographical feature found in many coastal regions is a **continental shelf**, the narrow ridge of a continent that is submerged under seawater. The sea shelf is shallow, its entire depth mostly within the photic zone. And it is biologically very productive (1.5%–2% global primary productivity). Primary production in the continental shelf takes place among an array of marine life, including seagrasses, algae, and corals.

Fig. 4.11

Hemispherical photography of winter roosting habitat, with ratio of the area of canopy to sky used to approximate LAI.

Source. S.B. Weiss, Wikimedia Commons

4.4 Terrestrial Biomes Are Determined by Abiotic Factors

By the end of this section students should be able to:

- Identify the major terrestrial biomes.
- Classify biomes on land by annual precipitation and temperatures.

There are two major groups of biomes: terrestrial and aquatic. The aquatic group is further separated into freshwater and marine biomes, based on saline concentration. The terrestrial group is separated into at least 10 major biome types. Classification of

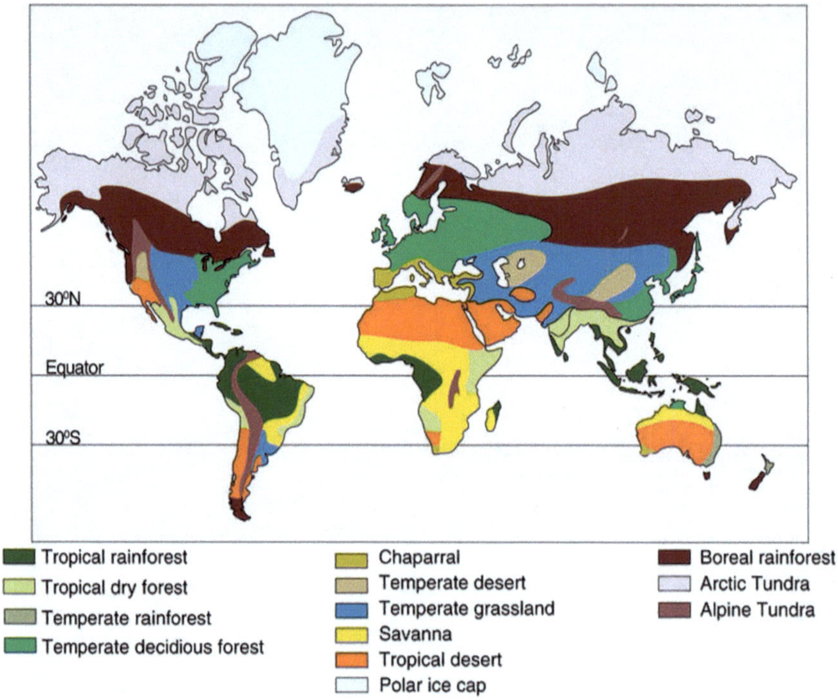

Fig. 4.12

World's major biomes are distinguished by their temperature and precipitation.

Source: CK-12 Foundation, under CC BY-NC 3.0

terrestrial biomes is based on its annual total precipitation and temperature range. If a biogeographer knows the degree of temperature variation within a biome on a daily and seasonal basis they could predict the type of vegetation. Geographically distinct areas can be defined as the same biome type, as long as they experience similar climates. But because a region's climate is partially defined by latitude biome locations in general show a horizontal banding pattern (**Fig. 4.12**).

We will now review eight of the major terrestrial biomes, noting the broad global regions in which they are found, degree of precipitation and temperature, including periodicity, and the diversity of plants and animal species that inhabit the biome. These include:

1. Tropical wet forest
2. Savannas (tropical grasslands)
3. Deserts
4. Chaparral
5. Temperate grasslands

6. Temperate deciduous forests
7. Boreal forests
8. Artic tundra

Tropical Wet Forest (0°–13° Lat)

Tropical wet forests, or tropical rainforests, are ubiquitous in a perimeter (0°–13° Lat) around the equator (**Fig. 4.13**). As expected the climate is hot and moist year-round. To be classified as a rainforest, the annual precipitation must be at least 1750 mm (69 in) with the rainfall distributed year-round. Vegetation in tropical wet forests form a dense canopy, with plant morphologies suited to capturing sunlight for photosynthesis by its broad leaves. The leaves are not deciduous (seasonal leaf drop) because there is little seasonal variation in the tropics. A forest morphology has three layers, with the uppermost tree canopy rising upward of 30 to 40 m with the tallest reaching 60 m. The canopy is inhabited by living plants (epiphytes) and vines forming tree communities among the trees. The middle layer is the understory, and its

Fig. 4.13

Tropical wet forests in Madre de Dios, Peru, near the Amazon River.

Source: OpenOregon under CC BY 4.0

humid space is filled with vines, shrub-like ferns, and palms that are adapted to limited sunlight for photosynthesis. The forest floor is most notable for its thick layer of decomposing matter, insects, and roaming animals. Here, only a few shade-tolerant plants (ferns, elephant ears) grow because the area receives only 1% to 2% sunlight (very high LAI). And even with rapidly decomposing matter the forest floor contains the least amount of nutrients of the rainforest layers. There is also minimal soil due to torrential run-offs and fast nutrient uptake by plants and insects.

Tropical zones have stable temperatures with consistent sun availability profiles year-round. The temperatures range from 20°C to 34°C (68 °F to 93 °F). In close proximity to the equator tropical ecosystems are not seasonal and don't experience long day/short night vs short day/long night cycles found elsewhere. The tropics have a constant cycle of equal length day and night, ideal for plant growth. This biome type is known for its high primary productivity (and NPP).

Tropical rainforests experience significant variation in annual precipitation, which ranges from 125 cm to 660 cm (50 in to 200 in). During wet months there may be periods of rainfall deluge up to 30 cm (11–12 in) rain. In drier months precipitation is typically less than 10 cm (3.5 in). Humidity can be as high as 80% due to the climatic factors that lead to heavy precipitation and total biomass, which releases more water into the atmosphere through transpiration.

Known for their high biodiversity the tropical wet forests have a greater number of species than elsewhere on the planet. Many species are not as yet discovered because of the unapproachable terrains in some segments of tropics. Nonetheless, scientists have estimated this biome type contains between 3 and 50 million species. This is about half of all species in the world! In the South American Amazonian rainforests there are on average 100 to 300 different species of trees per hectare (2.5 acres). A walk through the rainforest is always alive with a myriad of sounds, chatters, buzzing, and hums. Diverse species of trees and shrubs provide food and shelter. And many of its resident animals exist solely above ground, tens or even hundreds of feet suspended in air, because there exists a rich biomass community in the canopies.

Savannas—Tropical Grasslands (10°–20° Lat)

Savannas are tropical or subtropical grasslands, with scattered trees, located in flat areas between true deserts and the tropical rainforests (10°–20°NS Lat). Savanna grassland biomes are represented in quite large regions across South America, India, and northern Australia. In the African continent almost half of its land mass is classified as grassy savannas (**Fig. 4.14**). Savanna grassland annual temperatures remain generally constant year-round, ranging from 24°C to 29°C (75 °F to 84 °F). But the amount of rainfall is not a constant year-round, ranging from 10 cm to 40 cm (3.9 in to 15.7 in). There are two seasons: a wet summer and a dry winter. During its extremely dry season savannas are prone to fires because their wide expanse of grasslands is exposed to high winds with few trees and shrubs. Plants that live in the savannas must be adapted to heavy grazing, drought, and fires. They will have well-developed root systems and the capacity to quickly re-sprout. For the most part savanna grasslands

Fig. 4.14

Tree savanna. Tarangire National Park in Tanzania, East Africa.
Source: ProfessorX under Public Domain

are low-lying, no more than 152 m (500 ft) above sea level and exposed to sunlight all day with almost no leaf cover (low LAI).

Biodiversity is low in savanna grasslands because of harsh living conditions of long periods of drought juxtaposed against long periods of rains and flooding. And while the temperature remains fairly constant year-round, daily temperature fluctuation between day and night can be drastic, as much as a 15°C (60 °F)! The soil is generally poor and vegetation type is dominated by grasses. Other plants that inhabit savanna grasslands are sedges, legumes, and the sparse eucalyptus or acacia tree. Animal species show more diversity, though that depends on the geographic locations within the biome. Hoofed-animal grazers like zebras, giraffes, water buffalo, and kangaroos and several species of rodents, reptiles, and amphibians are typical inhabitants. A significant number of bird species (3000 species) and insects are also quite common in savanna grasslands.

Deserts (15–30° Lat, mostly)

Moving away from the equator are the deserts (15–30°NS Lat) in the latitudinal location where dry air returns back to Earth's surface from Hadley cell wind patterns (**Fig. 4.15**).

Fig. 4.15

Baja California Desert in the Cataviña region, south of Ensenada, Mexico.

Source: Tomas Castelazo under CC by SA 2.5

Deserts are known for being very dry and more extreme than savanna grasslands, both in their day-to-night temperature swings and the degree of moisture evaporation. Rainfall is typically less than 25.4cm (10 in) per year with an evaporation rate slightly above 25.4cm (10 in) per year. Desert lands cover 38% of Earth and are defined by having a higher evaporation rate than its precipitation rate, leading to dry, arid conditions. The characteristics of a desert, however, vary enough to be classified into one of four types:

- Hot/Dry: 23°NS Lat, includes the Tropics of Cancer and Capricorn
- Semiarid/steppe: below 23° Lat, includes arid mountain altitudes
- Coastal: slight humidity by proximity, more seasonality
- Cold: includes polar caps, low biodiversity

Hot/Dry desert temperatures can reach as high as 44°C to 50°C (111–122°F) with their heat-absorbing surfaces even hotter, 70°C (158°F). Perhaps the most famous, the expansive Sahara Desert, receives only 2cm (0.78 in) of rain over the year in most regions.

Semiarid deserts, also known as a steppe, are found in high mountain altitudes and include flat sage brushes. Their climates are less extreme than hot/dry deserts with long, hot summers and cool winters with some rainfall. Day/Night temperature ranges are wide, between 38°C and 10°C (100.4°F and 50°F). In the Australian Outback there are regions that are classified as semiarid deserts, with annual rainfall as low as 2 cm (0.8 in).

Coastal deserts are cooler and slightly seasonal with highs of 35°C (95°F) in summer and winter temperatures around 0°C to 4°C (32 to 39°F). Coastal deserts lack rainfall due to the rain shadow effect on the leeward side of the coastal ranges. The Namib Desert in eastern Africa stretches 2000 km, receives only 8 cm (3.14 in) of rainfall annually, and is known to be the oldest semiarid desert (55–80 mil years). The absolute driest place on Earth is found in another coastal desert, the Atacama Desert in Chile, which can experience decades without a drop of rain.

Cold deserts are mostly in high flat ranges and have cooler temperatures, ranging from 25°C to as low as –30°C (–22°F). Cold desert biomes include the Gobi Desert of Mongolia and the Oregon/Washington Desert. Cold deserts are not confined to the mostly 15°–30°NS Lat of other desert biomes.

Overall vegetation and animal diversity is lower in a desert biome because of the limited amount of rainfall and its sandy soils. But the desert climate creates a space for some unique life forms that developed specialized adaptations to the harsher environment. Plants are more likely to be annuals that grow quickly. To conserve water plant will develop deep roots, water-storing anatomical features, and reduced foliage to limit transpiration. Seeds are capable of remaining dormant for long periods, waiting for the occasional rains to germinate. Desert animals develop adaptations for the scorching heat with nocturnal behavior patterns like daytime burrowing or torpor strategies. Some species retain water by their waste management processes. Instead of excreting a dilute urine they precipitate a solid paste uric acid. This is the same strategy used by birds, which have low water intake.

Chaparral (30°–50° Lat)

The chaparral biome is sometimes called a Mediterranean forest because it contains semiarid expanses of scrub, the low-growing hardy shrubs, and small trees surrounding the Mediterranean Sea. But chaparral biomes are found in other locations between 30° and 50°NS lat. Geographic regions that are conducive to this biome type include Southern California, the southern coast of Australia, Southern Africa, and parts of South America. Chaparral biomes have mild weather with defined seasons including a hot, dry summer with less than 5 cm (2 in) precipitation and cooler, moist winters of 25 cm to 44 cm (10–17 in) precipitation. With fairly mild temperatures all year long, averaging 15°C to 30°C (59–86°F) in summers and 4°C to 20°C (39.2–68°F) in winter, it is known for sharp declines in temperature from day to nighttime. Often there are no clouds to trap daytime heat, causing the temperature to drop from 30°C down to 1°C (100 °F to 30 °F). The harsh terrain is not very fertile but the presence of broadleaf evergreens is possible, including the eucalyptus trees found in Southwest Australia chaparral.

Because of long summer droughts water remains a determining factor on what species survive in the chaparral (**Fig. 4.16**). Vegetation is varied and many plants will remain dormant during the summer months. The dense thicket of evergreen shrubs includes the small trees that are called chaparral (in California, anyway) and known by other names elsewhere. For example, chaparral shrubs are known as matorral in Spanish-speaking cultures and maquis around the Mediterranean. Vegetation must also be adapted to periodic fires, which shape the ecology of the Chaparral biome. Ashes left after a fire are rich in nutrients like nitrogen that help fertilize the soil and promote plant regrowth. Some plants are so well adapted to the fire season that they are known as pyrophytes, or fire-loving plants. Their life cycles will depend upon fires for seed germinating.

Sometimes a Chaparral biome will be formed after other disturbances in forests, for example, logging and overgrazing. Animals that thrive in chaparral biomes have less need for water and will be adapted to hot weather. For example, the jack rabbit holds ample blood supply in its long ears that permit loss of body heat directly to the atmosphere. And the kangaroo rat and bearded dragon limit their water excretion by a strategy of concentrating urine into a uric acid precipitate.

Fig. 4.16

Chaparral biomes are dominated by shrubs.

Source: Ricraider, English Wikipedia, Public Doman

Temperate Grasslands (23°–55° Lat)

Temperate grasslands, also known as mixed grasses or tallgrass prairies, are found throughout central North America. Sometimes subtropical savannas and temperate grasslands are considered part of the same biome type because they are both dominated by grasses. But there are a few key differences. Notably, temperate grasslands are prairies without the *sporadic* trees and shrubs that are found in subtropical savannas (**Fig. 4.17**). Temperate grasslands are located a little more distant from equatorial tropics and as such have less rainfall and more temperature fluctuation. This leads to their classification as temperate biomes. Another distinction is that temperate grasslands are more seasonal, with cold winter temperatures that are mostly dormant.

Annual precipitation averages 50 cm to 90 cm (20–35 in). The fertile soils are rich and dark, with a dense carpet of grass roots and rhizomes (underground stems) and they possess a carbon sink on par with the tropical rainforests. Grasslands are densely populated by a few species of grasses and a scattering of wildflowers like coneflowers and blue asters. Some prairies sustain herds of grazing animals like bison and a wide variety of smaller mammals, like the prairie dog, badgers, and rabbits.

Temperate grasslands contain unique biodiversity but are most important because of the overall effect they have on the global climate. The dense root structure of tall

Fig. 4.17

A northern prairie in the Manitoba Tall Grass Prairie Preserve.

Source: Krazytea under CC by SA 3.0

grasses is capable of storing an enormous amount of carbon below ground. There was a time when temperate grasslands were the largest carbon sink in the world, far greater than the carbon-storing capacity of the Amazon rainforest! The destruction of the American prairies has had devastating effects on their local wildlife, which were dislocated with many species going extinct or threatened with extinction. Today temperate grasslands are some of the most endangered biomes in the world, especially those identified as tallgrass prairies, of which only 1% to 4% exist today.

Interestingly, without an occasional fire to regenerate prairies the grasslands eventually succumb to scrub and dense forests. In land management it is a common strategy to suppress unwanted growth and restore the health of temperate grasslands through controlled burns so that the resilient grasses return quickly to dominate the landscape.

Temperate Deciduous Forests (23°–66° Lat)

Temperate deciduous forests are common wherever there are wide fluctuations in temperature representing all four seasons and a fairly consistent amount of rainfall all year long (75–150cm [29.5–59 in]). Temperate deciduous forests are found in mid latitudes (23° and 66° Lat) and are described as the most common biome in eastern North America. Temperate deciduous forests are also common in Midwest Europe and parts of Russia, East Asia, and, more southernly, in New Zealand and Chili. The temperatures range between –30°C and 30°C (–22°F and 86°F) and drop to below freezing temperatures in winters, leading to months of dormancy.

The moderate level of precipitation and cold winter snaps are optimal for the deciduous tree, which dominants this biome. Deciduous trees lose their leaves when the temperatures drops rapidly in the fall. New buds sprout again with the first warm days of spring (**Fig. 4.18**). Productivity is nonexistent in the dormant winter months. The annual decomposing leaves result in a fertile topsoil with a thick layering of humus (O and A horizons).

Overall, the temperate deciduous forests have high biodiversity and many of its trees (oak, maple, and evergreen juniper) are common across different biomes, known as **cosmopolitan**.

The ecological structure of a temperate deciduous forests is its four layers, not unlike the three-layered structure we identified in the tropical rainforests. The uppermost tree canopy towers over an understory of smaller, more immature trees or those that are more shade tolerant. Beneath the two layers of trees are shade-tolerant woody shrubs, and the bottom ground is composed of shade-tolerant grasses, ferns, mosses, and fungi. Because of winter leaf litter the ground of the forest is protected from soil erosion. Mixed among the decomposing leaf litter are many microbes and invertebrates, like the roly-poly (*Armadillidium vulgare*).

Boreal Forests (50°–60° Lat)

The boreal forest (Boreas is the Greek god of the Northwind) is also known as taiga, a Russian word meaning moist and swampy. It is populated with dense coniferous,

Fig. 4.18

Deciduous trees are a dominant plant in temperate forests.

Source: US National Park Service, Pubic Domain

or evergreen trees. Up to 30% of trees in the Boreal forests remain mostly still undisturbed. The area covers the terrain south of the Arctic Circle across a broad swath of Russia, Canada, and parts of Alaska and northern Europe. It is only found in the Northern hemisphere, between 50°N and 60°N Lat (**Fig. 4.19**).

The boreal forests are distinguished by their extreme cold climate, evergreen trees (pine, spruce, fir), and ability to trap freshwater (glaciers). Annual precipitation isn't very high, averaging 40 cm to 100 cm (15.7–39 in), and it typically falls as snow. The water content remains high in the boreal forests because it is too cold for evaporation to occur. The excess water can form swamps and bogs but more likely it will be frozen in lakes. The boreal biome represents the greatest source of freshwater on the planet. There are four seasons, with a very long winter, mild spring and fall seasons, and a very short summer. The growing season is only 2 to 3 months. Temperatures are extremely low, from −5°C to 5°C (23 °F to 41 °F), and even colder in places like Siberia, where average temperatures year-round are below freezing, −10°C (14 °F).

Biodiversity is understandably quite low due to a combination of the frozen climate and varying lengths of daylight. The soil is low in nutrients as there is little biomass to decompose and only a small population of microbes and invertebrates are present. Few photosynthetic plants exist in this harsh climate other than evergreens

Fig. 4.19

The boreal and conifer trees of the Adirondack Mountains in upstate New York represent the southernmost part of the world's taiga biome.

Source: R khot under Public Domain

with thick, waxy-coated needle-like structures. One of the only deciduous trees of the boreal forest, the larch tree, is found in Russia.

Compared to other forest biome types the boreal forests have a lower rate of productivity. One plant adapted by finding an alternative energy source. The carnivorous pitcher plant is able to lure and trap insects into its sweet floral parts and digest them with a strong acid secretion. Boreal trees are slow growing, long living, and quite tall with significant aboveground biomass. But in geologic time boreal forests are considered to be very young. Tens of thousands of years long ago (Younger Dryas) these lands were covered with sheets of polar ice. Ever since, encroaching and retreating glaciers have formed a complex, hilly landscape with deep crevices of waterlogged swamps and bogs.

There are large animals that thrive in the boreal forests including the moose, wild boar, woodland caribou, Siberian tiger, and bear. Their size, extra fat layers, and fur help them stay warm in the arctic cold. Some species hibernate, and one species, a wood frog (*Lithobates sylvaticus*), simply freezes its entire body through the coldest part of winter.

Arctic Tundra (60°–70° Lat)

The Arctic tundra is the northerly most biome (60°N–70°N Lat). There is an Antarctic tundra in a similar latitude in the Southern hemisphere but with only isolated land masses there isn't consensus yet to call the Antarctic region a tundra. The word *tundra* comes from one of the Scandinavian languages, as *"tunturia"* means treeless plain. The average temperatures in the Arctic Tundra ranges from –34°C (–30 °F) to 3–12°C (37–54 °F). Precipitation is neither variable nor plentiful, with an average of just 15 to 25 cm (6–10 in). Little evaporation is possible, even with the strong winds of the tundra, and water in all its forms is plentiful.

The difference in the length of day vs night is even more extreme in the tundra than in the boreal forests. The growing period is even shorter, as plants have approximately 10 to 12 weeks to complete their life cycles. At the peak of summer season it is completely daylight for the full 24 hours so that rapid plant growth is possible during this time.

Vegetation in the Arctic tundra is generally ground hugging, for example, mosses and lichen (**Fig. 4.20**). The underlying soil is in a state of **permafrost**.

Fig. 4.20

Shrubby low-growing plants dominate the arctic tundra in Sydkap in Scoresby Sund, East Greenland.

Source: Hannes Grobe under CC by SA 2.5

Species diversity, biomass, and NPP are all insignificant in the tundra. During summers, with melting of the top layer of the permafrost, there may be activity in the soil and bogs across the landscape. But neither water nor plant roots can penetrate the solid permafrost beneath, leaving the thin soil layer waterlogged and poor in nutrients. The reason there are no trees in the Arctic tundra is because of the presence of the permafrost.

Even under these most harsh, extreme conditions there are almost 2000 species that live in the Arctic tundra, many of which are transient migrants from the boreal forests. Seasonal herds of caribou and flocks of birds fly across the Arctic tundra. Fewer than 50 species are year-round residents, which include the polar bear and the muskox. Muskox is well adapted to the Arctic tundra, having been around since the time of the Wooly Mammoth (extinct approximately 10,000 years ago). The muskox developed a unique cold adaptation, helical nasal cavities, which warm the arctic air before it enters their bodies.

4.5 Desertification

Ongoing climate crises occur across terrestrial biomes and impact ecosystem sustainability, including its biodiversity, soil, and water quality. When conditions become severe enough it is possible for loss of an entire biome. According to the IPCC one of the greatest challenges of the 21st century is a phenomenon called **desertification**. Desertification occurs when fertile land is degraded to non-arable land, designated as arid or semiarid. The transition can happen after land mass movement and uplift alter wind patterns, which in turn alter the amount of moisture a region receives. Dr. Martin Williams, Earth Sciences, University of Adelaide, proposed that mountain uplift led to a slow decline of the once fertile Sahara Desert. Tectonic movement in the East African highlands (up to 19,000 ft altitude at Mt. Kilimanjaro) rose across the Great Rift Valley, approximately 3 Mya, diverting winds that once brought rainfall to civilizations along the Sahara. Another factor that triggers desertification is anthropogenic climate change. The end result is an expansion of deserts with limited access to water and a marked decline in soil quality suitable for vegetation and wildlife. Arid and semiarid lands have a high rate of evaporation and transpiration, potentially as high as 65% loss of their scarce rainfall. The region will also experience socio-economic repercussion for the human demographic. Most desert biomes (90%) are found in developing countries, where 2.7 billion people live.

Understanding the susceptibility of desertification is vitally important to protect Earth's limited resources. A region's terrain reverting from fertile to desert type includes declining water table and an alteration of its soil composition, becoming more compact with loss of its topsoil. Many interacting factors contribute to desertification, commonly involving human activity such as deforestation, overgrazing, overcultivation, and even global warming. Global warming is typically monitored by atmospheric, or oceanic, temperatures but the land itself warms. According to NOAA (National Oceanic and Atmospheric Administration), Earth's surface temperature since 1850 has warmed almost 2°C (35.6 °F). The rate was approximately 0.06°C (32.1 °F) per decade but has increased three times, to 0.20°C (32.4 °F) per decade since the 1980s. Furthermore, climate change leads to increasing frequency of abnormal weather, like droughts and floods, that impact the region's vegetation. Not unlike the Dust Bowl event, once plants die-off the soil will begin to erode.

Most wildlife cannot adapt in the short time spans in which desertification occurs. Local flora and its value as a carbon sink diminish with encroaching desertification. Another side effect of desertification is increasing atmospheric dust, estimated as a 25% global rise in dust emissions since the 1900s. Heavy dust emissions block solar radiation, which further reduces primary productivity. Increasing desert biomes go hand-in-hand with global warming triggers—one exacerbates the other.

Examples of contemporary desertification events include:

1. The Sahel (African region south of Sahara Desert) during the 1900s. Attributed to changing weather patterns with prolonged drought and overgrazing.
2. American Dust Bowl due to unsustainable land management and prolonged droughts led to rapid degradation of arable lands.
3. Pakistan, China, and Mongolia's encroaching desertification over the past two decades contributed to the rise in global dust emissions. Major factors are cited as climate change, overgrazing, and irresponsible mining.

Scientists project **drylands** (scarcity of water) will continue to increase with changing climate patterns and global warming, from its current 38% of global terrestrial lands to upward of 55% by the end of the 21st century. Areas expected to experience desertification include Southwest North America, Northern Africa, Southern Africa, and Australia (**Fig. 4.21**). Notably these latitudes border semiarid regions, which are also expected to expand in range.

Soil Restoration Is Part of the Global Response on Climate Action

The key to mitigating the potential starkness of arid landscapes is preempted by effective climate change responses. As the Dust Bowl demonstrates it is possible to reverse desertification processes by decisive government action designed to reverse soil mismanagement. Desertification and increased drylands are directly correlated with climate change events. Arid climates are net emitters of CO_2 because they don't have ample vegetation to sequester CO_2 emissions. The lack of water in the soil, or its evaporation into the atmosphere, means more extreme temperatures, more soil erosion, and less food production. According to the UN's drylands official, Luc Gnacadja, "Desertification is the greatest environmental challenge of our time" and

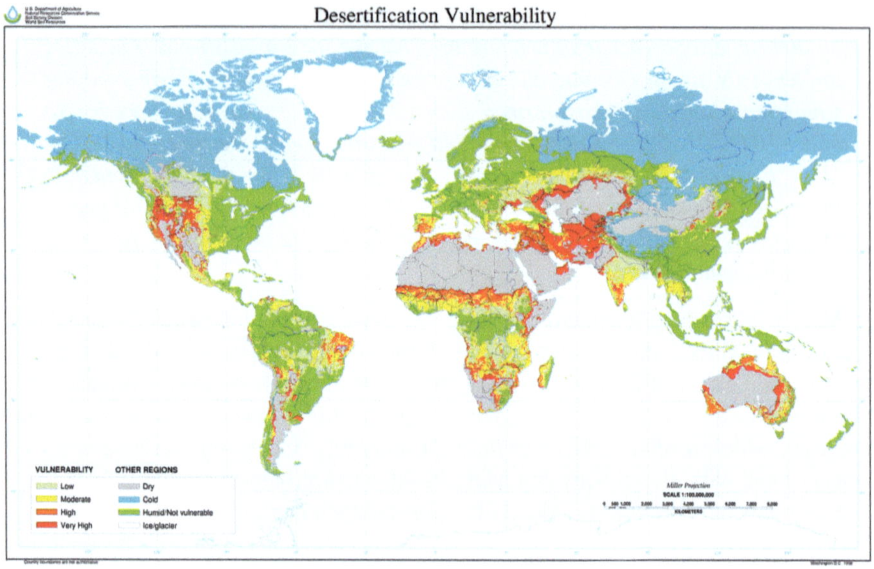

Fig. 4.21

Global Desertification Vulnerability Map. Areas in red represent the highest vulnerability to desertification. Areas in green are not vulnerable but may be susceptible to heavier flooding with climate change events.

Source: USFWS Arctic National Wildlife Refuge, Public Domain

"a threat to global wellbeing." The overarching measures to address global warming under IPCC guidelines must therefore also address the following:

- soil erosion
- vegetation loss
- overgrazing
- land mismanagement
- water overuse
- strong winds

How the Sahara Desert Is Turning Into a Farmland Oasis

In this regard many private and publicly funded organizations are working internationally to tackle desertification through installation of green walls (trees, shrubs) to block the impact of high winds, implementation of sustainable land use, and training on alternative farming techniques. In conjunction with this list of actions is the adoption of economic policies in surrounding regions to prevent the impoverished demographic from overgrazing their lands and to protect endangered crops during drought periods. The Global Environment Facility (GEF) has funded initiatives in Africa and China to successfully reverse the impact of poor soil management.

The expansive 3000 miles of the Sahara Desert resulted in extreme hunger, forced migration, and extreme climate events. However, Niger (80% lands within Sahara Desert) recently began restorative developments with 200,000,000 trees planted and new farmlands serving 2.5 million people. It is a remarkable endeavor considering the region only receives 16.5 cm (6.5 in) of rainfall per annum.

Glossary

Hibernacula **Cosmopolitan** **Drylands**

Citations

1. Jennifer A. Clack, Getting a leg up on land, Scientific American. Evolution (2005). Dec 5 issue. https://www.scientificamerican.com/article/getting-a-leg-up-on-land/
2. G.B. Rigg, B.S. Henry, On the origin of the gases in the float of bladder kelp, Am J Bot 22 (3) (1935) 362–365. https://doi.org/10.2307/2436362
3. G.E. Johnson, Hibernation of the thirteen-lined ground squirrel, Citellus tridecemlineatus. The rise in respiration, heartbeat and temperature in waking from hibernation, Biol Bull 57 (2) (1929) 107–129.
4. Johnson, C., et al. (2021). Long-term changes in occurrence, relative abundance, and reproductive fitness of bat species in relation to arrival of whitenose syndrome in West Virginia, USA. Ecol Evol. 11(18), 12,453–12,467. https://doi.org/10.1002/ece3.7991
5. C. Hartman, Novel inactivation of the causative fungal pathogen of white-nose syndrome with methoxsalen plus ultraviolet A or B radiation, PLoS One 15 (9) (2020) e0239001. https://doi.org/10.1371/journal.pone.0239001
6. G. Taylor, R.A. Eggleton, Regolith geology and geomorphology, Geol Mag 138 (6) (2001) 727. https://doi.org/10.1017/S0016756801216239
7. M. Schnepf, A history of natural resource inventories conducted by the USDA's Soil Conservation Service and Natural Resources Conservation Service (pp. 1–40), Soil and Water Conservation Society. Updated July 2016 by Patrick Flanagan (2008, September). https://www.nrcs.usda.gov/sites/default/files/2022-10/NRI_history.pdf
8. T. Cowan, Present-day greenhouse gases could cause more frequent and longer dust bowl heatwaves, Nat Clim Change 10 (2020) 505–510. https://doi.org/10.1038/s41558-020-0771-7
9. R.L. Chazdon, R.W. Pearcy, The importance of sunflecks for forest understory plants, Bioscience 41 (1991) 760–766. https://doi.org/10.2307/1311725
10. J.-P. Gattuso, Light availability in the coastal ocean: impact on the distribution of benthic photosynthetic organisms and their contribution to primary production, Biogeosciences 3 (2006) 489–513. https://doi.org/10.5194/bg-3-489-2006
11. E.O. Wilson, The Diversity of Life, WW Norton & Co, New York, NY, 1999.
12. L.C. Prentice, et al., A global biome model based on plant physiology and dominance, soil properties and climate, J Biogeogr 19 (1992) 117–134.

13. J.M. Storey, et al., Mitochondria and the frozen frog, Antioxidants (Basel) 10 (4) (2021) 543. https://doi.org/10.3390/antiox10040543
14. Designed for the Arctic: Muskox. National Park Service. Dec. 16, 2020. https://www.nps.gov/gaar/learn/nature/muskox-designed-for-the-arctic.htm
15. M. Williams, When the Sahara Was Green: How Our Greatest Desert Came to Be, Princeton Press, 2021. https://press.princeton.edu/books/paperback/9780691253930/when-the-sahara-was-green
16. T. Stanelle, et al., Anthropogenically induced changes in twentieth century mineral dust burden and the associated impact on radiative forcing, JGR Atmosphere (2014). https://doi.org/10.1002/2014JD022062
17. J. Han, H. Dai, Z. Gu, Sandstorms and desertification in Mongolia, an example of future climate events: a review, Environ Chem Lett 19 (2021) 4063–4073. https://doi.org/10.1007/s10311-021-01285-w
18. R. McSweeney, Desertification and the role of climate change, Resilience Environment (2019). https://www.resilience.org/stories/2019-08-07/desertification-and-the-role-of-climate-change/

CHAPTER

Aquatic Environments

5

Chapter Outline

5.1 Physical Properties and Constraints in Aquatic Species

By the end of this section students should be able to:

- Explain the role of the hydrologic cycle on aquatic systems.
- Discuss major abiotic factors as a constraint on aquatic life.
- Explain the importance of water's life-giving properties.
- Understand how gases are exchanged between bodies of water and the atmosphere.
- Make connections between atmospheric carbon and ocean acidification.

Abiotic Factors

In Chapter 4 we learned that gravity was an important constraint for terrestrial living. In submerged habitats the challenge of gravity is nonexistent. But there are plenty of other physical properties that require specialized adaptations. For example, in some waterways there are strong **flow forces** that impart drag and lift properties. Flow force becomes significant if the aquatic animal must adhere to surfaces against fast-moving currents and tides. This is the case for the populations of water climbing gobies and

Ecology for the 21st Century. DOI: https://doi.org/10.1016/B978-0-443-24031-7.00005-2

Fig. 5.1

Underwater life on a coral reef off the islet Klein Bonaire, Venezuela.

Source: TyBy Michal Strzelecki under CC BY 3.0

limpets that live at the intertidal zones (**Fig. 5.1**). And many mobile stream macroinvertebrates are sessile, spending their entire lives attached to substrate surfaces. They are governed by water forces imposed by stream flow conditions.

There are challenges regarding surface adherence in some terrestrial species but dry adhesion requires a different set of strategies. For example, dry adhesion principles of van der Waals forces and chemical bonding are the physical properties that allow a gecko to stick to and walk along a ceiling. These properties will not be helpful to animals submerged under water, as the forces would be too weak. Aquatic animals have evolved a diverse range of adaptive strategies for adhesion attachment, including glue, friction, and suction. A few mechanisms rely on mechanical principles, for example, the development of structures that resemble a hook, lock, or clamp. In shallow tidal seas, underwater plants are rooted to the ocean floor and other submerged species attach to coral structures (**Fig. 5.2**).

In general the major abiotic factors are the same for aquatic and terrestrial species. The three key abiotic factors include:

- solar radiation
- temperature
- precipitation (for most bodies of water, not an issue)

The Water Cycle

The water cycle, known also as the **hydrologic cycle**, explains the global movement of water (**Fig. 5.3**). The global hydrologic cycle depicts the various water reservoirs (bodies of water) and exchanges that take place with the atmosphere as it converts between its liquid water phase and its gaseous phase (water vapor).

Fig. 5.2

Limpets survive strong tidal movement by suction using their muscular feet with a layer of adhesive mucus.

Source: Ella Patenall under CC BY 2.0

Solar radiation is the driving force behind the hydrologic cycle. As the sun heats up Earth liquid water converts to water vapor through **evaporation**. Recall that water vapor in the atmosphere proceeds through **condensation** into cloud formation and eventually falls as **precipitation** (Chapter 2). The rate of rainfall passage through soils varies with types of terrain, slope, and soil composition. Once the soil is saturated, **water runoff** occurs and makes its course into gullies, streams, and larger waterways. Water uptake by plants returns to water vapor in the process of **transpiration**. The water that isn't taken up by plants or filtered into waterways will seep into the soil horizons until it reaches an impervious layer of clay or rock. There is collected as **groundwater**. Water that reaches the streams, rivers, and wetlands will eventually flow to the coastlines, where transitions occur between freshwater from terrestrial land and the salty marine water environments. Available surface water can evaporate directly into the atmosphere, dependent on temperature and the amount of water vapor already present in the atmosphere. This completes the movement of water through the hydrologic cycle. Take a moment and review **Fig. 5.3** to be sure you can trace the water cycle and fill in the spaces on a blank image.

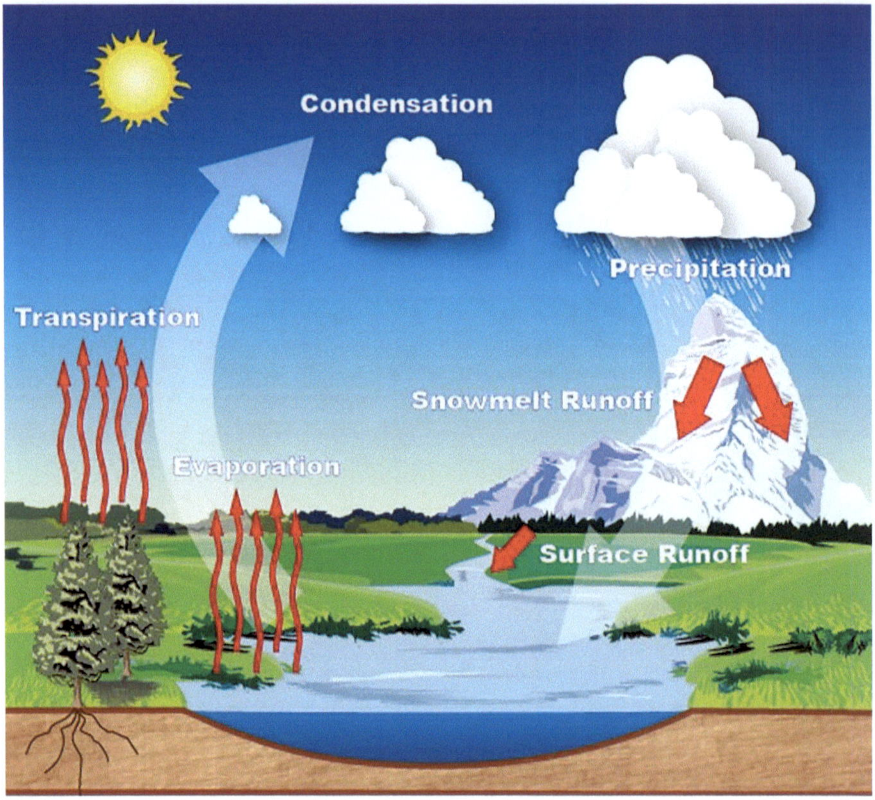

Fig. 5.3

The hydrologic cycle.

Source: Anishct, Public Domain

Just over one-half of Earth's freshwater is trapped in polar ice caps, glaciers, or groundwater (NOAA). The remaining (1.2%) is collected in lakes, marshes, and rivers. Water is also present in soil, living things, ground ice, and permafrost (**Fig. 5.4**).

The Life-Giving Properties of Water

In our world almost every physiological process requires water as a fundamental constituent. Across the planet, the dominant environment is aquatic, with much of it in the form of saline water, or marine water. For aquatic animals reliance on the specific properties of water will define their existence, beginning with the distinction of those species adapted to live in freshwater vs marine water. A limited number of species are capable of living in both environments. According to the Florida Fish and Wildlife Conservation Commission (FFWCC), the gulf sturgeon (*Acipenser oxyrinchus desotoi*) are large prehistoric-looking fish that spend their summers in the

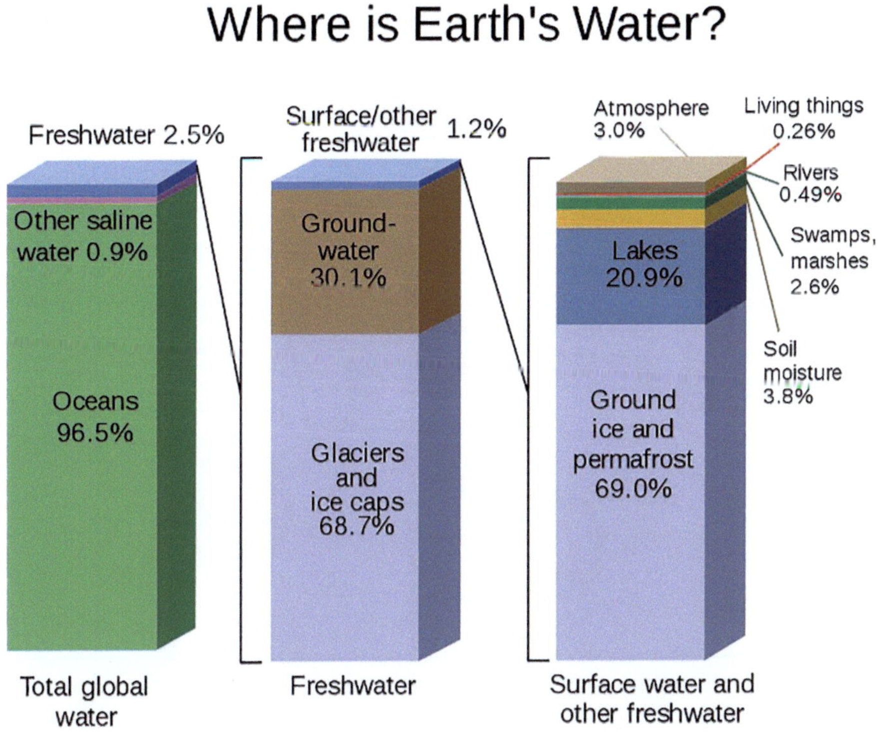

Fig. 5.4

Water distribution on Earth.

Source: US Geological Society (USGS), Public Domain

freshwater deltas of Louisiana and Florida but then spend winters in the marine habitats of the Gulf of Mexico. Another example is the Pacific Salmon, which are born in rivers but spend their adult lives in the open oceans, only returning to freshwater to spawn, which completes their life cycle.

The properties of water help define all aquatic environments and are an important constraint on the evolution of aquatic animals. The unique physical properties of H_2O are caused by its molecular chemistry. Two hydrogen atoms bond to the oxygen in a way that polarizes the molecule. Different angles across the molecule have either positive or negative charges, which in turn lead to interactions with neighboring charged molecules. It's the reason why ionically charged salts dissolve in water. Water is a covalently bonded molecule with the oxygen holding more tightly onto the bond electrons formed with hydrogen. As oxygen holds more electron character it becomes slightly more negative than its bonding partner, hydrogen. Since the hydrogen atom receives less time with the bonding electron it ends up being slightly positively charged. This is the reason for the polarized nature of water molecules (**Fig. 5.5**). Altogether the molecule is neutral because there are

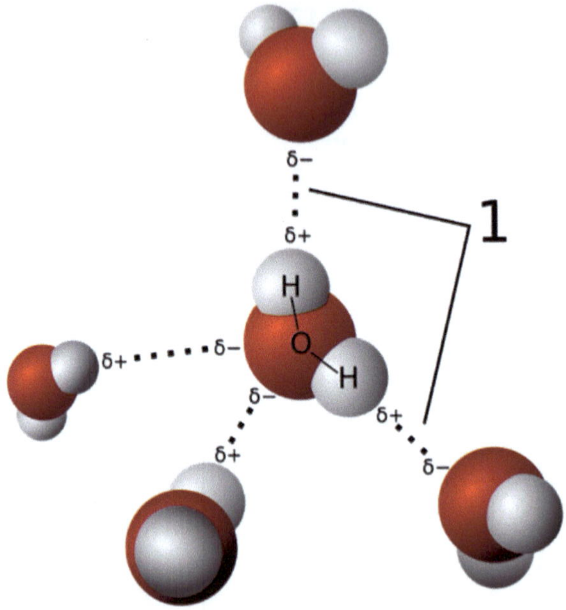

Fig. 5.5

Unique hydrogen bonding based on polarity found in the H_2O molecule.

Source: Qwerter under CC BY-SA 3.0

equal number of electrons and protons, there are pockets of negative charge and pockets of positive, which create an opportunity for more chemistry. An additional layering of bonding interactions, called hydrogen bonding, surrounds the molecule in addition to its covalent bonds. They are weak bonds but so important. Any water-loving molecules (those with a charge) can dissolve into water, releasing positive and negatively charged ions (solutes) that interact with the hydrogen bonds. Water is a dynamic solvent, forming weakly interactive hydrogen bonds that are easily broken, reestablished, and then broken again.

The unique properties of water are absolutely required for life on Earth to continue. No other material on Earth can fulfill these requirements nor substitute for water to provide life-giving benefits to organisms and their environments. Some of the most important properties of water are:

- ***High Specific Heat Capacity*** A standard value defined as the number of calories required to raise the temperature of 1 gram of water by 1°C. Water's specific heat is 1, and all other liquids are measured relative to water's specific heat capacity. As it turns out water has the capacity to store a large amount of energy, which gets absorbed in the form of heat, without raising the temperature. It is because of this property that the maritime influence brings milder winter temperatures, cooler summers, and less temperature variation along coastlines. In the environment organisms near large bodies of water are better protected from the cooling effect of seasonal climate. In the body, just like in bodies of water, water's high specific heat capacity works to prevent wide fluctuations in body temperature during activities that raise the metabolic rate.

- ***Cohesion and Viscosity*** Because of the weak electrical charges of hydrogen bonding water molecules tend to stick to one another, giving water the property of cohesion. Slightly more energy is required to break the hydrogen bonds surrounding water molecules. Viscosity is the measure of the force required for an object to pass through a liquid. The more viscous a material the more frictional resistance is present. The frictional resistance of water is 100 times greater than air and accounts for an adaptation in water-dwelling animals—a streamlined shape to counteract friction drag.
- ***Density*** Water's high specific heat capacity also means that larger quantities of heat must be absorbed before changes occur in its chemical phases (liquid, gas vapor, or solid ice). When the temperature drops to freezing, water's hydrogen bonding properties form a lattice structure with large open spaces, "locking" the hydrogen bonding into a stabilized form. The frozen water molecules are spread out *further* and therefore occupy more space than water. Because the molecules are less concentrated, ice is less dense than liquid water (**Fig. 5.6**). Ice floats over liquid water for this reason.
- ***Cohesion and Surface Tension*** In a body of water the cohesive properties of water (forces of attraction) surround each water molecule with hydrogen bonding. At the surface, however, there is a much weaker force of attraction between water molecules and air. The molecules of water at the surface tend to be dragged slightly inward, creating a taunt surface in which very light objects can land and not be submerged—an interesting adaptive trait that pond insects, like water striders and water spiders, developed to walk across water without breaking the surface accounts for this property of water. To accomplish this engineering feat the insect will spread its limbs in a manner that distributes its weight without penetrating the water's surface (**Fig. 5.7**). Tension at the air–water interface can impede a large moving body as it has to break inertia to overcome both gravity and surface tension. And on land it is the property of surface tension that forms dew into droplets.

While most materials become denser as they cool the opposite is true of water. If that were not the case, just imagine the scene at a pond or the ocean as temperatures drop and ice forms across the surface. If ice were denser than liquid water it would sink to the bottom, leaving a new top layer of water exposed to the freezing temperatures, which next freezes and drops. Eventually the entire body of water would become a solid cryosphere! The lighter ice, due to formation of a lattice structure by hydrogen bonding, ultimately keeps aquatic life insulated from the freezing atmospheric temperatures.

Liquid water is a dense medium, 860 times denser than air, and the reason water has a buoyancy affect. Buoyancy is an advantage to aquatic animals because it omits the need to counteract the effects of gravity. However, water's density is far less

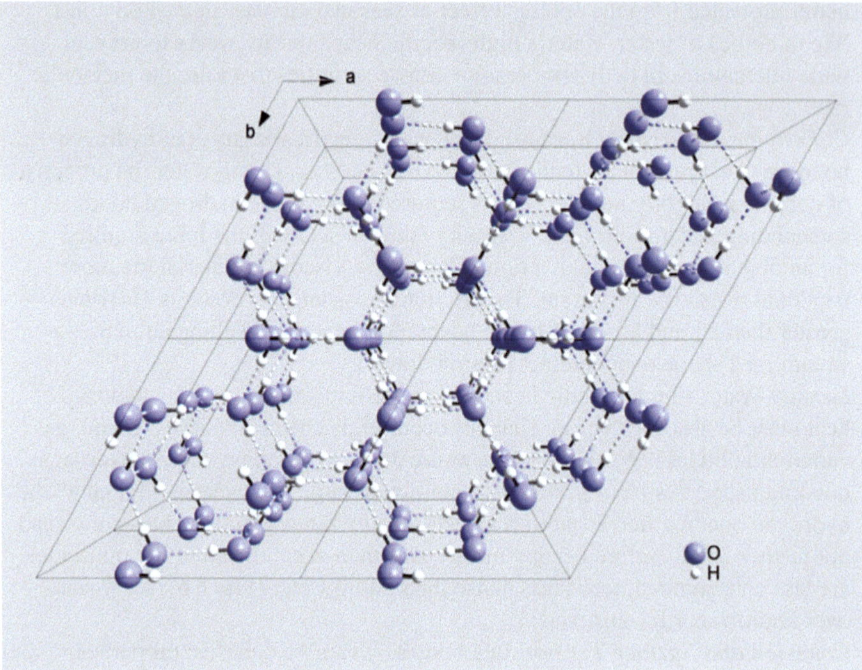

Fig. 5.6

Crystal structure of ice, using Diamond 3.0.

Source: Solid State, CC BY-SA 3.0

favorable with increasing ocean depths. Now the density of water becomes a constraint due to increasing barometric pressures. The barometric pressure of surface water is 1 atm (atmosphere). At a depth of 5000 meters, the barometric pressure of water is 500 atm, which is 500 times more than surface water. Life in deep oceans necessarily required the evolution of adaptations to thrive in this harsh environment. According to the Deep Seal Conservation Coalition (DSCC), it is estimated that 500,000 to 10 million species successfully inhabit deep oceans.

Aquatic challenges that arise from living in deep oceans are extensive, including high pressure, total darkness, limited oxygen, extreme cold, and inadequate nutrients. The main reason that barometric water pressure is an impediment is because of the air pockets in the anatomy of animals. For example, the lungs of beaked whales limit them to swimming down to a depth of 10,000 ft but this can only happen because of a few unique adaptations. Air is affected by barometric pressure and can depress yet water does compress. The beaked whales actually divert all the gases in their lungs into other compartments of the body (blood and muscles) for the dive. Below 3000 m up to about 7500 m depth (approximately 10,000 to 25,000 ft) any residing aquatic animal doesn't have air pockets. Instead they develop adaptations to

Fig. 5.7

Insects can "walk on water" due to water's properties of cohesion and surface tension. Media Tumblr

Source: Media Tumblr, Public Domain

high barometric pressure by production of molecules. For example, Hada snailfish (*Pseudoliparis amblystomopsis*) produce trimethylamine oxide (TMAO). This is a hydrophilic substance that prevents distortion and compression of cellular proteins and other molecules under intense external pressure. Below 7500 m the majority of deep-sea creatures have not been explored. Among those identified, microorganisms have membranes so specialized that they cannot maintain their integrity except under those extreme barometric pressures.

Abiotic Factors Are Present in Aquatic Gradients

The quantity of light, temperature, and gases (oxygen and carbon dioxide) all vary across water depths. The angle at which light reaches the surface of water will determine how much gets reflected back into the atmosphere and how much light energy gets absorbed into the water. At the equator, direct solar radiation is absorbed to the greatest degree while the low angles of solar radiation that reach the poles mean the least light energy is absorbed. Since more solar radiation is reflected back to the atmosphere at the poles, or higher albedo, the waters are much cooler.

Daily and seasonal changes in light availability must be taken into account when determining the amount of light available to conduct primary productivity across bodies of water. Once solar radiation enters the water medium it will be scattered and dispersed according to the number of objects and solutes present to obstruct its path. In addition, the molecules of water themselves can directly absorb light from wavelengths greater than 750 nm (nanometers; red and infrared radiation). As solar radiation penetrates the surface and travels through the water depth less and less of the spectrum will be visible. And at the furthest depths of the photic zone only blue wavelengths are still present. Full spectrum is the ideal light quality.

The distribution of species in waterways is directly affected by the quantity and quality of light availability. Therefore any limitations on light represent a constraint and require an adaptation. Producers might develop an adaptation for low light conditions by altered pigments to harvest light waves. Or, when there is no photosynthetic capacity consumers might develop other means for carbon and energy acquisition. Among the diverse survival adaptations for deep water prey or predator are the following: larger eyes, the capacity for bioluminescence to enhance visibility, and loss of all skin pigmentation except for the color red. Red will camouflage animals since most deep-sea creatures have lost their ability to see red.

The rate at which oxygen gas from the atmosphere dissolves into water follows gas diffusion laws. The rate of oxygen's diffusion will depend upon the water's solute concentration (salinity), pressure, and temperature. **Diffusion** refers to the ability of a solute to move from a region of its higher concentration to a region of lower concentration. In this example oxygen is the solute moving from high to low concentration. In aquatic gas exchange there will be a net transfer of two metabolically important gases (oxygen and carbon dioxide). Gas exchange occurs at the surface of water, along its concentration gradient, with oxygen generally moving in the direction of air to water, and carbon dioxide (often dissolved in water as carbonic acid) moving from water back to the atmosphere in its gaseous form. CO_2 is continually generated in waterways as a metabolic by-product of cellular respiration. The concentration of dissolved oxygen (DO) in water varies. DO is increased in areas of high photosynthetic activity as it is the main by-product. DO decreases wherever animal density is high because oxygen will be consumed during cellular metabolism (respiration). Once oxygen enters water from the atmosphere it generally diffuses slowly. But diffusion rates can be increased by aeration or fast movement of water. For example, stream turbulence, tides, or currents all have the effect of increasing the amount of oxygen in waterways.

The rate of oxygen diffusion from the atmosphere can also be increased with any of the following: cooler temperatures, decreased salinity, or increased barometric pressure. All three have the effect of increasing oxygen's *solubility* in water. Conversely, the following conditions decrease oxygen's solubility in water: warmer temperatures, increased salinity, or a drop in gas pressure. The presence of a solid layer of winter ice will also impede diffusion of oxygen from the atmosphere. The seasonal variance in oxygen solubility is another constraint in aquatic species. Similar to hibernation, some species will reduce their metabolic activity during winters to offset the demands on oxygen consumption. And since temperature, pressure, and salinity levels all vary widely across time and space, different habitats support different combinations of biodiversity richness. Unlike terrestrial environments that for the

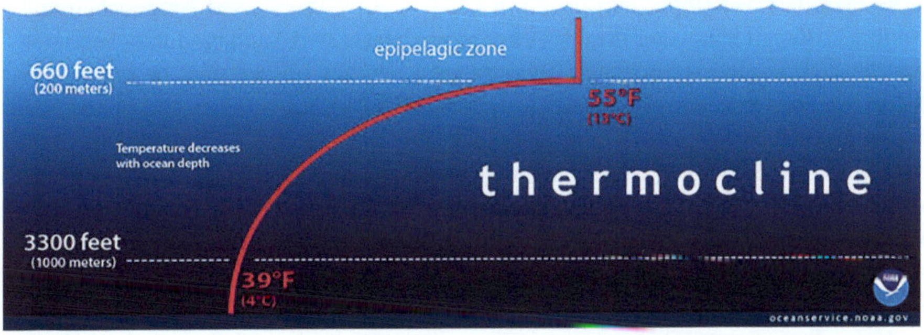

Fig. 5.8

Seawater temperature profile showing thermocline rapid drop in sea temperature between epipelagic zone and deep water of mesopelagic zone.

Source: Sbyrnes321, Public Domain

From 1000-4000m depths, water temperature remains constant; National Oceanic and Atmospheric Administration (NOAA), Public Domain

most part have unlimited access to oxygen, aquatic life is dependent upon oxygen as a limited resource. This is the case even at saturated levels of dissolved oxygen. Thus oxygen accessibility remains a major constraint to aquatic life.

The third abiotic factor in waterways is temperature. In oceans a zonation pattern is present with a mixed upper layer and a thermocline temperature gradient that drops quickly. The warmed water at the surface reflects the incoming and outgoing solar radiation that is distributed through currents, wave action, and winds. This surface mixing and heat distribution occur only until around the 200 m (660 ft) depth (**Fig. 5.8**). Below this surface layer is the region called the **thermocline**, which displays a rapid decline in vertical temperature to about 1000 m (3300 ft). The thermocline serves as a transition zone between the mixed upper layer and the stable, cold waters of deep oceans. The depth and strength of the thermocline in oceans and lakes may vary seasonally. For example, in the polar regions the thermocline is nonexistent because surface temperatures are so cold that the entire water column stays at the same temperature. And in the tropical zones the thermocline is mostly permanent with a stable warm surface water.

Freshwater lakes and pond have layered or stratified temperature and oxygen levels under similar conditions to oceans. Diffusion occurs at the water's surface. Deeper waters do not have access to either solar heat or oxygen. Yet decomposer organisms that reside in lake bottom sediments still need to consume oxygen for their respiratory metabolism. Spring and fall turnover are important events occurring in ponds and lakes that help with dispersal of oxygen, temperature, and nutrients as vertical columns of waters are mixed. Seasonal changes in the thermocline are possible wherever ecosystems have sufficient variation in solar radiation and temperature (e.g., temperate zones).

It is the seasonal temperature fluctuations, found in temperate zones, that cause lake turnover (topic introduced in Chapter 9, "Nutrient and Biogeochemical Cycling"). Recall that water is most dense around 4°C (39 °F) and gradually becomes less dense as temperature moves away from 4 degrees (either rises or falls). The denser waters drop beneath less-dense waters (phenomenon in lakes and ponds, not deeper oceans). In summers and winters lakes and ponds maintain a distinct thermocline with limited access to nutrients and oxygen in their benthic zone. The surface ice of temperate climates will insulate underlying water and animals when air temperatures drop below 4°C (39 °F). With the fall and spring weather changes lake and pond waters experience their lake turnover, pushing deep water upward along with any nutrients dissolved.

Note: Turnover events are typical of smaller lakes and ponds but not large bodies of water like the Great Lakes, where the bottom water never fully mixes with the surface layer. Also, turnover is a phenomenon of climates with seasonality and does not occur in tropical zones, which won't experience sharp temperature changes year-round.

Temperature is an abiotic factor that impacts flowing stream ecology as well. It may have pockets of increased or decreased solar radiation, for example, areas under the shade of a tree canopy. But stream ecology does not experience a thermocline or turnover.

Carbonic Acid Is the Ocean's Buffering Agent

The biogeographic distribution of aquatic organisms is also dependent on the environmental range of pH concentrations. Environmental water will range from a pH of 2 to 12. However, the average pH of seawater is 7.5 to 8.4. Freshwater ranges from pH 6 to 8 (USGS). Most aquatic organisms survive close to neutral pH with varying tolerance limits, but for the most part they cannot sustain life under pH 4.5. One explanation for this constraint is that aluminum will leach from nearby rocks and river sediment under acidic conditions. Aluminum is highly toxic to aquatic animals. Whenever there is an excess of CO_2 a phenomenon known as **ocean acidification** occurs. With rising greenhouse gases in the atmosphere the diffusion rate of CO_2 from water to atmosphere slows down and oceans retain more CO_2, making the water more acidic. This also creates an acidic environment that is not conducive for shell-building species.

The relationship between atmospheric CO_2 levels and ocean acidity occurs as increased CO_2 dissolved in water converts into carbonic acid (**Fig. 5.9**). The full chemical equation is CO_2 to carbonic acid to bicarbonate ion, known as the carbonic acid cycle. This is a reversible chemical reaction with bicarbonate serving as a buffer against large fluxes in pH. The carbonic acid cycle also serves as a buffering agent in the human body. In human blood biochemistry this buffer is used preserve pH in a narrow range of 7.35 to 7.45 pH. A condition of acidity results when the buffer system has been overridden and can no longer take up excess hydrogen ions (H+).

The term pH represents the measure of concentration of H+ and is expressed on a negative logarithmic scale. A solution's pH is determined by the concentration

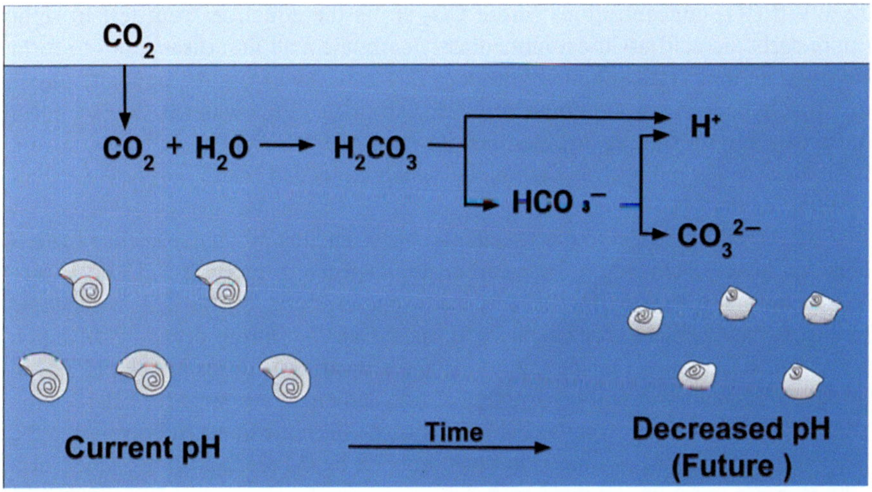

Fig. 5.9

CO_2 dissolves to form carbonic acid (H_2CO_3), breaks down to bicarbonate (HCO_3^-), and again to hydrogen ion (H^+) and carbonate ions (CO_3^{2-}). Higher concentrations CO_2 is acidifying and shifts equilibrium toward bicarbonate. Loss of free carbonate negatively impacts calcifying organisms.

Source: Elizajans under CC BY-SA 4.0

of free H+ ions compared to its concentration of free hydroxyl ions (OH−). If you add together free H+ ions with OH− ions you get water molecules (H_2O). As water molecules disassociate the free ions can be absorbed in chemical reactions by attaching to different molecules or they remain suspended in the solution. Higher free H+ means the solution is more acidic (low pH value). Higher free OH− means it is more basic (alkaline) and has a higher pH value. An equal concentration of free H+ and OH− ions results in a neutral solution, pH of 7, on the log scale from 1 to 14. Pure water of pH 7 is rarely found as some disassociation of ions will be present. In a buffering system excess free ions are scooped up rather than affecting the solution's pH. This is the reason that oceans retain their relative pH range of 7.5 to 8.4 and freshwater pH of 6 to 8. In **Fig. 5.9** the arrows show excess CO_2 moving in the direction from atmosphere to water, converting to carbonic acid, then bicarbonate and then carbonate while releasing free H+ ions into the solution. This direction *acidifies* the water if there isn't enough bicarbonate to scoop up excess hydrogen ions.

Let's look at the relationship of CO_2 to ocean acidity from the perspective of an equilibrium shift. Equilibrium in chemical reactions is dependent on reactant and product concentrations, according to Le Chatelier's principle shifting from higher to lower concentrations. In aqueous solutions CO_2 gas quickly reacts with water molecules (**Fig. 5.9**). The carbonic acid equation shift depends upon dissolved CO_2 concentrations (more CO_2 shifts the equation from left to right (more carbonic acid). Almost immediately carbonic acid first dissociates to form bicarbonate ions (HCO_3^-) and hydrogen (H^+) ions. And from the opposite direction free hydrogen ions combine with free carbonate in the water (CO_3^{2-}) to form more of the buffering agent, bicarbonate (a shift from right to left). But going toward the left requires free carbonate, which is needed for the process of shell building.

$CaCO_3$ is the material of hard skeletons, coral, and the protective shells of **calcifiers** (mollusks, clams, etc.). Free carbonate is needed to produce $CaCO_3$. At seawater's normal pH range the carbonic acid equation shifts left to right, with ample free carbonate available for calcifiers to make $CaCO_3$. During ocean acidification atmospheric CO_2 rises outside the norms and seawater pH drops below pH 7.5. In this environment there will be insufficient free carbonate available for calcification and in the worst scenario shells begin to dissolve. Since the industrial revolution the average pH of surface marine water has dropped by 0.1 pH, which accounts for a 30% increase in acidity (climate.gov).

In aquatic systems it is limestone rock that leaches calcite and aragonite (crystals of $CaCO_3$) into waterways. The crystals are dissolved into free carbonate and calcium and made available for the aquatic calcifiers to form their structures. Limestone is one of the major reserves for carbon and will be covered in more detail in Chapter 9, "Nutrient and Biogeochemical Cycling."

5.2 Marine Biomes

By the end of this section students should be able to:

- Distinguish between ocean stratification and what type of species can exist in each.
- Identify the zonation patterns in open ocean.
- Be familiar with the effects of water temperature on nutrient mixing.
- Discuss the high productivity of the neritic realm.
- Name at least two foundation species in marine biomes.

The ocean is a continuous body of salt water with mostly dissolved sodium chloride (3.5% NaCl) and other mineral salts with some decayed biological matter. The physical diversity of the ocean has a significant influence on plants, animals, and other organisms (**Fig. 5.10**). Within the ocean biomes there are subgroups based on climate and water depth:

- pelagic realms
  - divided by water depth into epipelagic, mesopelagic, bathypelagic
- coastal realms
  - divided into neritic realms (coral reefs and mangroves in tropical zones; sea kelp forest biomes in temperate zones) and transitional realms (estuaries and intertidal biomes)

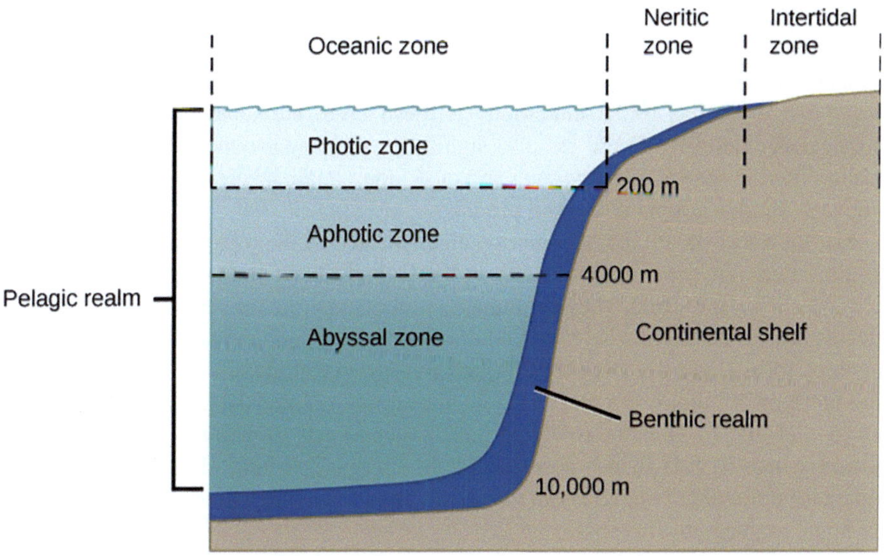

Fig. 5.10

The ocean is divided into different zones based on water depth and distance from the shoreline.

Source: James Rosindell under CC BY-SA 4.0

Unique Water Properties According to Its Depth

Amount of sunlight and temperature are major abiotic factors in all biomes. The relationship between light to the medium of water is unique. Water absorbs light waves to a different degree than traveling through air. And after a certain depth no light waves are able to penetrate. The depth where light penetrates is called the **photic zone** and beneath that is the **aphotic zone**. Primary productivity by photosynthetic organisms only occurs in the photic zone. Because sunlight is also the source of solar radiation in the form of heat, the temperature in bodies of water will be affected by access to sunlight. Heat exchange in water occurs more slowly than it does on land, leading to the maritime influence (Chapter 2). The function and survival of marine systems is dependent on thermal properties of water. Solar radiation also has a major impact on global climate and weather patterns from global current patterns (e.g. the Gulf Stream).

The Neritic Zones (Coastal)

Coastal biomes are part of the **neritic realm** and constitute the range of waterways from intertidal zones to the continental shelf. In this realm some light penetrates to the bottom of the continental shelf, meaning coastal biomes are mostly photic and photosynthetically active. Because of currents and wave action the aquatic environment is well oxygenated. And since the depth of the continental shelf is generally not more than 200 m (660 ft) it is low pressure with a stable, mild temperature. The neritic realm is productive and biodiverse, providing most of the aquatic diet of humans. Current estimates are that global coastal oceans contain almost 30,000 species of fish, more than 10,000 cnidarians (jellyfish, coral, etc.), and more than 50,000 crustaceans (shellfish). There are also significant primary producer phytoplankton, floating marine seaweed (*Sargassum sp.*), and other small species at the base of the food web, animal zooplankton, and protists.

Sea kelp forests are massive marine algae located in the rocky ocean shelves of neritic waters, growing predominantly in the United States along the Pacific Coast, from Alaska to California. They are found in other temperate higher-latitudinal zones such as the Tasman Sea, where we find the weedy seadragon (**Fig. 5.11**). Kelp communities form along the coastlines in proximity to ocean upwelling which delivers rich nutrients to supply its massive perennial growth. Giant kelp reach heights of 30–65 m (100–213 ft) as the towering underwater forests grow quickly towards the surface where its leaf blades access sunlight energy. Sea kelp also require cooler waters for optimal growth rates (up to 4 in/day) and form into underwater forests that are home to many diverse species.

Fig. 5.11

Giant kelp forests and weedy seadragons in the Tasman Peninsula, Tasmania.
Source: TomCatX, Public Domain

Fig. 5.12

In the kelp forests just off of the rocky cliffs of La Jolla, California, you can hear a symphony of fish sounds that reaches its peak just after dusk. (Photo: Camille Pagniello).
Source: California Sea Grant under CC BY 2.0

The kelp forest represents one of the most diverse marine habitats. The structure of the kelp forest is tiered, with multiple layers of understory lying beneath vertical columns of dense brown macroalgae species. The giant kelp (*Macrocystis pyrifera*) and bull kelp (*Nereocystis leutkeana*) form an overarching canopy (sometimes called the nursery zone) enveloping thousands of different marine species (**Fig. 5.12**).

Biodiversity of sea kelp forest includes invertebrates, fish, marine mammals, and birds. The nursery zone houses plankton and larvae, which grow easily in its protected warmth of sunlight. The stalk of sea kelp shelters medium-sized animals, like snails, sea stars, and soft spine sea urchins. The kelp itself is anchored to the rocky coastlines by **holdfasts**, where octopus might be hiding. The holdfasts grow vertically using air-filled bladders, called **pneumatocysts**. The kelp forest habitat even provides protection and shelter for larger animals, like sea otters, harbor seals, and even whales, which frequent the forests to feed on its species-rich habitat. During stormy weather kelp forests are particularly useful as a refuge from the force of strong winds and waves.

Sea kelp are known as an important **foundation species**. Without its presence the kelp forest community at large becomes threatened. Studies on the herbivore sea urchin, which grazes on giant sea kelp, have illustrated this fact. When the sea otter, a natural predator of the sea urchin, diminishes population densities of sea urchin explode. High densities of sea urchin can completely destroy a kelp forest. This has a negative impact on the entire biodiversity of its underwater forest community.

Coral Reefs

Coral reefs are even more biodiverse than the tropical rainforests and represent another type of foundation species. Coral reefs are mostly found in warm coastal, photic zone waters within 30°NS Lat. The $CaCO_3$ reef structures that make up the multicolored coral are built by small animals, known as **coral polyps** (phylum Cnidarian) (**Fig. 5.13**). Coral-building polyps live in a symbiotic relationship with a community of microscopic dwellers. The calcium carbonate coral structure provides housing and protection but the colorful pigments are produced by its co-inhabitants, photosynthetic unicellular algae (zooxanthellae). In this beneficial relationship the algae provide nutrients in exchange for housing. Photosynthesis requires pigment molecules to capture a range of solar radiation from the light spectrum. Zooxanthellae possess a wide array of pigments to capture varying radiant energy transmitted through the shallow waters.

An interesting side note, coral polyps are the only animal besides humans that build residential structures visible from space (Coral Reef Alliance). The Australian Great Barrier Reef, a massive 2000-mile-long reef structure, is found just off the northeastern coast of Australia. Like all coral reefs it is located in shallow waters of only 60 m (200 ft) or less. Many smaller coral reefs developed over thousands of years in shallow tropical waters, sometime surrounding islands or forming their own **atoll** (enclosing a lagoon).

The coral found along eastern coastlines tend to be more diverse. This is because of the direction of currents. The most biodiverse coral reefs are located in the Indo-Pacific waters, with fewer species found in coral reefs located in the Caribbean, Chile, and Florida.

Coral reefs are foundation species because of the large communities that inhabit the structures. There can be over 4000 fish species swimming through the coral reefs, feeding on coral polyps, small invertebrates (crypto fauna), seaweed, or predation of other smaller fish. The current health status of coral reefs is at risk and will be covered in a later section.

The Pelagic Realm (Deep Ocean)

Open ocean biomes of the **Pelagic Realm** are divided by water depth. The word "pelagic" is derived from Ancient Greek *(pélagos)*, meaning "open sea," and is an imaginary column of water extending from the upper surface of the sea almost to the bottom (**Benthic Realm**). Seawater covers 70% of Earth's surface, with the majority of that being deep waters (below 200 m). With increasing depth from the surface, pressure increases while temperature, DO, nutrients (iron, magnesium, and calcium), and light quality all decrease. In the open ocean there is an abundance of **plankton** as a food base for larger animals like whales and dolphins. Plankton are microscopic plants and animals at the bottom of the aquatic food chain. Open oceans are less productive than coastal waters because the necessary nutrients needed for photosynthesis are scarcer. When dead organisms fall to the ocean bottom, upwelling is prevalent in coastal waters but not open waters. The

Fig. 5.13

$CaCO_3$ skeletons of coral organisms form the coral reefs.

Source: Mattole Restoration Council under CC BY-SA 3.0

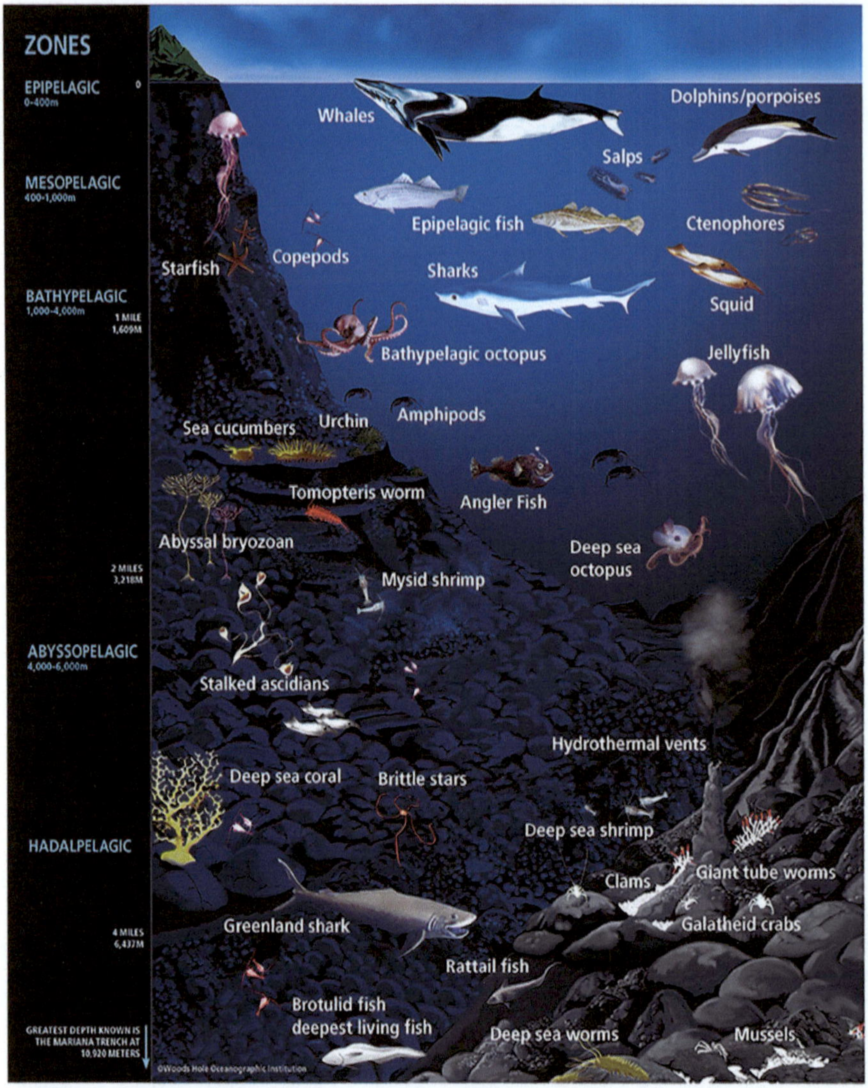

Fig. 5.14

Ocean animals (not to scale) in depth-defined ecological habitats.

Source: Haiku Deck under CC BY 4.0

open ocean is separated vertically into distinct pelagic zones: epipelagic (photic), mesopelagic (twilight), bathypelagic (midnight), and abyssopelagic (lower midnight) (**Fig. 5.14**).

Most photosynthesis occurs in the **epipelagic realm** and consequently most aquatic animals reside here, such as plankton, jellyfish, sharks, and dolphins. The epipelagic zone is within the photic zone and includes depths to 200 m.

The **mesopelagic realm** is in the uppermost, broadly defined aphotic zone from 200 to 1000 m. Some of the most abundant organisms in the aphotic zone are sea cucumbers (phylum Echinodermata). Heterotrophic bacteria (nonphotosynthetic) are abundant in the mesopelagic (twilight zone), which does not have adequate sunlight for productivity. Mesopelagic residents have large eyes, including swordfish and bioluminescent creatures like hatchetfish (family Gasteropelecidae). Bioluminescence is a useful adaptation used for communication, luring prey, or to disguise silhouettes from predators. That is the use for bioluminescence in hatchetfish that glow from their underbellies. The adaptation allows the hatchetfish to blend into the lighter waters above so that their predators lurking below cannot see them.

At the **bathypelagic realm** begins the truly aphotic zone and where the ocean becomes far less diverse. It is populated by many bioluminescent creatures, squid, angler fish, and various organisms that feed off of debris (detritus) falling from above. Within the mesopelagic and bathypelagic realms is a layer or zone recently discovered, known as the **Deep Scattering Layer**. Here deep-sea species (squid, crustaceans, and fish) follow a daily evening ritual of traveling toward coastal waters for feeding before returning to the protection of the murky depths of the ocean (NOAA). The migration occurs only in the safety of the night, as these species are adapted for no sunlight.

In the deeper parts of the ocean the **Abyssal realm** begins at depths ranging from 3000 m to 6000 m (9860 ft to 19,680 ft) and is in perpetual darkness. Its name comes from the Greek word for "bottomless sea." The abyssal realm is directly beneath the bathypelagic realm, beginning at whatever depth the water reaches 4°C (39 °F). The abyssal realm is not just freezing water, it is also highly pressurized (reaching 11,000 psi) with very low nutrient content. In the abyss the ocean is very still, not being subject to the action of waves and currents far above. A variety of invertebrates and fishes can be found in the abyss as well as many hydrothermal vents. Since ocean depth averages 4000 m (13,123 ft), most oceans have an abyssal realm. Given that oceans cover 60% of Earth's surface, they are the most expansive biomes, with an estimated 300,000,000 square km (115,000,000 mi). Since the abyssal zone extends to the ocean's benthic floor, it is actually more nutrient rich than the ocean layers above it.

The abyssal realm is comprised of two pelagic regions, the first of which is known as the **abyssopelagic realm** (once thought to be the bottomless ocean), where extremely high pressures require further adaptations among the limited distribution of species that can survive its challenges. Temperature increasingly drops with increasing water depth though remains above freezing. Of the few aquatic species found in the abyssopelagic are anglerfish, squid species, echinoderms like the basket star, marine arthropods like the sea spider, and other species that are transparent and eyeless. In hydrothermal vents chemosynthetic bacteria utilize hydrogen sulfide and other minerals emitted from the vents for energy utilization and serve as the base of the food chain for the entire abyssal zone. Occasional deep-sea trenches, specialized regions known as the **hadopelagic realm**, occur in 2% of the ocean. Nearly three times the depth of the average sea floor depth, it is characterized by its total darkness and pressures of up to 6000 pounds per square inch. The **Marianas Trench** is more than 11,000 m deep (36,000 ft) and considered to be the deepest place on Earth. The hadopelagic realm is largely unexplored due to difficulty operating submerged vessels for cameras and/or humans to spend time in this region.

The **benthic realm** comprises the layer of water present just above the ocean's bottom. The benthic zone is not unique to marine waters. It includes any column of water in contact with the riverbed, lake bottom, or ocean floor. Depending on the waterway, its depth can be just a meter below the surface (e.g., shallow rivers) or thousands of feet in deep oceans. A wide range of diversity can be present in the benthos, depending upon the composition of the water above it (saline, freshwater, mix). The benthic floor is comprised of sand, silt, and dead organisms, making it nutrient-rich from collecting debris and particulates from the layers above. Species diversity can include fungi, sponges, sea anemones, marine worms, sea stars, fishes, and bacteria, depending on depth to the ocean floor.

5.3 Water Movement and Transition Zones

By the end of this section students should be able to:

- Describe the unique properties of intertidal zones.
- Explain why an estuary can be classified as a transition zone.
- Explain why extreme challenges of estuary environments can create more diversity.

Water flow and velocity will determine the nature of aquatic life and are abiotic factors in aquatic species. Winds impart frictional forces on the water surface, creating waves or ripples. In large bodies of water ocean currents like the Gulf Stream bring warm tropical waters northward and southward, generating oceanic **upwelling**. Similar to lake turnover, the deeper, colder, and poorly oxygenated waters are brought to the surface by this water movement. And comparable to lake turnover they mix nutrients, oxygen, and water temperature, making them available where organisms reside.

The gravitational pull of the moon and sun create Earth's tidal movements. This, in turn, has a large influence on aquatic life. An intertidal zone is the area that lies between high and low tides. Like estuaries, it is an environment of extremes. In high tides the area is bathed in water. At low tides it is exposed to desiccation and atmospheric temperatures. The temperature at intertidal zones may fluctuate as much as 10°C to 38°C.

The Littoral Zone (Intertidal)

Anywhere that the ocean meets land is an **intertidal zone**. The intertidal zone is part of the **littoral zone**, which comprises any marine and freshwater meeting land (**Fig. 5.15**). This area is exposed to the air during low tide and immersed in sea water

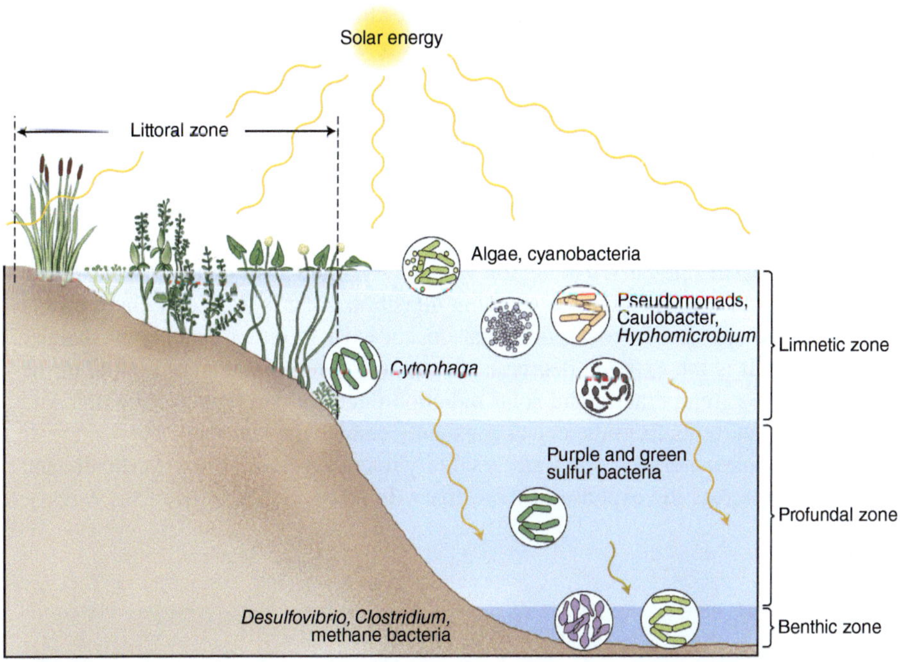

Fig. 5.15

The intertidal zone is a rich assemblage of biodiverse and productive species.

during high tides. It can have a sandy or muddy beach. It can be rocky without any beach. It is an incredibly variable environment because of the forces of tides and intermittent dehydrating effect of air and sunlight at low tide. When exposed to the atmosphere the temperatures can become extremely hot or freezing cold. There can be pockets of trapped salts from tidepools of retreating seawater. After evaporation the area would be even saltier than coastal waters. Intertidal plants, phytoplankton, and algae must be capable of adapting to their environment. The flora may be permanently submerged or transiently submerged, depending on the water depth. The zone extends to approximately the continental shelf in marine environments and the depth of photosynthetically active waters in lakes.

For any organisms, like crustaceans, to inhabit the shores they must be adapted to the pounding action of waves without dislodgment and their exoskeletons must be tough enough to prevent desiccation. In the high-tide region of the intertidal zone species that flourish include anemones, barnacles, crabs, limpets, mussels, sea stars, and snails with green algae. The low-tide region is most similar to coastal waters, with limited primary productivity and more diversity of aquatic animals (sea cucumber, snails, shrimp).

The littoral zone is an actively investigated area by ecologists who study how species survive challenges like interspecies relationships and competition. The study of intertidal species of rocky coastlines separates high tide (submerged only in high tide, mostly saline), middle tide (equally underwater and exposed to the air), and low tide (mostly underwater, almost normal saline) resident species.

Estuaries

From mountain streams water flows into rivers and follows a pathway toward the coastline and flows into regions where freshwater and saltwater mix, known as diluted (brackish) saltwater. The region where a river and river delta meet the ocean is known as an **estuary** (**Fig. 5.16**). This transition zone is a unique environment and presents aquatic life constraints that do not exist in either entirely freshwater or saltwater. It is the extreme degree of fluctuating conditions that is unique to the transition zone. Temperature and solar radiation fluctuate daily and seasonally. And the tides and currents from the ocean constantly change the landscape between river mudflat and rich waterway. The water's salinity fluctuates widely, both vertically and horizontally, across the expanse of the estuary. In less-active regions of the estuary,

Fig. 5.16

View of Rio de la Plata Estuary taken by a NASA astronaut. On the right is Buenos Aires and on the left is the city of Montevideo. The waters are brown where sediments are carried down from its rivers (Paraná and Uruguay), changing to blue as it mixes with marine waters.

Source: eol.jsc.nasa.gov, Public Domain

away from moving water, vertical separation is still achieved as lighter incoming freshwater will layer above the denser, high-solute seawater.

In regions with incoming tides freshwater and saltwater are mechanically mixed to form even vertical salinity. This is called **tidal overmixing**, creating a homogeneous solution from top to bottom. Horizontally a salinity gradient runs from the mouth of the river, across the estuary, to the edge of the sea. Physiological adaptations to survive in this unusual ecosystem require adaptations to continuous, rapid changes in salinity. One behavioral adaptation might be restriction of an organism's movement to just the regions where salinity levels are relatively constant. Oceanic species might be restricted to only moving inward at times when the flow of freshwater is lowest and salinity is highest. Overall, estuaries are low in biodiversity, with fewer animals successfully thriving in their harsh environmental conditions.

Estuaries are protected from the full force of waves by various boundaries like reefs, mudflats, or barrier islands. This creates a unique habitat for species that wouldn't survive the wave action but can handle a gradient of salt concentrations. You may already know of many estuaries under a different name, for example lagoon, bay, inlet, or sound. Chesapeake Bay and Long Island Sound are estuaries. The salinity of estuaries is a variable and a significant abiotic factor. The rate of flow of the freshwater sources determines salinity overall. But during certain times of the day it is the incoming tides that modify the salt concentration.

Estuaries are productive for all or some of an organism's life. Many juveniles of crustaceans, mollusks, and fish begin their lives in the shallow pools, freshwater, salt marshes, or microhabitats of estuaries (mud or sand flats, rocky shores, and wood swamps) (**Fig. 5.17**). The buildup of organic matter in estuaries can be as high as in a forest.

With the constant changes in salinity, estuaries create a very specific physiological challenge for plant and animal species. Estuarine plants tend to be **halophytes**, which have the ability to remove excess salts by water filtering systems in their roots, not unlike adaptations of aquatic fish to remove excess salts. Typical estuarine animals are mussels, clams, crabs, lobster, and other shellfish and both sea and shore birds will frequent the abundant food sources of the transitional zones.

5.4 Freshwater Biomes

By the end of the section, you will be able to:

- Describe freshwater ecology as distinct from marine ecology.
- Explain the differences between wetlands and intertidal zones.
- Explain how flow forces are a major abiotic factor in stream ecology.
- Give an example of adaptations to freshwater biomes.

Freshwater biomes with flowing waters, for example, rivers and streams, are known as **lotic**. Freshwater biomes with standing water, for example, lakes, ponds,

Fig. 5.17

Bloody Marsh, Georgia. Salt marshes and mud flats are havens for bird life.

Source: Bubba73 under CC BY SA 3.0

and wetlands, are known as **lentic**. The value of freshwater ecology cannot be overstated as they are natural resources for drinking water, irrigation, sanitation, and wildlife.

Freshwater wetlands are waterlogged terrestrial environments because the soil is permanently or periodically saturated with water. Wetlands have shallow bodies of water that don't vary in depth as they do in lakes and ponds. Wetland plants are rooted in a benthic soil with vegetative growth (leaves, stems, and flowers) above the water's surface. Wetland subgroups include marshes, swamps, bogs, mudflats, and salt marshes (**Fig. 5.18**). Note that these subgroups are recognized in the section above on intertidal zones. The distinction is whether the aquatic biome is a true freshwater habitat or has a saline component for at least some of the time, in which event it is classified under intertidal zone. Water ecology characteristics of wetlands include their type of vegetation and soil. **Hydrophytic vegetation** is suited to saturated soils, for example, duckweed, cattails, water lily, and hydrilla. **Hydric soils** are anaerobic at its surface due to continuous saturation and flooding.

Freshwater marshes and swamps will have especially slow-moving water. Bogs are depressions of stagnant, oxygen-depleted water that is trapped with no water flow entering or leaving and poor percolation from having a clay bottom. Bog water is also very acidic, as little decomposition can occur without oxygen and organic acids

River of Grass, Homestead, Florida

Florida Everglades National Park

Fig. 5.18

Salt marshes and estuarine mangrove forests of the Florida Everglades National Park, Homestead, Florida.

Source: Geograph Bot under CC BY 2.0

will begin to build up. The resulting anaerobic and acidic soil is much less conducive to plant growth.

Lakes and ponds vary in size and can be as large as thousands of square kilometers. In temperate zones a thermocline will participate in nutrient and temperature distribution. Light penetrates the photic (littoral) zones and phytoplankton (algae and cyanobacteria) are the primary producers at the bottom of the food web. The upper layer is fairly diverse. In the area just beneath the surface are zooplankton, such as rotifers and small crustaceans, representing its primary consumers. Usually fish are abundant in this layer, which is also called the **limnetic zone**. And finally, the benthic zone of lakes and ponds contains nonphotosynthetic bacteria feeding off dead organisms that sink to the bottom. The species present in lakes are generally more limited than other bodies of water, as they tend to be more isolated. In some climates a pond may be present only seasonally, returning for a few months every year.

For primary productivity the main limiting nutrients in lakes and ponds are nitrogen and phosphorus, needed for protein and DNA synthesis of living organisms. These macronutrients are also limiting in terrestrial soils. While essential to water ecology, there is nonetheless the risk of excessive leaching of nitrogen and phosphorus from sewage or agricultural runoff. This may result in a downward spiral of water pollution and overgrowth of algae (**algal bloom**). Excessive algal growth leads to excessive oxygen consumption, depleting its availability for aquatic animals. Algal blooms eventually spread so densely that they reduce light penetration in water, now hindering photosynthesis and its oxygen generation (**Fig. 5.19**). Once the surface water has become aphotic and anaerobic plants and fish alike experience mass die-offs. This condition is called **eutrophication**, which is neither exclusive to freshwaters nor a rare process. Eutrophication is a topic that is explored in more detail in Chapter 14 under Drivers of Biodiversity Loss.

Fig. 5.19

Algal overgrowth on Lake Erie, 2017.

Source: bunnygoth under CC BY-ND 2.0

Rivers and streams are continuously moving bodies of water that originated at their **source water**, or headwater (likely in a mountain), which will eventually make their way downhill to a river, delta, ocean, or lake. Some of the world's largest rivers are the Nile River in Africa, the Amazon River in South America, the Yangtze River in Asia, and the Mississippi River in North America.

In stream ecology the amount and strength of water flow is an abiotic factor affecting species diversity. Other abiotic factors vary along the length of the river or stream, as it widens and collects more nutrients traveling toward a final destination. When water is fast moving there will be less silt accumulation, making the water clearer. Productivity may be limited to rocks covered with algae or river plants when the current is swift. The nutrient content may vary seasonally, such as organic material falling into rivers or streams with the fall leaves. Some animals have adapted to flow forces by anchoring themselves to substrate by suckers, for example, leeches (phylum Annelida). Freshwater rainbow trout of phylum Chordata prefer fast streams and are well adapted to spawning in head waters before traveling to the open ocean to complete their lifecycle. This migratory pattern is similar to salmon lifecycle.

As the river or stream heads toward a tributary and joins other streams, its width increases and current begins to slow down. Calmer water deposits more sedimentation and a different ecology emerges, with phytoplankton and burrowing species like worms (phylum Annelida) and insects (phylum Arthropoda) now present. Waterfowl and different fish species, adapted for slower moving water and richer nutrients, will also be spotted further downstream.

5.5 Global Decline of Coral Reefs

After reading this section students should be able to:

- Develop an understanding of factors that led to decline of coral reefs.
- Relate the importance of coral reefs to oceanic life.

Coral reefs represent only 1% of the world's marine environment yet are home to 25% of all marine species, including fish, sea snakes, marine iguana turtles, and algae. Large aquatic animals, including sharks, rely on the coral reef habitat to hunt their prey. Photosynthesis and carbon fixation take place in coral reefs by symbiotic zooxanthella algae, which make their home within the tissues of the reef-building corals.

The health status of coral reef fisheries is a good indicator of environmental stress. The death of a coral reef suggests a much darker diagnosis for the region's marine health. Hypersensitive to climatic change, the coral reef is the colloquial "canary in the coal mine" with respect to the vulnerability of water ecosystems.

Climate change poses a major threat to the viability of coral reefs. The Global Coral Reef Monitoring Project has estimated there has been a 14% loss of coral reefs between 2009 and 2018. Destruction of coral reef includes the disintegration of its $CaCO_3$ architecture by increasing ocean acidification (recall the carbonic acid cycle). With global warming ocean temperatures are raised, increasing the amount of CO_2 that enters the oceans and lowering the average ocean pH (acidification). The buffering action of the carbonic acid cycle during acidification leads to more carbonates being unavailable for the $CaCO_3$ structures of the coral reef. Furthermore, excessive heat in the ocean expels the zooxanthellae symbiont from the coral reef. Ecologists are able to monitor the degree of loss of symbionts by measuring the amount of coral reef bleaching. Coral polyps are heterotrophic, dependent on their symbionts to acquire their energy, and do not survive whenever loss of their symbiotic relationship is prolonged. Climate scientists believe that an increase in temperature of just 1°C to 2°C would be the tipping point at which many of the world's coral reefs will no longer be able to recover.

Climate change is only one factor driving global decline in coral reefs. Approximately 55% of coral reefs around the globe are threatened, which dynamic fishing and overfishing exacerbate. The fishing method known as **dynamite fishing** destroys delicate coral tissue and wreaks havoc on the architecture of coral colonies around the globe, making them more vulnerable to wave action and tropical storms. Dynamite fishing is a form of "blast fishing" and is now illegal in most nations. But despite government crackdowns dynamite fishing continues unabated in regions of Southeast Asia, including the Coral Triangle, and Tanzania. Dynamite fishing was first introduced as an economical fishing method during World War I when European soldiers began using grenades to quickly gather seafood. Researchers have demonstrated that one small dynamite is capable of blasting a crater the size of a meter in diameter. With repeated blasts, it becomes impossible for the coral reef to recover.

Human population growth is an indirect driver of global decline in coral reefs. The increase in pollution from agricultural run off and overfishing of popular fish species alters the aquatic food web. Lower trophic level species that feed on coral polyps grow in overabundance without their predators from the upper trophic levels. This disruption in the food web balance results in decimated coral polyp populations.

Ecologists are aware that once the reefs are gone the fish will disappear as well but many fishermen do not understand this relationship. Decline of coral reefs poses a serious threat to coastal economies in terms of commercial fishing and tourism alone. The death of coral reefs has an estimated value of $375 billion per year (some estimates are as high as 500 billion pounds). With loss of coral reefs, erosion of the continental shelf from powerful wave action becomes amplified. This is because coral reefs protect the benthic realm and shorelines analogous to how rooted trees prevent soil erosion.

Coral reefs are foundation species that need protection. And scientists are taking broad measures to save coral.

Glossary

flow forces
deep scattering layer
source water
dynamite fishing
eutrophication
algal bloom
foundation species
water runoff
groundwater
epipelagic
mesopelagic bathypelagic
abyssopelagic
hadopelagic
atoll
hydrologic cycle
hydrogen bonding
cohesion
viscosity
high specific heat capacity

Citations

1. A.M. Smith, The role of suction in the adhesion of limpets, J Esp Biol 151 (1) (1991) 151–150. https://doi.org/10.1242/jeb.161.1.151.
2. J. Staughton, Why do deep-sea fishes not get crushed by pressure on the sea floor? Science ABC (2023). https://www.scienceabc.com/nature/animals/deep-sea-fishes-not-get-crushed-pressure-sea-floor.html.
3. V.J. Merten, et al., Deep-sea predatory-prey dynamics revealed by biologging and eDNA analysis, Frontiers in Ocean Observing (2023). https://doi.org/10.5670/oceanog.2023.s1.31.
4. T.A. Blewett, et al., Physiological and behavioural strategies of aquatic animals living in fluctuating environments, J Exp Biol 225 (9) (2022). https://doi.org/10.1242/jeb.242503.
5. "Temperature of Ocean Water." Windows to the Universe. University Corporation for Atmospheric Research. 2001-08-31. Archived from the original on 2010-03-27. Retrieved 2019-12-27
6. G.K. Meyer, et al., Impact of Climate Change on Sensitivity of Lake Stratification: A Global Perspective: Water Resources Management in the Face of Climatic/Hydrologic

Uncertainties, Springer Netherlands, 1996.225–270 10.1007/978-94-009-0207-7_9 ISBN 978-94-010-6577-1

7. A. Botté, M. Zaidi, J. Guery, et al., Aluminum in aquatic environments: abundance and ecotoxicological impacts, Aquat Ecol 56 (2022) 751–773. https://doi.org/10.1007/s10452-021-09936-4
8. P.W. Atkins, The Elements of Physical Chemistry, 3rd ed., Oxford University Press, 1993.
9. O. Sulpis, E. Jeansson, A. Dinauer, et al., Calcium carbonate dissolution patterns in the ocean, Nat Geosci 14 (2021) 423–428. https://doi.org/10.1038/s41561-021-00743-y
10. Mae Rennick, et al., Detrital supply suppresses deforestation to maintain healthy kelp forest ecosystems, Ecology (2022). https://doi.org/10.1002/ecy.3673
11. M. Charette, Walter Smith, The Volume of Earth's Ocean, Oceanography 23 (2) (2010) 112–114. https://doi.org/10.5670/oceanog.2010.51 hdl.1912/3862
12. (2017). Hatchetfish cloaking strategies unveiled. *Nature*. 545, 137. https://doi.org/10.1038/d41586-017-00641-7
13. J.V. Gardner, A.A. Armstrong, B.R. Calder, J. Beaudoin, So, how deep is the Mariana Trench? Marine Geodesy 37 (1) (2014) 1–13. https://doi.org/10.1080/01490419.2013.837849
14. M.P. Angelo, "Understanding the world of benthos: an introduction to benthology," chap 1. Ecology and Biodiversity of Benthos, Walag (2022) 1–19. https://doi.org/10.1016/B978-0-12-821161-8.00002-7
15. L.M. Nordlund, et al., Intertidal zone management in the western Indian Ocean: assessing current status and future possibilities using expert opinions, Ambio 43 (8) (2014) 1006–1019. https://doi.org/10.1007/s13280-013-0465-8
16. T.J. Flowers, T.D. Colmer, Plant salt tolerance: adaptations in halophytes, Ann Bot 115 (3) (2015) 327–331. https://doi.org/10.1093/aob/mcu267

CHAPTER

Natural Selection and Adaptations 6

Chapter Outline

6.1 Darwin and Natural Selection

By the end of this section students should be able to:

- Provide a historical timeline for preevolutionary thought.
- Compare essentialism and enlightenment in terms of species alterability.
- Garner an appreciation of 19th-century discoveries that defined the concepts of evolution.
- Distinguish between natural selection and descent with modification.
- Name three principles of nature that underlie natural selection.

Ecology for the 21st Century. DOI: https://doi.org/10.1016/B978-0-443-24031-7.00006-4

Before Darwin

Evolution by **natural selection** describes a *mechanism* for how species change over time. That species change has been observed and debated since ancient times. The topic of evolving species can be found throughout recorded history, in writings from Greeks, Romans, Chinese, and even medieval Islam. However, it was during the 17th century that two opposing ideas were firmly established and debated in the scientific realm: the inalterability of species in **Essentialism** (Aristotle's school of thought) and the "new age" alterable, nonstatic **Enlightenment** (variability within species). By the 19th century, the world's leading naturalist, **Alexander von Humboldt** (1769–1859), had successfully argued that across geographic regions there are distinctly different plant and animal populations, even when their environments appear to be similar. Recall from Chapter 2 that it was von Humboldt who introduced the concepts of biogeography.

By the end of the 18th century the Scottish naturalist **James Hutton** (1726–1796) had proposed that geological changes occurred in the landscape gradually by the accumulation of small changes over long periods of time. This hypothesis contrasted with the predominant view of the era: that Earth's geology was as constant as the inalterability of its species. Hutton's theories on gradually changing landscapes were ridiculed by **Georges-Louis Leclerc Comte de Buffon** (1769–1832, known as **Baron Georges Cuvier**) (**Fig. 6.1**).

Cuvier happened to be the world's leading naturalist and a strong proponent of extinction by catastrophic events over *short* periods of time. Cuvier was the curator of the National History Museum in Paris, where his global influence inspired and mentored many of the leading scientists as the world entered into the golden era of science (1800s). His groundbreaking work as an exemplary anatomist working with fossils established the fields of comparative anatomy and paleontology. The emerging field of paleontology sought to explain discrepancies in **extant** species and the strange fossils being discovered, which provided evidence that species could indeed go extinct. Extinction was a radical view that refuted Essentialism's static view of the natural world as God's creation. Even as openminded as Cuvier was to the topic of extinction he rejected the newly emerging concept that evolving changes occurred in *living species* over time. Cuvier spoke of a **Lost World of Fantastic Beasts** (e.g., the massive mammals like mastodon and woolly mammoths). Cuvier's vision accepted evidence for cyclical periods of new creation and destruction of life forms. He concluded these events occurred as a result of global extinction events, specifically catastrophic floods, and land mass movements. But paradoxically his perspective remained incapable of accepting evolution of species. The fantastic beasts lived and then died off.

A few contemporary scientists of the mid-1800s proposed alternatives to Cuvier's **Catastrophism**. For example, **Jean-Baptiste Lamarck** published a book on the mechanism for evolutionary change in which he argued that acquired characteristics could be modified in an individual by its environment through the use or disuse of one of its appendages during its lifetime and that this modification could

Fig. 6.1

Georges Cuvier.

Source: Georges Cuvier, public domain

be inherited by its offspring, which would in turn lead to a **Transmutation** of the species. Lamarck and other preevolutionary theorists were ridiculed heavily by the highly revered Cuvier. The harshness of his criticism of Lamarck served as a deterrent to other naturalists speculating or hypothesizing about gradually changing species. While no serious discussions took place within the science community during the first half of the 19th century, soon to enter the debate would be two naturalists, **Charles Darwin** and **Alfred Russel Wallace**, and their groundbreaking publications on evolution. (**Fig. 6.2**).

The Pattern and Process of Natural Selection

The mechanism for evolution of species was independently developed by Charles Darwin and Alfred Russel Wallace during the mid-1800s. Both explored the natural world by traveling to the tropics, including the Galápagos Islands west of Ecuador, to observe and propose an explanation. Darwin traveled between 1831 and 1836 to

Fig. 6.2

(a) Alfred R. Wallace. (b) Charles Darwin wrote scientific papers on natural selection that were presented together before the Linnean Society in 1858.

Sources: Fig 6.4a: London Stereoscopic and Photographic Company, Public Domain; Fig 6.4b: Messrs. Maull and Fox, Public Domain; Darwin and Wallace images modified by author

South America, Australia, and the southern tip of Africa. Shortly thereafter (between 1848 and 1862) Wallace traveled to Brazil where he collected insects from the rainforest and the Malay Archipelago. In the Galápagos Darwin observed species of organisms on the different islands that were similar, appearing to be the same species yet with distinct differences. He documented many different species of ground finches among the islands with unique beak shapes (**Fig. 6.3**). The species displayed a gradation of beak sizes and shapes. He also observed that among all variations of Galápagos ground finches they still resembled a related finch species he'd identified on mainland South America. Darwin proposed that the various island ground finch species all originated from the mainland ground finch and were modified by its continued existence on the respective islands (over time). The naturalist had also observed and measured the beaks of each island ground finch, which variation suggested they would be suited to a specific type or size of food. For example, seed-eating finches were found to break seeds using their stronger, thicker beaks, and insect-eating finches were able to stab at their prey using a much narrower, spear-like beak.

Fig. 6.3

Darwin's ground finches. (a) Variation in ground finch beak shape found on the Galápagos Archipelago, as sketched in "Voyage on the Beagle." (b) Contemporary photographs of the same genus, species.

Sources: Fig. 6.3a: Kiwi Rex under CC By SA 4.0; Fig 6.3b: Voyage of the Beagle, Public Domain; Kiwi Rex and Voyage of the Beagle images modified by author

Recall from Chapter 1 that investigative approaches using the scientific method include deciphering a pattern and process. Wallace and Darwin had observed specific patterns in species variation occurring across geographic ranges. The two naturalists then independently came to the same conclusion of a process that could explain how and why such changes were occurring. Darwin gave his hypothesis a name: **natural selection**. Natural selection is also known by the phrase **survival of the fittest**. Natural selection states that there will be prolific reproduction in individuals with favorable traits within its specific environmental challenge. This favorable trait will impart a selective advantage, which in turn will lead to an overall change, or modification, of the species over time (i.e., those who survive to reproductive age can contribute offspring with their favorable traits). Under a different environmental stressor a different set of individuals within the population would possess the most favorable traits (**Fig. 6.4**). They would then likely survive longer and become more prolific toward producing their future generations. Darwin's natural selection hypothesis was a logical explanation for what was observed in nature.

Darwin also described giant tortoises in the Galápagos Archipelago with much longer necks compared to tortoises found on nearby islands. His Galápagos tortoises could use their longer necks to reach the foliage during times of drought. He described this species as evolving under a selection mechanism to increase its survival odds. During long periods of drought, tortoises with shorter necks would die off as they couldn't access their food resources (leaves). Consequently it would be the long-necked tortoises reaching reproductive age and more likely to pass on their long-necked trait to offspring. If long periods of drought conditions became

Fig. 6.4

(a) Saguaro cacti (Carnegiea gigantea) can absorb up to 750 L H2O during the rare rainfall deluges in the arid desert biomes of Mexico and the southwest United States. (b) Andean cold-blooded lizard (Potamites montanicola) are unusual lizards with nocturnal lifestyles to survive in the 1570–2100 m altitudes of the Andean mountains.

Source: Gentry George, USFWS, Public Domain

the norm in a particular environment only long-necked tortoises would be representative of the tortoise population. In contrast the same tortoise species living in a region experiencing significant rainfall for productive foliage growth would not need to select for this particular modification. The selective process, when applied to reproductive success, is known as **descent with modification**. In summary, Darwin's name for evolution was *descent with modification*, and the process (mechanism) he attributed to bringing about the modifying elements was *natural selection*, aka survival of the fittest.

The theory of evolution was an advanced concept compared to how Lamarck described the process of transmutation a century earlier. Transmutation proposed species evolve up a ladder of complexity by intent. Evolution proposed the process of change is random and without design, simply coping with current conditions. Lamarck's concept of changing species did not consider a common ancestor; he proposed the changes made new species, created by spontaneous generation. By the time the theory of evolution was published Darwin and Wallace had worked out the mechanisms with well–documented research data. Within a very short time the scientific community accepted its explanation on how organisms changed over time.

Descent With Modification

According to Darwin natural selection is the inevitable outcome of three principles of nature:

1. *Characteristics of organisms are mostly inherited, descending from parent to offspring*. In the mid-1800s no one knew the mechanism of inheritance

but it was a commonly accepted understanding based on observation. *Note:* Mendel's genetic experiments were ongoing around the same time but the significance and implications for inheritance were still unknown.

2. *Far more offspring will be produced than are able to survive*. The capacity for reproductive success in a generation's individuals will outstrip availability of resources to support the members of all species, which in turn will lead to competition for resources.
3. *Individual offspring of a species vary in regard to their characteristics and those variations are inherited*. As demonstrated in the example of the tortoise under drought conditions both Darwin and Wallace reasoned that some offspring would have inherited characteristics more favorable to compete for limited resources in their specific environment. Some offspring will have greater chance of survival to reach reproductive maturity. Individuals that did not possess a favorable variation (i.e., suitable attributes under the presiding environmental conditions) will be less able to compete and survive to reproductive maturity. This inequitable success rate will in time lead to a sustained change in the species at its *genetic level*. Whenever a change reaches the genetic level of a species evolution has occurred. This is the process that Darwin called **descent with modification**. *Note:* Darwin was unaware of genes or DNA but the changes in species were observed in their form and behavior.

In his book "On the Origin of Species" Darwin outlined in considerable detail his arguments for evolution (i.e., descent with modification by the process of natural selection). Indeed Charles Darwin and Alfred Russell Wallace both cited a broad range of evidence for evolution, drawn from animal husbandry, biogeography, geology, morphology, and embryology. However, one of the best-known examples was demonstrated in Darwin's Galápagos finches.

Alfred Russell Wallace was a fan of the more well-known Charles Darwin. During his explorations in 1958 Wallace took to a feverish state in Ternate (Indonesia) when he first composed his understanding of survival of the fittest passing on their traits. He immediately wrote to his peer Darwin, who had been developing a similar theory for the previous 20 years. Wallace's contribution to evolutionary concepts began to be overshadowed after Darwin's "Origin of Species" was published the following year. But the two men did not entirely share the same views on species, as Wallace believed the human intellect was not subject to the laws of natural selection nor could be explained by it. He was a spiritualist and held there was a distinction between humans and animals that the human spirit did not extinguish even upon death. Wallace's spiritualist views were generally not supported by his scientific contemporaries, but the realm of spiritualism is not subject to scientific investigations either. Wallace was also one of the first members of the scientific community to raise concerns over the environmental impact of human activity.

6.2 Processes and Patterns of Evolution

At the end of this section students should be able to:

- List other mechanisms of evolution besides natural selection.
- Give a definition of adaptation.
- Distinguish between convergent and divergent evolution.
- Describe homologous and vestigial structure.
- List other evidence of evolution and discuss misconceptions about the theory of evolution.
- Differentiate acclimation from adaptation.

For natural selection to occur there must be present some **variation** among individuals in a population. For these differences in individuals to continue to the next generation they must have a genetic basis. In other words, different traits between individuals due to nutritional status are not carried to the next generation. Descent with modification is the heritable passing of changes in the genetic code from parent to offspring. In natural selection *survival of the fittest* gives direction to an otherwise random descent with modification process. *Survival of the fittest implies pressure is being placed on the population, most notably environmental stressors.*

Genetic migration, also known as gene flow, is a mechanism of evolution leading to distinct populations of the same species expressing different characteristics because of slightly different genetic codes. As individuals migrate and breed among themselves their young population will mix and change the genetic profile in offspring. *Genetic migration occurs by a process not under a survival of the fittest pressure.*

Genetic drift is a random change in the number of occurrences of a particular variation of a gene's phenotype. For example, in a group of equal proportions of green and brown beetles if the majority of brown beetles happened to be feeding on the side of a tree closer to their predator (a bird, for example), the number of brown beetles will be drastically reduced overall, and the population will now have more green beetles. The interaction between prey and predator in this case was a random event that wiped out a significant proportion of brown beetles from the beetle gene pool. It does not fall under natural selection because the environment didn't *favor* the traits for green over brown beetles. Rather the brown ones just happened to be in the wrong place at the wrong time. Genetic drift can also happen at the molecular level when genes are duplicated (another random event), which increases its frequency in the gene pool. *Genetic drift occurs by a process that is random, not under a survival of the fittest pressure.*

Mutation is a random process that changes the genetic profile. Offspring inherit an altered genotype after random mistakes in gene replication or when large pieces of chromosomes break and move around. The genes impacted may now code for proteins with a slightly different function. Mutations are mostly spontaneous, although some are induced by mutagenic agents, like X-ray or UV light waves. In a population of beetles feeding on a green leaf a mutation in one individual that produces a green pigment instead of the usual brown pigment will have a new advantage as the green

Fig. 6.5

Ancon sheep.

Source: Archives of LIFE magazine, 1947, Public Domain

color provides camouflage against its predators. The brown beetles will continue to be easily picked off while the green beetles live long enough to produce many more offspring carrying their green mutation. In time the green beetles will overtake the brown beetle phenotype in the population.

One of the earliest known records of a gene mutation occurred in 1791 when a sheep farmer, Seth Wright, noticed he had a lamb with unusually short legs. Seeing the advantage in herding sheep that were too short legged to jump his low stone walls, he kept breeding for the desirable trait, eventually developing the Ancon Sheep species (**Fig. 6.5**).

Charles Darwin wrote about the Ancon sheep (of common sheep *Ovis aries)* in his "Origin of Species" as he considered the breed a perfect example of **macroevolution**, even though the Ancon sheep is more appropriately an example of **microevolution**. While those terms had not yet been attributed to the phenomenon Darwin was arguing correctly that small changes within a population over a relatively short time period will ultimately add up to larger changes over time. This is the origin of new species. However, Darwin was erroneously arguing that the existence of Ancon sheep supported his theory that offspring can inherit characteristics independently from their parents. Modern genetic analysis reveals the short-leg phenotype mutation to be a homozygous recessive trait, meaning both parents must carry one copy of the hidden mutated gene. It is masked in the parents because two copies of recessive traits are required for the phenotype to be expressed. *Animal husbandry and breeding of traits does not fall under a selective pressure survival of the fittest mechanism.*

Genetic diversity enters a population through random gene mutations (like the Ancon Sheep) and also through sexual reproduction. When two parents reproduce their combined genotype creates a unique combination and phenotype in each of the offspring. The ultimate strategy of sexual reproduction is to increase the pool of alleles in a population so that some individuals within the species will survive under any given number of environmental stressors. In nature new combinations of genes from random mutations, migration, genetic drift, and sexual selection all provide important variation within the gene pool that lends a selective advantage for the species survival. It is **natural selection** that makes certain changes are passed on to subsequent generations. *because those traits that were the most suited to survive in the current environmental pressures will direct the species' adaptation.*

Adaptations Increase Fitness

An **adaptation** is any genetic trait developed to help an organism survive in its present environment. Populations of species become *adapted* to their environment when a change in their genetic variation over time increases or reinforces their "fitness" to the environment. Adaptations can be morphological or behavioral or anything that would increase the species fitness. For example, the characteristic spotted patterns of the snow leopard took many generations to evolve. The spots help the leopard hide from its prey by blending into its environment more effectively. The thick fur on snow leopards is another adaption for survival in cold climates. Adaptations are traits found within the gene pool of the *entire* species' population in the particular geographic region. Whichever individual members express the adaptation will have the selective advantage. For example, bacteria have adapted to most environments by synthesizing their own nutrients. One of the best examples is the bacterium that inhabits the human intestine, *Escherichia coli*. Since the supply of nutrients varies with the host's diet, the bacterium must develop adaptations to a less-restrictive diet to survive. For example, being able to produce the metabolic pathway and enzyme, lactase, which breaks down lactose from milk as its carbon source.

Whether or not a trait is favorable will always depend on the specific conditions under which the species exists. For example, if a plant grows in a moist climate, it will not need a mechanism to conserve water. Producing larger leaves under these conditions would be advantageous because greater leaf surface area will increase rates of photosynthesis and energy production. In a moist environment, therefore, we could anticipate traits for producing larger leaves will be selected for. However, in dry climates, larger leaves do not provide any selective advantage. When there's a shortage of water large leaves present a disadvantage as they are a drain on water reserves. Can you name any large-leafed plants in a dry desert biome? Not likely! And if we consider thousands of years of gradual climate change toward a drier environment we can anticipate that the direction of natural selection in plants would shift to smaller leaves. In summary, adaptation depends upon conditions that are most favorable for the species' survival.

Divergence and Convergence of Species Evolution

The evolution of species has given rise to a seemingly endless degree of variation in form and function. The direction of evolution can also lead to duplicative modes of form and function. These pathways are known as divergent and convergent species

Fig. 6.6

(a) The common snap dragon (*Antirrhinum majus*) and the (b) geranium (*Pelargonium spp.*) vary in appearance yet share a basic morphology as flowering plants.

Source: Images by author

evolution. When evolution results in groups of organisms from the same common ancestor *diverging* into distinctly different species it is called **divergent evolution**. An example of divergent evolution is found in the reproductive organs of flowering plants, which share a common ancestor and basic anatomy. Flowering plants have adapted over time to different climates and types of pollinators present, resulting in an enormous diversification of their reproductive strategies (**Fig. 6.6**).

It is possible that similar phenotypes may evolve independently in distantly related species. For example, the capacity of flight has evolved independently in bats and insect species. Each developed analogous structures as their adaptation to perform flight. However, they have no common ancestral origins and wings evolved from very different original structures. This phenomenon is called **convergent evolution**. There is no homology between the evolutionary pathways in bats and insects.

Evidence for Evolution

There is compelling and extensive evidence for evolution and natural selection. Biologists have documented a record of ongoing evolution at every level of organization in living systems, from molecular to organismal to populations and communities. Darwin wrote at length in "On the Origin of Species" about the patterns he had identified in nature that were consistent with evolution. Since those golden years of biology of the mid-1800s our understanding has become broader and more in-depth. In the same era of Darwinism Mendelian genetics was in its infancy but the two would eventually be integrated into a modern synthesis of the theory of evolution, called evolutionary genetics.

1. **Fossils** constitute a form of physical evidence that organisms from the past are not the same as organisms existing today. Fossils reveal evolutionary progression in species because it is possible to determine the age of the specimen. Through the fossil record scientists tell the story of Earth's past and its evolution of form over millions of years. *Homo sapiens* is the only surviving member of a once diverse genus, *Homo*, which is classified with other genera in the more inclusive *Hominidae* family (includes *Pongo, Pan, Homo* and *Gorilla*). However, in horse lineage only one genus, *Equus*, is still extant in the once highly diverse genera in the horse family, *Equidae*. Fossils from diverse specimens of *Equus* have been recovered from all continents except the Arctic and Antarctic and date back to 5 Mya. The morphology of early *Equus* horses had striped bodies like modern zebras, short donkey-like heads, and stiff, upright manes (**Figs. 6.7a, b**). The genus *Equus* today includes only the horse (*Equus ferus*), zebra (*Equus zebra*), and donkey (*Equus africanus asinus*). It is unknown why the horse disappeared altogether from North America nearly 11,000 years ago, although it is suggested that the indigenous peoples ate them instead of domesticating them. Historical records show that approximately 3000 years ago humans in other parts of the world began cultivating the horse for agriculture, transportation, and warfare. *E. ferus* was reintroduced to North America by the Spaniards in the 15th century.
2. **Anatomy and embryology**—the presence of similar structures across related species provides evidence for evolution. The appendicular skeletal bones in dogs, bats, birds, and whales share a similarity in their overall construction (**Fig. 6.8**). These species also have the same ancestral origins. Over time evolution led to modifications in the shape of appendages in each of these species as different functions were needed, which is an example of divergent evolution. The overall layout shows homology. For example, the wings of a bat and the flippers of a whale are **homologous** because the structures originated from forelimbs of common ancestral mammalians. Through divergent evolution one appendage evolved for flight and another for swimming.

   In the white landscape in the Arctic, the arctic fox and the ptarmigan grouse provide unique evidence for convergent evolution. These are very different species and not related by a common ancestor. Yet they developed a similar seasonal color change (from dark to white) as camouflage to avoid being eaten by winter predators. When structures evolve independently, under selective pressure to serve a similar function in their environment, they are **analogous structures** (**Fig. 6.9**). An example of analogous structures by convergent evolution occurs between the organs of flight in bats and birds. They do not share a recent common ancestor. There are sufficient dissimilarities between

Fig. 6.7a

(a) Fossil hominids. (b) Artist's rendition of extinct species of genus *Equus*.

Source: Heinrich Harder under Public Domain

Fig. 6.7b

George Catlin's bison hunt. Fierce wild horses roamed North America and were integral to Native American Indian culture. To conquer the Indian the US Calvary resorted to repeated massacres of entire Indian horse herds—a testimony to the toughness of the breed and superiority in battle.

Source: George Catlin, WikiVisually

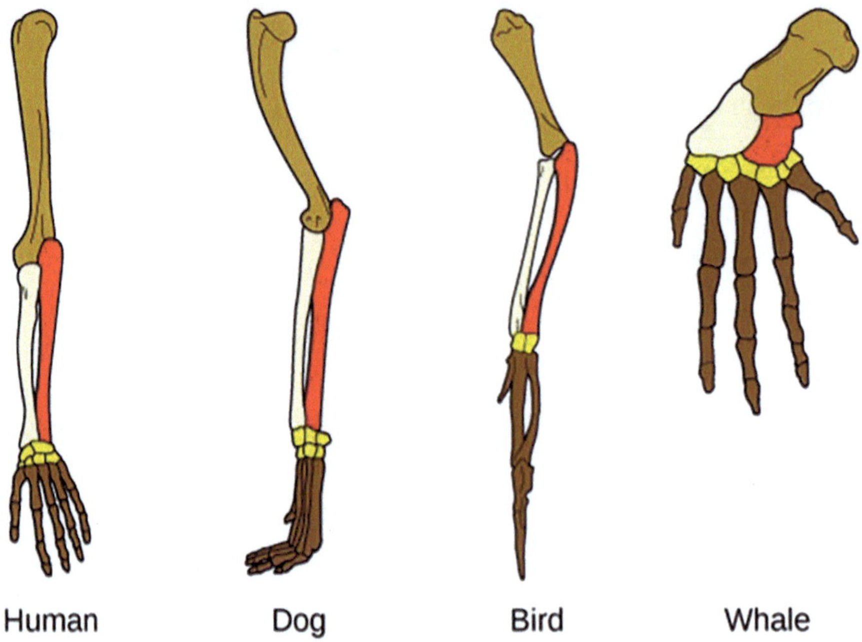

Fig. 6.8

Similar construction of appendages indicates organisms share a common ancestor.
Source: Волков Владислав Петрович under CC BY SA 4.0

their wings to indicate they were not inherited from a common ancestor that already had wings. The wings on birds have feathers across their forelimb, whereas bat wings have flaps of unfeathered skin that stretch across their fingers and forelimb. These are analogous structures because their evolutionary origins are distinct. However, the presence of a *forelimb* structure is still considered a homologous structure inherited from a four-legged common ancestor to the bat and bird.

3. **Embryology** is the study of the developmental anatomy of an organism. The process of embryology is highly conserved, as even the slightest change in gene expression will have broad-scale implications on the growing individual. Cellular and molecular events occurring during this critical period of development provide evidence of relatedness among widely divergent groups of organisms. Structures that may be absent in species may actually have been present in the species' embryonic form. This is evident in vertebrate embryos, including humans, which exhibit gill slits and tails at some point in their early embryonic development. Those structures disappear in the adult forms of the terrestrial groups but are maintained in adult forms of aquatic groups, like fish and amphibians.

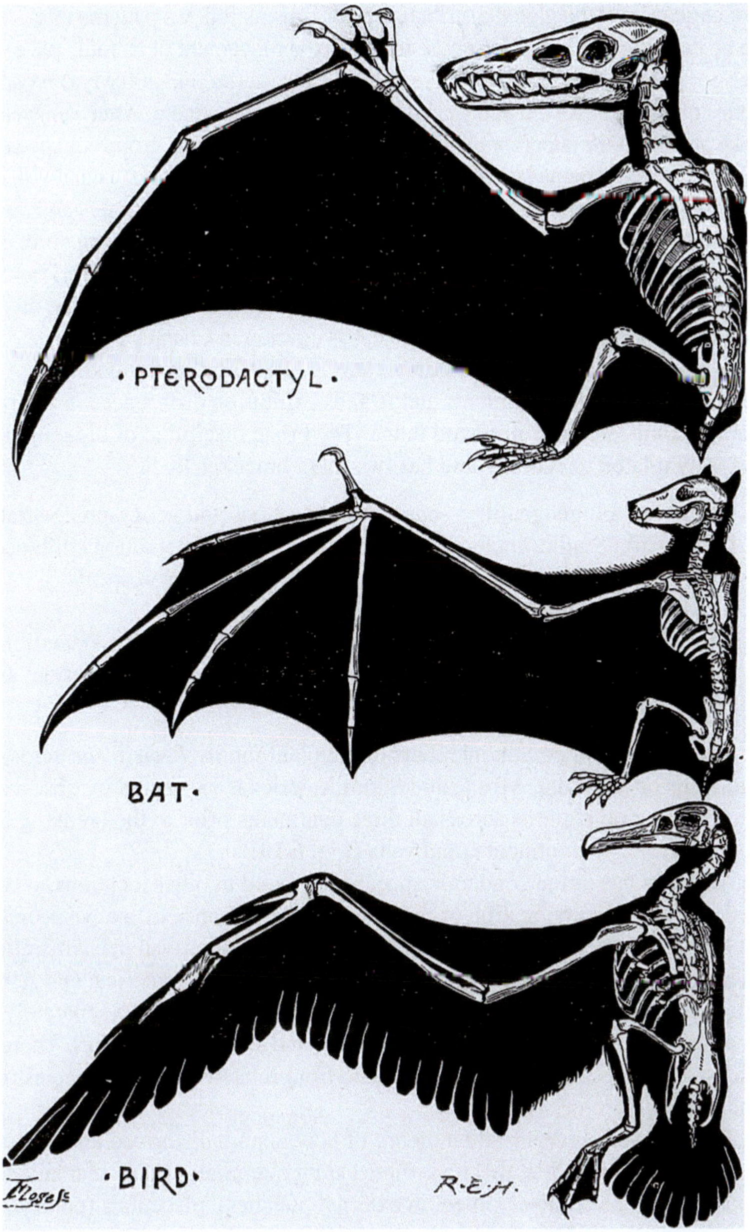

Fig. 6.9

Analogous structures for flight occur in the vertebrate wings of (a) pterosaurs, (b) bats, and (c) birds, which all evolved independently.

Source: John Romanes under Public Domain

4. **Biogeography**—the global distribution of organisms follows patterns that can be explained by evolution consistent with the movement of tectonic plates over geological time. Prior to the breakup of Pangaea (around 200 Mya) broad groups of species evolved and spread across the supercontinent. After Pangaea broke apart into the supercontinents Laurasia and Gondwana, groups of species began to evolve separately in specific regions of the planet. This is because they were separated from each other by oceans or other geographic barriers and the environment was distinct. Further stages of separations occurred in the north, as smaller northern continents broke off from the Laurasia and in the south continents formed out of the Gondwana (**Fig. 6.10**). Biogeography is a subdiscipline of geography and its relationship to climate was covered in Chapter 2.

   A biogeographer examines Earth's surface based on its divided land masses, primarily by continents and islands exhibiting differences in their average composition of flora and fauna. The broad discipline of biogeography is closely related to ecology and has two main branches.

   - **Ecological biogeography**—current distributions and geographic variation in diversity. Studies include how biotic and abiotic interactions influence species' distributions and the unique interactions between species (predation and competition).
   - **Historical biogeography**—examination of continental drift, glaciation, evolutionary lineages reconstructing the origin, dispersal, speciation, and extinction of species.

   The presence of certain members of the plant family *Proteaceae* across the continents of Australia, Africa, and South America is explained by observing their distribution patterns across all three continents prior to the breaking up of the southern supercontinent Gondwana (**Fig. 6.10**).

   Australia has unique endemic species not found in other locations, particularly its diversification of marsupials. The presence of many species not found anywhere else is typical for isolated island ecosystems. Australia was once a part of Gondwana, connected to India, South America, and Africa. After the supercontinent land mass split species begin to evolve separately, according to its distinct natural environment and selective pressures. There is extensive evidence for species divergence from related mainland species of which it has become separated.

   But let's delve deeper into a theory of how mammals thrived and evolved in Australia. Worldwide, there are three extant mammal groups: marsupials (embryos complete development in external pouches), placentals (embryos develop internally), and monotremes (embryos deposited in eggs). Paleontologists have used fossils from the Cretaceous period to explain why marsupials thrived even as placentals died off in Australia. During this time period South America, Antarctica, and Australia still abutted each other, with a forest beltway connecting Australia and South America (Fig. 6.10). The fossil record showed all three mammal types present in both continents.

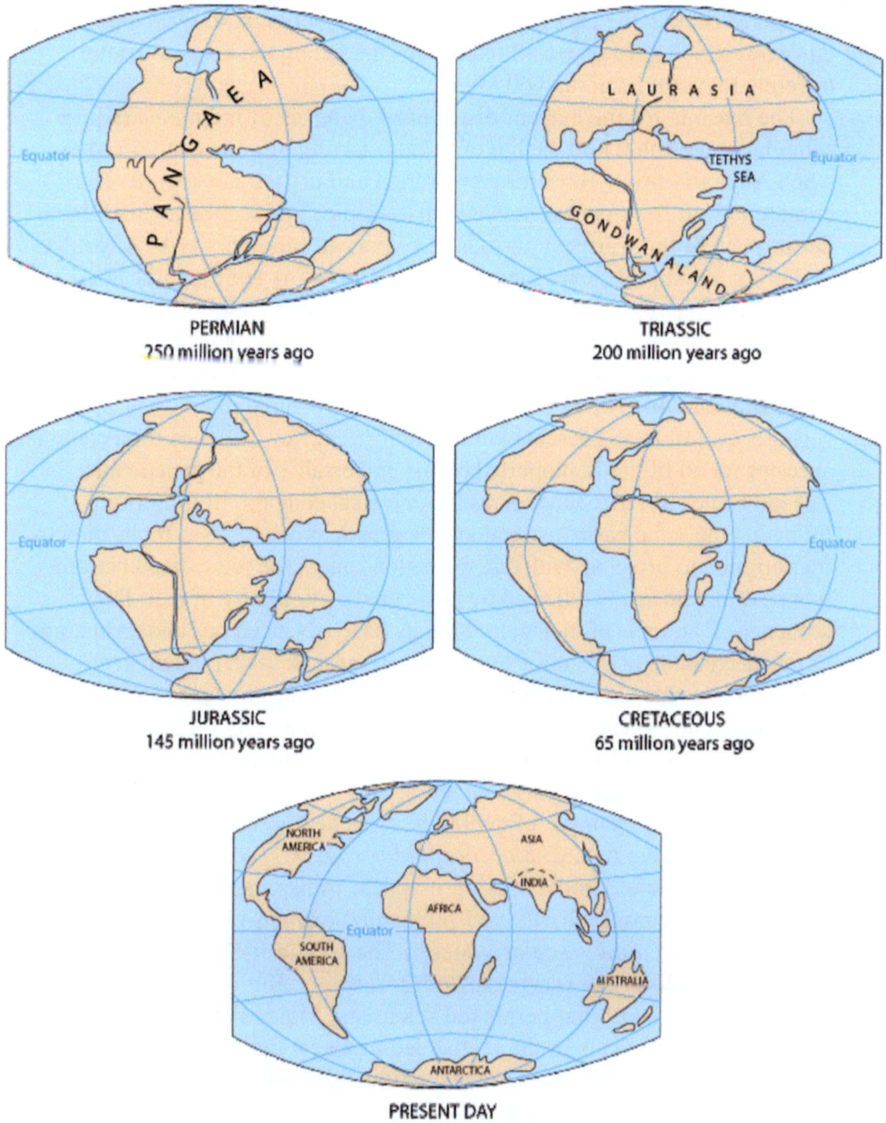

Fig. 6.10

Continental drift to current geographic distribution.

Source: Historical Perspective, USGS, Public Domain

According to Australian paleontologist Tim Flannery, during this period only low-energy monotremes like the platypus thrived in Australia while high-energy placentals only thrived in South America. The marsupials have intermediate energy requirements and are still present in both continents today (though with far greater diversity in Australia). The earliest known marsupial was traced back to the ancestor of a tiny Chilean opossum, *Dromiciops gliroides*, which gave rise to all marsupials, including the kangaroo, after traveling to Australia around 60 Mya.

The reasons for the current distribution patterns of mammals require some biogeographic knowledge of events starting around 40 to 45 Mya, when Australia broke free from Antarctica and drifted northward. The climate cycled for a period between lush tropics and icehouse conditions. During one of the warmest periods (23 Mya) all biodiversity in Australia, including marsupials, exploded. And by 15 Mya the continent began drying out and has continued to cool ever since. The lower metabolic demands of marsupials thrived in this new climate; however, many species had gone extinct, including early placentals with their high metabolic requirements.

5. **Molecular biology**—the molecules of life are structured in a way that supports evolution and descent with modification. Key evidence for a common ancestor for all of life is supported by the universality of the genetic material. The molecules of DNA function as the universal code for all lifeforms, following the same basic instructions for genetic processivity—replication to transcription to proteins, with rare exceptions only. The relatedness between groups of organisms is reflected by the degree of similarity of their DNA sequences. The tools of molecular biology have advanced our understanding of the mechanism of evolution as new functions arise with random gene duplication events, gene mutation, drift, or natural selective pressures. After a gene duplication event two distinct proteins with distinctly different functions may result. And if the duplication event proves beneficial it will increase variation in the species' gene pool.

An interesting example of evolution by gene duplication events is found in the Antarctic ice fish, *Champsocephalus esox*. *C. esox* has colorless blood and a scaleless body, making it appear transparent under water. *C. esox* lacks hemoglobin, the red-pigmented oxygen-binding protein present in most cardiovascular systems. The fragile-looking *C. esox* ice fish seemed unlikely to survive in subzero waters. But it was thriving, which was a mystery that perplexed scientists. Molecular analysis revealed gene duplication events in the Icefish led to a new gene that produced an antifreeze glycoprotein. Further modifications enhanced the ice fish's survival as it evolved over time.

Sometimes solving one question leads to another. For example, what were the environmental conditions that led to a beneficial antifreeze adaptation in the ancient Icefish? Around 25 to 22 Mya the Antarctic currents changed, thermally isolating the Southern Ocean from the warmer subtropical currents from the north (around the time Australia was experiencing one of its heat waves). This time period is believed to be when the bulk of the aquatic inhabitants in the Antarctic, including *C. esox*, were put under pressure to evolve or go extinct. By random genetic events, the ice fish was one of the few organisms that did not go extinct and was free to colonize into many more niches with limited competition for resources or predation.

6. **Misconceptions of evolution**—within two decades of the printing of "On the Origin of Species" the theory of evolution was universally accepted by most scientists, particularly younger biologists. The mechanism, descent with modification, took longer to be accepted. But by the early 20th century descent with modification had been the subject of intense investigations and was also accepted across scientific circles. This theory has been difficult to conceive for the layman, however, especially when it is confused with origins of life. So let's review the causes for misconceptions about what evolution actually is and how it works.
7. **Evolution is not just another theory**—sometimes simple semantics is the reason for confusion about the theory of evolution. Critics might unknowingly emphasize the everyday usage of "theory." But they are using the word "theory" in the common vernacular. When we state that we have a theory about how actors are selected as winners of the Oscar awards, that is really akin to making a guess or stating an untested hypothesis. But when scientists talk about a "theory" it has an entirely different word meaning. A scientific theory is based on an enormous body of investigative research and repetitive and rigorous testing with fully verified explanations for observations about the natural world. It does not represent a good guess or suggested explanation. It is an accepted hypothesis within the scientific community and remains valid until evidence can be found that disproves it. There are relatively few scientific theories but among those that have passed the rigors of scientific scrutiny are the theory of gravity, the theory of relativity, germ theory, cell theory, the Big Bang theory, and, of course, the theory of evolution. Each of these scientific theories describes well-defined phenomenon about the world we live in.

   Critics of evolution might imply that there is little evidence to support evolution, that there is still ongoing debate in the scientific community, or that tests still need to be done to "prove" it. None of these claims have credence, any more than it would make sense to question the validity of the theory of gravity.
8. **Individuals do not evolve**—today we understand the molecular level of life and that evolution is basically a change in genetic composition within a *population* over time. The molecular level of variation is known as the differential alleles for each gene. Individuals themselves may change to a degree over their lifetime, developmentally (infant to puberty to adulthood) and to some extent in response to environment or choices (for example, consuming a high-fat diet vs a leaner Mediterranean diet or living a lifestyle that is high stress vs one that is low stress). Developmental processes are programmed by the genes acquired from birth and coordinate with timed environmental cues. These events are not due to changes at the genetic level. Unhealthy environments may bring about changes in an individual, even leading to genetic mutations. A key distinction for evolution is whether the variation is heritable, or universally experienced by the entire population.

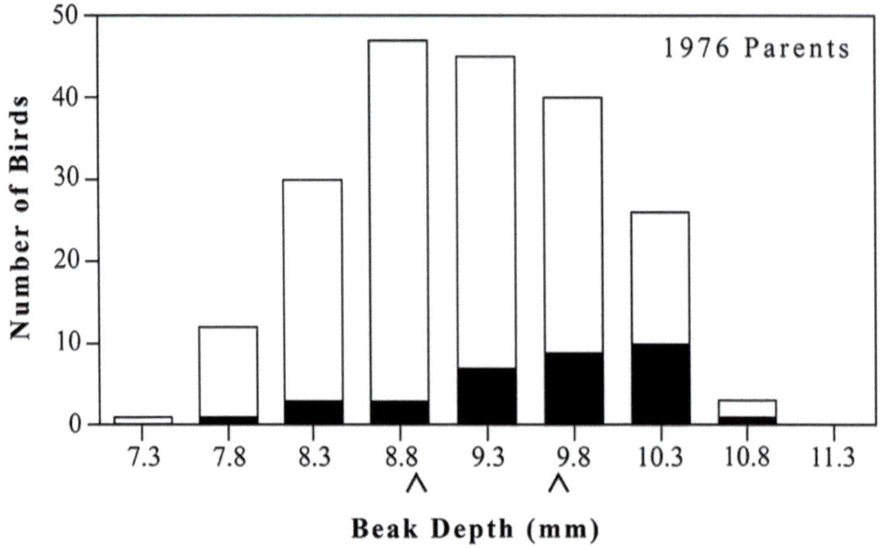

Fig. 6.11

In 1976 a study on the distribution of beak depths in medium-ground finches (*Geospiza fortis*) on Daphne Major (white bars) included survivors from a 1977 drought (black bars).

Source: HHMI Biointeractive under CC BY 4.0

For any heritable characteristic, when thinking about its potential for *evolution* it's a good idea to consider the average characteristic variation within the entire population over a period of time. Using Darwin's ground finches as an example, natural selection drove a change in its **bill** size. But the measurable value did not represent any individual finch's bill. Rather, the average bill size among all individuals in the population was measured over different seasons and indicated a change as the result of evolution. It's important to note that under the pressures of natural selection not all individuals will survive. And those that do survive do not possess a bill that can change in size across the seasons. Within the data collected over the seasons, however, there will appear an abundance of new individuals that contributed to the shift in average bill size (**Fig. 6.11**).

Acclimatization

Acclimation explains changes that can be sustained for short periods of time (hours to weeks) during an organism's lifetime. The purpose of acclimation is for organisms to be able to respond to environmental changes (temperature, photoperiod, pH, etc.) using their genetic tool kit. Organisms can accomplish some adjustments to their behavioral patterns, biochemistry, or even morphological traits based on their gene's ability to modify its expression. In other words the genetic capacity was already present. It is possible to adjust to high-altitude living because an individual's genes encode oxygen-transporting proteins capable of modifying their oxygen-binding affinity. The adjustment may take months or longer to achieve the best outcome but it occurs in an individual's lifetime. The ability to acclimate is in contrast to **adaptation**, which allows organisms, over generational time, to adjust to their environmental constraints by enhancing species' evolutionary fitness. The evolutionary change is observed in the gene pool as offspring become better adapted through descent with modification.

Evolution Cannot Explain the Origin of Life

It is a common misunderstanding that evolution provides an explanation for the origin of life. The theory of evolution does not assume to explain the origin of life. Evolution provides an explanation for the diversity of life and how existing populations change over time. There are many theories on the beginnings of life, including origin of the first cells, but they are separate and distinct topics studied by physicists, biologists, and other disciplines. The origin of life on Earth occurred a very long time ago and that makes it a particularly difficult problem to resolve. Biologists believe the presence of life began in a primordial form that later evolved as the conditions for more complex life forms became possible on Earth. This explanation falls under cell theory and disproved spontaneous generation. Questions on what happened once living organisms were present would be difficult to answer without a basic understanding of evolutionary processes.

Evolution Is Not Goal Directed

On two counts it is misleading to state that "*organisms evolve* in response to a change in their environment." Individual organisms do not evolve, nor is species evolution an intentional undertaking by individuals. When there is a change to the environment it disproportionately affects some individuals in the population more favorably. For the change to be favorable will depend on their particular phenotypes (e.g., variable alleles mentioned under "Individuals Do Not Evolve"). Any resultant change in the population's gene pool occurs only if the favorable characteristics were already present. In other words the variation that natural selection can work must already be present.

As an example we can look at a population of bacteria exposed to repeated dosages of an antibiotic. The application of this antibiotic to a bacterial population will eventually select for a population of bacteria with acquired resistance to that specific drug. Antibiotic resistance is an evolution caused by a gene variation that conferred resistance. The resistance variant gene did not arise in the population as the result of applying antibiotics. The presence of resistance was already present in the gene pool. The gene variation may have been rare, or a low-frequency occurrence. Applying the antibiotic began to kill off bacterial cells without the resistance gene. The antibiotic application was selected for those individuals that were resistant. With less competition, they are now able to thrive and multiply as the predominant genetic variant in the population. With the advent of broad-spectrum antibiotics populations of multidrug resistant bacteria have flourished.

Bacteria are a good example of evolution in action but do not suggest that evolution is goal directed. The bacteria did not strategize to overcome an antibiotic threat to their survival. Nor did the species become "better" as they evolved. It's simply a matter of adapting to the changing environment to maximize survival of the species. In the biological definition of survival it is a measure of reproductive fitness, carrying on as a species while a particular environment changes. Evolution's goal is not about making the biggest, fastest, smartest, or most complex species. Recall that preevolutionary thought considered species to have a goal toward complexity (Lamarck).

We now know that specific characteristics evolve in a species as a function of (1) the variation present in the gene pool and (2) the conditions within the environment. Both gene variation and environmental conditions are constantly changing in a nondirectional, nonplanned way. And certainly, while one trait is the most fit for its environment at any given point in time that same characteristic may prove to be fatal at another point in time. When a finch or other bird evolves to a characteristic bill shape optimized for cracking seed, if and when the environment changes, that bill morphology may prevent access to its new food resources. What was once a beneficial trait is now preventing fitness for the environment.

Thus while natural selection gives a *direction* to descent with modification (unlike random genetic changes) that direction serves no goal-orientated purpose and can be altered, simply in response to newly developing selective pressures.

6.3 Organization of Life on Earth

By the end of this section students should be able to:

- Display a comprehensive understanding of the classification system.
- List the different levels of the taxonomy.
- Describe and interpret a phylogenetic tree.

The evolutionary history of organisms and their relationships to other organisms or groups is called phylogeny. **Phylogeny** is a way to organize how species have evolved from earlier species and shows the degree of relatedness between existing or extant species. The phylogenetic data provides information on shared ancestry that can be illustrated in the root and branching patterns of a phylogenetic tree.

Phylogenetic Trees

The **phylogenetic tree** is a diagram that shows connections that occur among organisms and the evolutionary history that links them. The most universal phylogenetic tree is known as the **tree of life** and is comprehensive of all evolutionary relationships among known extant and extinct species. In **Fig. 6.12** the drawing of Charles Darwin's tree of life is compared to the complexity of a comprehensive modern phylogenetic tree. A phylogenetic tree is useful to hypothesize on evolutionary history that is still under investigation.

Each species can be traced backward along the "trunk" of the phylogenetic tree toward its ancestors. Branchpoints indicate where the species broke off from other lineages that share its common ancestry. To create a phylogenetic tree many sources of evidence and data are collected toward understanding how life evolved over time.

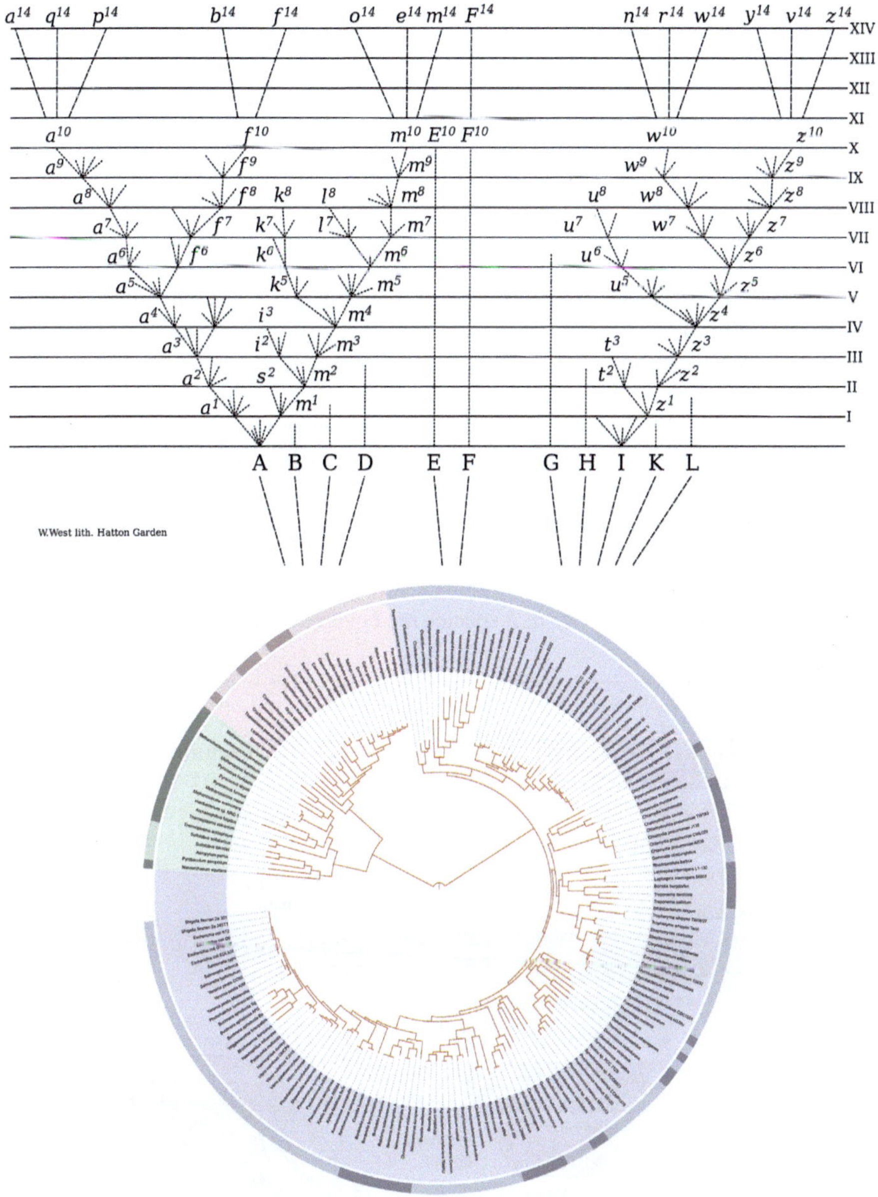

Fig. 6.12

(a) Tree of life in Darwin's Origin of Species. (b) Phylogenetic tree of life, 2006, showing relationship between species according to their genomes with universal ancestor of all life on Earth in center. Color blocks represent domains: pink is for *Eukaryota;* blue for bacteria; and green for archaea.

Source: On the Origin of Species, Public Domain

Sources of evidence include the fossil record, comparative anatomy, and DNA analysis. The collection of work requires combined contributions from the disciplines of paleontology, anthropology, and evolutionary genealogy.

The Levels of Classification

Taxonomy is the scheme of classification for living and extinct organisms constructed into an internationally recognized classification system. In this hierarchal system organisms are placed into large groupings based on easily identifiable similarities and then into more and more exclusive groupings. The taxonomic classifications are Latin, named according to the Linnaean system after its inventor, Carl Linnaeus. Linnaeus was a Swedish botanist, zoologist, and physician who began the daunting task of naming flora as genus and then more exclusively by species according to his binomial system. The binomial system was later adopted for all fauna.

Scientists identify the three large groupings as domains: Bacteria, Archaea, and Eukarya. Each domain is then divided into **kingdoms**. This is followed by **phylum, class, order, family, genus**, and, finally, the most exclusive of all, **species** (**Fig. 6.13**). Subspecies and strains are sometimes identified within a species because of slight alterations, providing the individuals still qualify as the same species. Rules for being classified as a member of a species are covered under speciation.

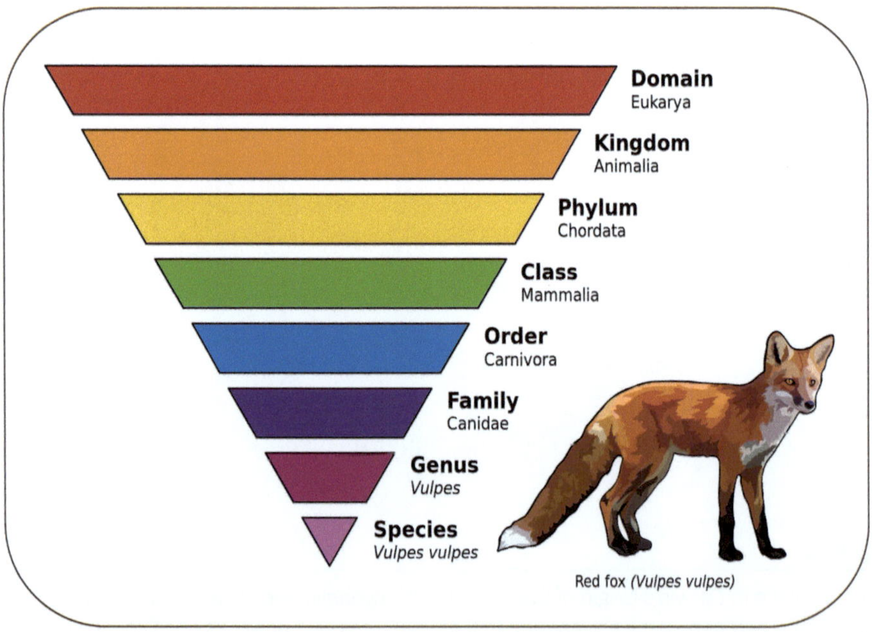

Fig. 6.13

Taxonomic classification system.

Source: Annina Breen under CC BY SA 4.0

The kingdom Animalia is within the domain of Eukarya. If we were to follow the red fox using taxonomy, its full name includes eight terms: **Eukarya, Animalia, Chordata, Mammalia, Carnivora, Canidae**, ***Vulpes*****, and** ***vulpes***. Note that the only category not capitalized is species and that only genus and species are italicized. These most exclusive categories fall under the Linnean **binomial nomenclature** for scientific names (e.g., *Vulpes, vulpes*). Each taxonomic category is called a **taxon**. For example, *V. vulpes* is in the broader, more inclusive taxon, Chordata. It is also in the more exclusive taxon, Carnivora.

Red fox is the common name in English vernacular for this animal species. In French the common name is Renard roux. One advantage to using a taxonomic classification system is that it applies a universal language, typically Latin or Greek, recognized and used internationally. One of 30 different subspecies may be identified for *V. vulpes*. For example, the North American subspecies are called fulva and can be written *Vulpes v. fulva*. The subspecies variation is typically due to geographic or behavioral isolation but not sufficient to be classified as a speciation event.

Following the taxonomical classification system, the human species share the larger inclusive groups (Animalia, Chordata, and Mammalia) with the red fox and other Canidae like the dog. The human species split phylogenetically after the taxon Mammalia. Chimpanzee species are closer to humans along the evolutionary tree, splitting phylogenetically after a subdivision of the taxon family (subfamily Homininae). At this point humans and chimpanzees evolved divergently. One did not evolve from the other. The two lineages separated from their common ancestor by speciation events in a long, drawn-out process of evolutionary changes occurring between 12 Mya and 5 Mya (**Figs. 6.14a, b**).

Human	Fox	Chimpanzee
Kingdom Animalia	Kingdom Animalia	Kingdom Animalia
Phylum Chordata	Phylum Chordata	Phylum Chordata
Class Mammalia	Class Mammalia	Class Mammalia
Order Primates	Order Carnivora	Order Primates
Family Hominindae	Family Canidae	Family Hominindae
Genus Homo	Genus Vulpes	Genus Pan
Species sapiens	Species vulpes	Species troglodytes
	Subspecies fulva	

****Some family rank classifications were later subdivided into subfamily, then tribe, then subtribe.***

Fig. 6.14a

Taxonomical comparison of three class Mammalia organisms.

Source: Image by author

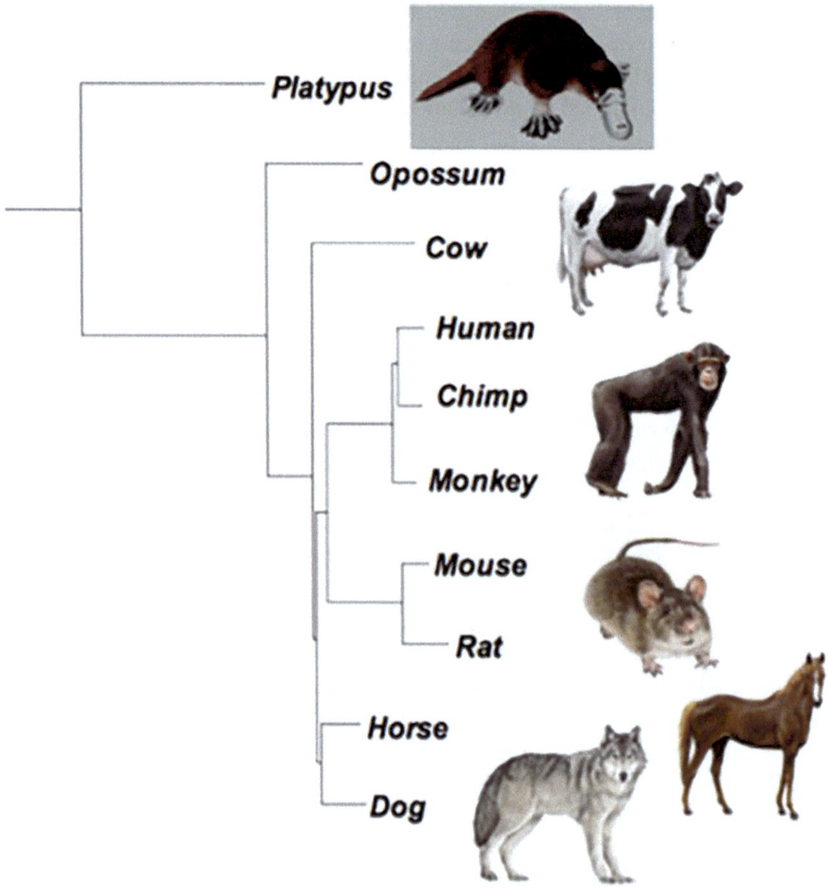

Fig. 6.14b

Phylogenetic trees illustrate level of relatedness between species.

Source: Lumen Biology under CC BY 4.0

6.4 Speciation

By the end of this section students should be able to:

- Explain how a species is identified as being distinct from another species.
- Describe the variables that lead to new speciation.
- Know the difference between allopatric and sympatric speciation.
- Describe modes of dispersal like adaptive radiation.

Although all life on Earth shares universal similarities in their genetic codes, organisms that are not of the same species do not have opportunities to combine their genetic information. Successful sexual reproduction resulting in offspring occurs in members of the same biological species.

Species and the Ability to Reproduce

A **species** is defined as a group of individual organisms that interbreed and are capable of producing fertile, viable offspring. This is how scientists distinguish one species from another. Different species do not successfully mate in a manner that produces fertile offspring and, therefore, no gene flow will occur between them. Individual members of the same species share compatible characteristics (reproductive organs and mannerisms) that develop from their genetic makeup and, as such, permit their reproductive fitness.

A species' appearance is not sufficient to determine its ability or inability to mate. This is demonstrated by the interbreeding of domestic dogs (*Canis lupus familiaris*). Thoroughbred dogs can display a wide range of phenotypes in size and morphology and still be capable of interbreeding to produce viable puppies. The mixed-breed puppies will be fertile and mature to sexually reproduce because they are all part of the same species. An interesting hybrid between wolves and coyotes, called the Eastern coyote, was first spotted in the Great Lakes but now is seen to be moving eastward (**Fig. 6.15**).

Fig. 6.15

An eastern coyote spotted in West Virginia.

Source: http://www.ForestWander.com under CC BY SA 3.0

(a) (b)

Fig. 6.16

(a) African fish eagle is similar in appearance to (b) American bald eagle, a different species.

Source: Nigel Wedge; USFWS, Public Domain

Alternatively, phenotypic similarity is insufficient for breeding success as entirely different species may appear to be similar. In **Fig. 6.16** the American bald eagle (*Haliaeetus leucocephalus*) and the African fish eagle (*Haliaeetus vocifer*) share many of the same characteristics. These eagles are both classified within the same genus but they are members of different species, and they are incapable of mating.

The collection of all variants of genes in a species comprises its gene pool. Earlier we covered inheritance, the genetic passing on of traits. Wherever variations occur within a gene pool they can be passed on to the next generation. This process occurs along one of two main pathways: asexual reproduction or sexual reproduction. In sexual reproduction mating must be successful, and the offspring acquires a mix of genetic material from its two parents.

A genetic change can only be passed on asexually if the reproducing cell already possesses the altered trait. In the process of sexual reproduction the altered trait would need to be present in one of its sex cells, called **gametes** (sperm or eggs). The odds are not 100% that a trait will be passed on to offspring with successful fertilization. This is because each parent only contributes half of its genetic diversity during the fertilization process. Sexually reproducing organisms only pass on traits that are present in their fertilized gamete cells. Because only heritable traits have the opportunity to evolve, reproductive strategies play a key role in how sexually reproducing populations can change over time.

New Species

Any group of interbreeding individuals that produces viable offspring meets the biological definition of a species. Asexually reproducing species obviously do not fit under this definition. There are a few exceptions, for example similar species in which mating and a hybrid offspring are possible (e.g., horse and donkey). But for these their offspring (e.g., mule) will not be viable *and* fertile, and thus they remain classified as distinct species.

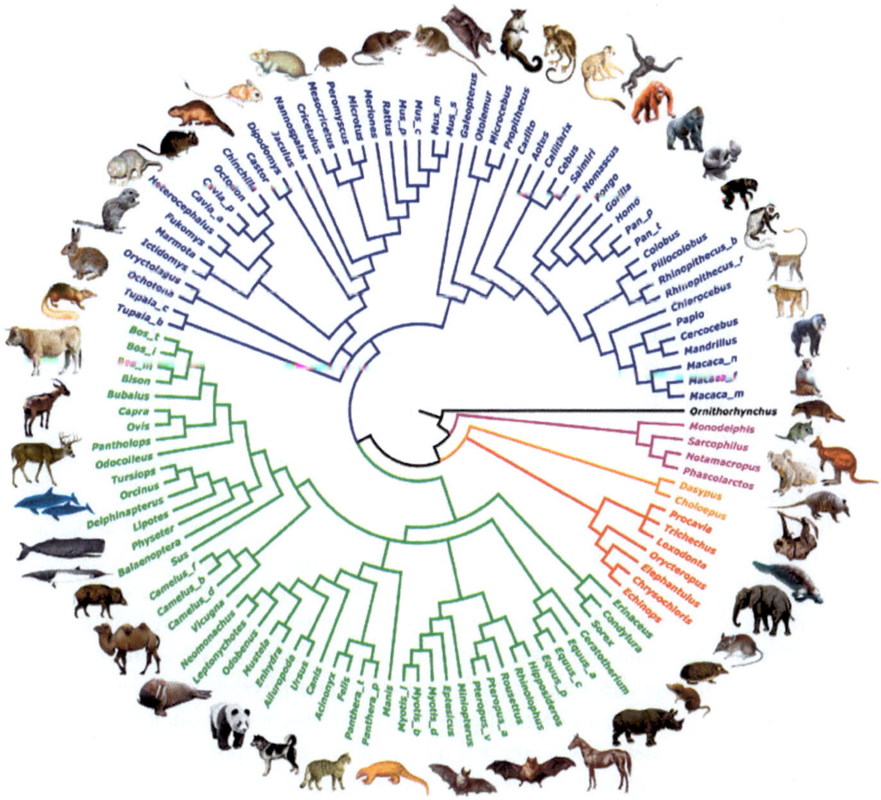

Fig. 6.17

(a) Darwin's "On the Origin of Species" speciation events leading to biological diversity and (b) modern-day phylogenetic of species relatedness in mammals. Combined Elsevier and Darwin's On the Origin of Species, Public Domain 978-0-12-822562-2.

Scornavacca, C., Belkhir, K., Lopez, J. et al. (2019). OrthoMaM v10: Scaling-up orthologous coding sequence and exon alignments with more than one hundred mammalian genomes. Molecular Biology and Evolution. 36 (4), 861 862. doi: 10.1093/molbev/msz015. PMC 6445298. PMID 30698751). The major clades are colored: Marsupials (magenta), Xenarthrans (orange), afrotherians (red), laurasiatherians (green), and euarchontoglires (blue). SEE: https://en.wikipedia.org/wiki/Mammal#/media/File:OrthoMaM_v10b_2019_116genera_circular_tree.svg

Speciation events occur whenever two distinct species split from one original species. The divergence into two separate species occurs if two populations of one species evolve along separate pathways and become incapable of interbreeding successfully (viable, fertile offspring). In Darwin's "On the Origin of Species" he sketched his vision of the process of speciation as a branching event on the tree of life (**Fig. 6.17**).

Speciation events will follow one of two pathways: allopatric or sympatric speciation.

- **Allopatric speciation** (allo- = "other"; -patric = "homeland") occurs when geographic separation of populations leads to subsequent, distinct evolutionary pathways from what was once a single parent species.

- **Sympatric speciation** (sym- = "same"; -patric = "homeland") occurs when the parent species remain in the same location but something causes gene flow to stop long enough between the populations that they begin to follow different evolutionary pathways.

Continue exploring speciation by reviewing the Orca Whale case study, which addresses whale speciation and the debate on whale ecotypes.

Allopatric Speciation

Gene flow is the movement of alleles (variation in genes) through sexual reproduction within a species. In a geographic range gene flow within a population is continuous and relatively homogeneous because they are free to move about and mate. The frequency of an **allele** occurring is randomly distributed across the distribution range as long as there are no barriers to prevent its appearance (gene flow) in new offspring. When barriers prevent populations from mating freely some alleles will no longer be present, or inherited in the offspring of the now-separate populations. If geographic barriers remain for a long enough period each population evolves and adapts to its specific environment. They begin to follow different trajectories for allele frequencies. Because mutations happen spontaneously and some allelic frequencies are based on survival of the fittest, the two gene pools will in time look quite different. This difference will be expressed in the species morphology, behaviors, and physiology. Environmental conditions that may impose differential constraints on the separate populations include climate, resources, predators, and competitors. All are selective pressures that will lead to divergent populations with a different set of adaptations.

Geographic barriers that lead to allopatric speciation include rivers, mountains, and similar areas such as any area that prohibits clear flow of alleles, including migration to new areas with an inability to return, or unintentional transport across an ocean to an island.

Along the west coast of the United States two separate subspecies of spotted owls (*Strix occidentalis)* exist. The spotted owl is a large bird, dappled with white feathers and lacking tufted ears. But the Northern spotted owl (*Strix occidentalis caurina*) has genetic and phenotypic differences from its close southern relative, a paler, larger Mexican spotted owl (*Strix occidentalis lucida*) (**Fig. 6.18**). Note that they are listed as the same species but different subspecies. This tells us immediately that they are capable of mating successfully to produce viable, fertile offspring.

The longer the time period that *S. occidentalis caurina* and *S. occidentalis lucida* inhabit distant ranges, the more likely that speciation events will occur. This is logical because as the distance increases the biogeography begins to differ, with environmental factors having less in common than if the subspecies lived in closer proximity. For the two spotted owls the climate is cooler in the north than it is in the south. The ecosystem biotic factors differ, as do hunting habitats and prey choices. These variances were sufficient to lead to distinct differences in the subspecies and suggest an eventual speciation.

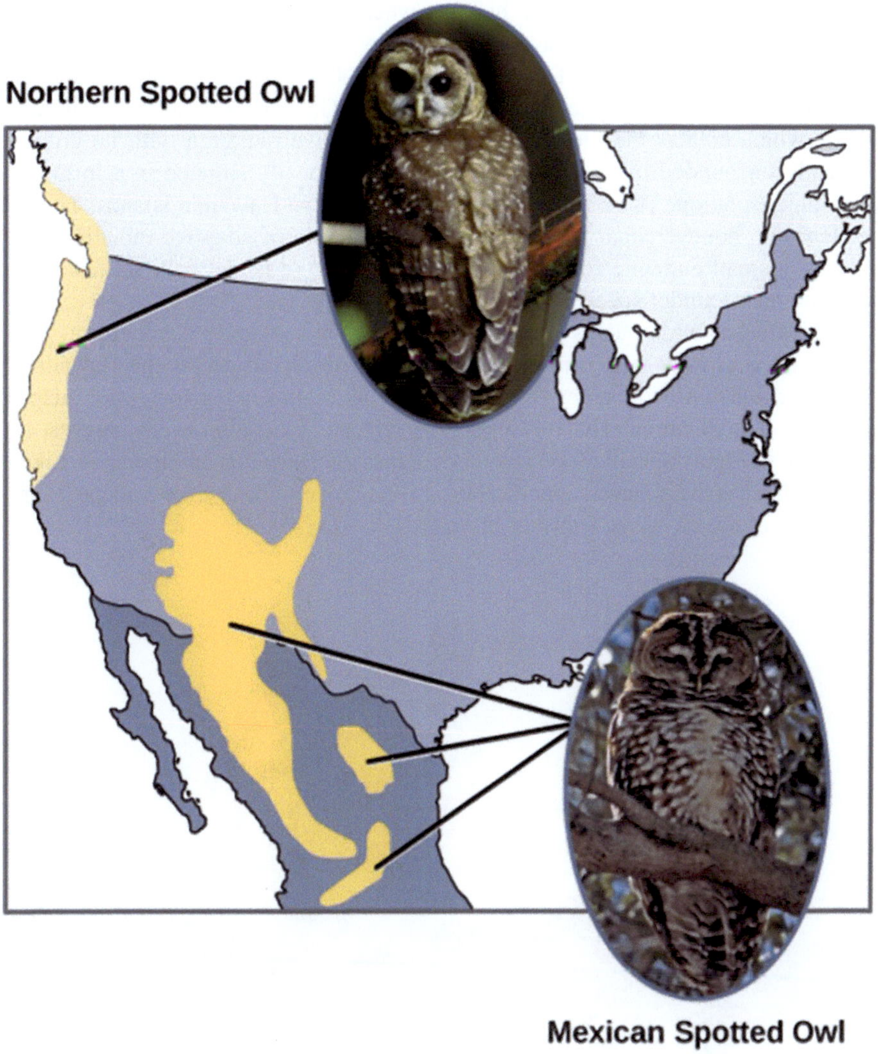

Fig. 6.18

Northern spotted owls and Mexican spotted owls are geographically separated by allopatric speciation but can still mate.

Source: J. and K. Hollingsworth, et al under CC BY-NC-SA 3.0

Adaptive Radiation

In some cases a population of one species disperses throughout an area, with small groups of individual members settling into a distinct niche or isolated habitat. The ecological niche will have different constraints, leading to different lifestyles and speciation events. The conditions may lead to **adaptive radiation** if several populations are dispersed to niche habitats where rapid, subsequent adaptations occur that can be traced back to a single point of origin. The original species "radiates" into several new species. Darwin's ground finches were the result of adaptive radiation from a founder species on the mainland. Galápagos is an island archipelago, which creates the ideal conditions for adaptive radiation with its group of islands surrounded by water. The conditions are ideally situated in a form of geographic isolation. There is a similar geography in the Hawaiian Islands. There the Hawaiian honeycreeper provides a similar example of adaptive radiation. In this case a small endemic forest bird evolved across the island habitats from its ancestral finch founder species (**Fig. 6.19**).

Honeycreeper birds (genus *Carpdacus*, in the finch subfamily Carduelinae) are small birds with beak sizes and plumage variation observed across the Hawaiian islands. Acquisition of food resources is one of the main driving forces for adaptation and diversification. The Hawaiian honeycreeper speciation events suggest a prevalence of highly specific food resources across the Hawaiian archipelagos. Like Darwin's finches the honeycreeper displays a wide range of beak shapes adapted for availability of nectar, seeds, fruits, or insects.

Fig. 6.19

Adaptive radiation in Hawaiian honeycreeper birds from a founder species includes different beak sizes.

Reproduced from Pratt, H. D. (2005). The hawaiian honeycreepers: Drepanidinae. OUP Oxford.

Adaptive radiation occurs primarily through allopatric speciation, although sometimes it occurs by sympatric speciation, which is thought to have been the driver for cichlid rapid speciation in the freshwater lakes in East Africa.

Sympatric Speciation

Divergence can occur even when no physical barriers are in place to separate individuals that are living and reproducing in the same habitat. In this case a randomly mating population begins to reduce its gene flow. This process of speciation is called sympatric speciation; the prefix "sym" meaning same, thus "sympatric" means "same homeland". In contrast "allopatric" means "other homeland." Sympatric speciation is the rarer form but a number of mechanisms for sympatric speciation have been proposed and studied. In the absence of geographic isolation some members of a species may begin to mutate, forming a new entry into the gene pool, which begins to isolate its population to a specific niche.

Sympatric speciation was first studied in the egg-laying habits of the apple maggot. This was the result of changes in the plants that female apple maggot flies selected to lay their eggs. For hundreds of years, the predecessor of apple maggot flies laid their eggs only on native North American hawthorn fruit. When the domestic apple tree was introduced by immigrants some populations of the maggot flies began to lay eggs on the apple fruit instead of large Hawthorne fruit. Female maggots generally choose to lay their eggs on the same type of plant they hatched on, and males tend to look for mates on the same type of fruit they grew up with (**Fig. 6.20**). The hawthorn maggot flies ended up mating only with other hawthorn maggot flies and apple maggot flies ended up mating only with other apple maggot flies. In time this behavioral pattern led to a reduction in gene flow between the populations based on their mating preference on different fruit trees. The apple-feeding race and the Hawthorne-feeding race are both taxonomically classified as *Rhagoletis pomonella* but the changes between the two races of maggot fly constitute an early step in the emergence of a new species.

Reproductive Isolation

Reproductive isolation is the inability to interbreed and forms a barrier to gene flow. This can result through the process of sympatric or allopatric speciation. When populations begin the process of diverging genetically and phenotypically they will eventually lose the capacity to successfully reproduce. If individuals from the two species were to be brought back together mating would be less likely. And if mating were still possible offspring would be nonviable or infertile. New behaviors and physiological processes that lead to speciation create barriers that ultimately maintain the integrity of the new species as distinct.

Prezygotic barriers vs postzygotic barriers. When sex gametes (egg and sperm) combine in a fertilizing event the zygote is formed. Reproductive isolation can occur either before or after a fertilizing event has taken place between two distinct species.

Fig. 6.20

Sympatric speciation in Hawthorne Maggot fly.
Source: Image courtesy of J.K. Clark, Understanding Evolution. Museum of Paleontology. 22 August 2019

In the mechanism of a **prezygotic barrier** the formation of a fertilized egg will be prevented. This can be due to incompatible timing of mating or alterations in mating structures. In the mechanism of a **postzygotic barrier** after successful zygote formation there will be an unsuccessful implantation, miscarriage, or sometimes live birth of a sterile offspring.

Temporal isolation is a type of prezygotic reproductive barrier that occurs due to incompatible breeding schedules (**Fig. 6.21**). Eventually their gene pool becomes isolated from the other populations in a form of reproductive isolation events leading to speciation.

Behavioral isolation is also a type of prezygotic reproductive isolation and occurs when specific mating behaviors prevent reproduction from taking place. The male firefly mating behavior depends upon specific light patterns to attract females. If the female does not recognize the light patterns from another species she will not mate.

Plant species are also capable of forming prezygotic barriers. In flowering plant diversification highly specific interactions with pollinators attracted to its floral structures are required for successful reproduction. The coevolutionary pathway ensures its pollen reaches and fertilizes another flower of the same species. Among the floral structures that ensure same species fertilization events are tunnels through which the pollinator must access its nectar, and that varies widely in length and diameter. The pollinator is an important intermediary in the plant's reproduction and coevolves with structures perfectly suited for its role in accessing and transmitting the pollen.

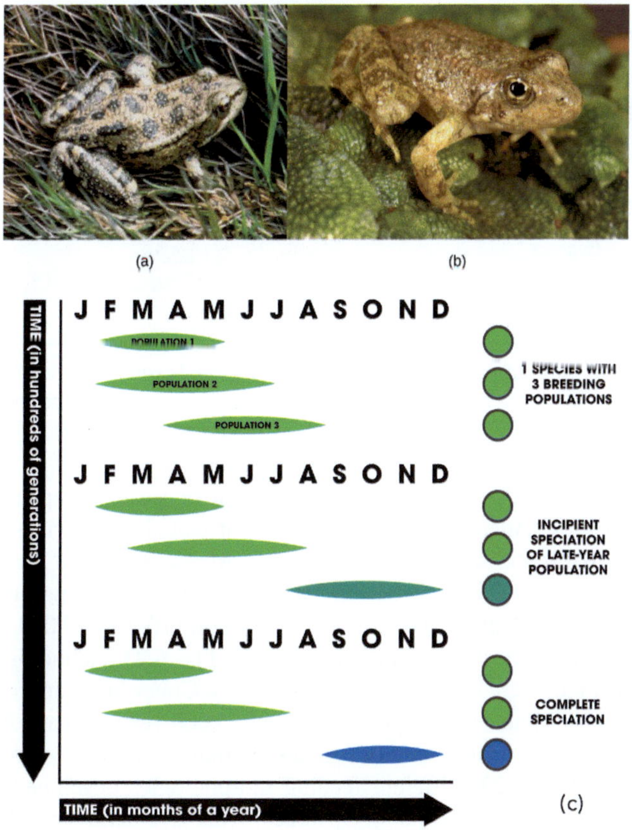

Fig. 6.21

Temporal reproductive isolation in frogs shown in (a) *Rana aurora* and (b) *Rana Boylii.* Temporal isolation is depicted in (c) whereby breeding seasons for populations of one species have shifted over time.

Source: Modification under CC BY SA 4.0 of Mark R. Jennings, USFWS, Public Domain

This strategy prevents the plant from being randomly cross-pollinated by an unrelated pollinator carrying pollen from a different plant species (**Fig. 6.22**).

Habitat isolation plays a role in influencing speciation events by ongoing dynamics in the habitat. For example, a species of lake fish may be undergoing robust population growth and fierce competition for food. Under pressure to find food sources a subgroup of the species may discover a new food resource unused by the other fish. If the new food source is found at a different depth of the lake that would impart another degree of habitat isolation. Over time those feeding on the second food source would interact less and less with individuals in the main group. This self-imposed isolation means less opportunity to breed together. As the offspring continue to inhabit the area of their parents, the subgroup experiences reduced gene flow with the origin group.

Fig. 6.22

Flower coevolution with pollinators (a) wide-cupped foxglove flowers are adapted for bees. (b) Tube-shaped flowers are adapted for the long proboscis of hummingbird moths.

Source: Kurt Stüber under CC BY SA 3.0 and Yusuf Akgul under CC BY SA 4.0

Habitat isolation falls under sympatric isolation because the species are still in the same location. Adaptive radiation by habitat isolation describes the highly diverse cichlid fish found within Lake Victoria in Africa. The large lake has many unique habitats but no geographic barriers. Therefore it is not allopatric speciation that caused the rapid adaptive radiation and speciation (10,000–15,000 years ago). Another example of sympatric isolation is shown in **Fig. 6.23**, cichlid fish diversification in a Nicaraguan crater lake, Laguna de Apoyo. Crater lakes are formed from volcanic eruptions. Laguna de Apoyo was formed 22,000 years ago and is a favorite tourist destination with benthic sulfuric gasses keeping its waters warm. In nearby Lake Apoyeque (**Fig. 6.23**) different feeding morphologies developed quickly to allow cichlids to feed from different food sources. What is remarkable is that this lake is only 1800 years old, and just 100 years ago it was stocked with one single population of cichlids! Cichlids appear to be primed for diversification at the genomic level, possessing rates of gene duplication far higher than other lake-dwelling fish. Lake cichlids represent a great experiment in evolution and natural selection.

Adaptations Reflect Trade-Offs and Constraints

Variation in temperature, precipitation, and the seasons produces a wide array of biodiversity in unique terrestrial habitats (Chapter 3). For aquatic life variations in pH, salinity, oxygen availability, and resulting zonation define diverse habitats for marine and freshwater animal life (Chapter 4). Whatever combination of environmental conditions prevail at any given time within the habitat will present its own unique set of **constraints** on survival and reproduction. John Maynard Smith, a post-modern British evolutionary biologist and geneticist, once remarked, "If there were no constraints on what is possible, the best phenotype would live forever, would be impregnable to predators, would lay eggs at an infinite rate, and so on." Fitness for the environment determines which species are most successful in populating the habitat. Natural selection pressures for the most favorable phenotypic traits will, over time, reflect changes in the genetic pool that are specifically adapted to the

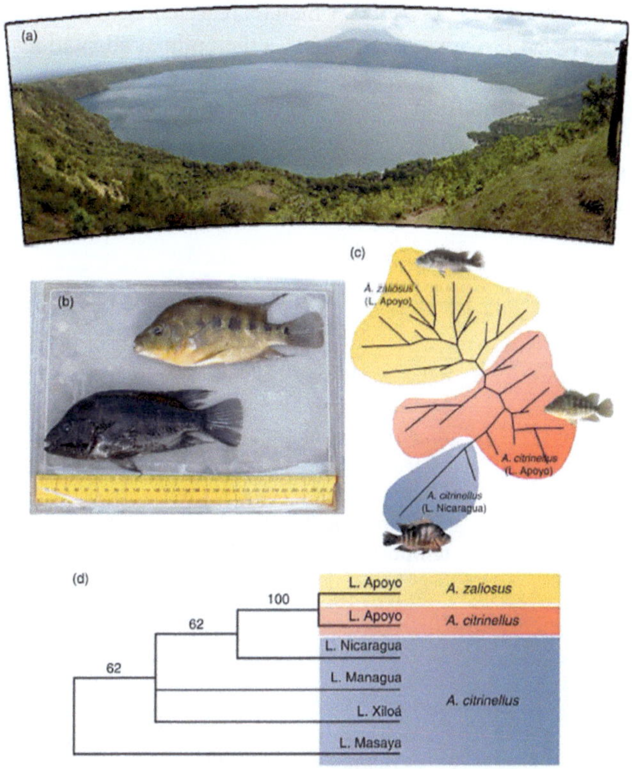

Fig. 6.23

Laguna de Apoyo, Nicaragua, cichlid fish sympatric speciation in Amphilophus zaliosus and A. citrinellus. Phylogenetic tree shows cichlid diversity in the larger nearby crater lakes, Nicaraguan and Managua, which are connected by waterways.

Images courtesy of Axel Meyer

habitat. However, the fittest phenotypic variation within a species may not be suited to another set of conditions. Those same features may now pose a limitation on its survival and reproduction. Changing conditions can occur in the same location in different time spans or in the same time spans but in different locations.

A limitation on species is a function of **trade-offs** that have been imposed by the constraints. Adaptive trade-offs were emphasized by Charles Darwin in his medium ground finch studies on the Galápagos Islands. In his work he demonstrated that ground finch access to their seed resource was determined by beak size vs seed size and hardness. The difference in diet (small or large seeds) was a direct function of beak size, which reflects an anatomical trade-off (small finches have small beaks, large finches have larger beaks). Finches with smaller beaks cannot crack larger, harder seeds. Finches with larger beaks can access a larger range of seed types. But a small seed diet would not be efficient for larger finches, partly due to their greater metabolic needs. Larger beaks require larger overall body, larger skulls, beak musculature, and body mass, all of which increase energy demands. In times of seed scarcity the larger birds would not possess the optimal adaptations for survival and would probably experience a die-off. A basic trade-off exists between morphology and diet and this relationship constrains evolutionary adaptation for the medium ground finch in its unique habitat.

In the next chapter we will discuss trade-offs and constraints that are unique to survival in a terrestrial or aquatic environment.

Glossary

Essentialism
Enlightenment
extant
taxon
founder species
ecotypes
nomenclature
catastrophism
transmutation
macroevolution
microevolution
Lost World of Fantastic Beasts
temporal isolation
behavioral isolation
habitat isolation
acclimation
adaptation

Citations

1. T. Dobzhansky. *Nothing in biology makes sense except in the light of evolution, Am Biol Teach 35 (3) (1973) 125–129*. https://doi:10.2307/4444260
2. *Lee, Jeffrey, &* Von Humboldt, Alexander. *(2014). The Encyclopedia of Earth*. Archived *from the original on 27 September 2015*. Retrieved 26 September 2015
3. American Museum of Natural History. *(2000)*. James Hutton: the founder of modern geology. Earth: Inside and Out.
4. Elizabeth Kolbert, The lost world: annals of extinction, *The New Yorker*, 2013.
5. Lamarck, J.B. *(1914). Zoological Philosophy. London*. https://en.wikisource.org/wiki/1911_Encyclop%C3%A6dia_Britannica/Evolution
6. C. Darwin, On the Origin of Species: A Facsimile of the First Edition, Harvard University Press, 1859. 540 pgs.
7. G. Bergman, Ancon sheep: a now disproven example of macroevolution, Riv Biol 98 (3) (2005) 435–448. PMID: 16440280.
8. R.F. Kimball, J. Huxley, The great biological generalization: evolution: the modern synthesis, The Quarterly Review of Biology 18 (4) (1943) 364–367. https://doi.org/10.1086/394682
9. E. Johnson, Horses part of Indigenous cultures longer than Western historians thought. National Science Foundation, *Science Matters*, 2023. https://new.nsf.gov/science-matters/horses-part-indigenous-cultures-longer-western
10. I.C. Cuthill, A. Székely, Stevens, in: S. Merilaita (Ed.), Animal Camouflage: Mechanisms and Function, Cambridge University Press 2011, p. 50.

11. https://theconversation.com/yes-eastern-coyotes-are-hybrids-but-the-coywolf-is-not-a-thing-50368. Accessed 2.22.2024.
12. D. H. Erwin, Climate as a driver of evolutionary change, Curr Biol 19 (14) (2009) R575–R583.
13. D.G. Jenkins, R.E. Ricklefs, Biogeography and ecology: two views of one world, Philos Trans R Soc Lond B Biol Sci 366 (1576) (2001) 2331–2335. https://doi.org/10.1098/rstb.2011.0064
14. I. Sanmartín, Historical biogeography: evolution in time and space, Evo Edu Outreach 5 (2012) 555–568. https://doi.org/10.1007/s12052-012-0421-2
15. B. Lamont, Out of Africa: linked continents, overland migration and differential survival explain abundance of Proteaceae in Australia, Perspect Pl Ecol Evol Syst 62 (2024).
16. T.F. Flannery, The Gondwanan origin of Tribosphenida (mammalia), Alcheringa (2022). https://doi.org/10.1080/03115518.2022.2132288
17. M. Nilsson, U. Arnason, P. Spencer, A. Janke, Marsupial relationships and a timeline for marsupial radiation in South Gondwana, Gene 340 (2004) 189–196. https://doi.org/10.1016/j.gene.2004.07.040
18. F.V. Seersholm, et al., Rapid range shifts and megafaunal extinctions associated with late Pleistocene climate change, Nat Commun 11 (2020) 2770. https://doi.org/10.1038/s41467-020-16502-3.
19. B.F. Minhas, et al., Novel mitochondrial genome rearrangements including duplications and extensive heteroplasmy could underlie temperature adaptations in Antarctic notothenioid fishes, Sci Rep 13 (2023) 6939. https://doi.org/10.1038/s41598-023-34237-1.
20. C.D. Jiggins, J.R. Bridle, Speciation in the apple maggot fly: a blend of vintages? Trends Ecol Evol 19 (3) (2004) 111–114 https://doi.org/10.1016/j.tree.2003.12.008 PMID: 16701238
21. A. Meyer, The extraordinary evolution of cichlid fishes (2015). https://www.scientificamerican.com/article/the-extraordinary-evolution-of-cichlid-fishes/

CHAPTER

7 Adaptation to the Environment

Chapter Outline

7.1 Plant Adaptations

By the end of this section students should be able to:

- Explain how light availability impacts photosynthetic activity and plant adaptations.
- Describe how photosynthesis is the primary exchange between plants and the atmosphere.

Ecology for the 21st Century. DOI: https://doi.org/10.1016/B978-0-443-24031-7.00007-6

- Describe mechanisms for plant adaptation to nutritional deficiencies and temperature.
- Contrast the process of carbon uptake between aquatic and terrestrial plants.
- Give an example of acclimation vs adaptation in response to temperature variation.
- Relate adaptations as trade-offs between rates of growth and tolerances to environment.

Primary producers require water, solar energy, and a carbon source to perform photosynthesis. All organisms require mineral nutrients for their growth and development. To acquire necessary nutrients for growth and survival plants and animals developed a variety of adaptations across a wide array of environmental conditions and stressors. Sometimes that even includes coevolutionary strategies between animal and plant species (**Fig. 7.1**).

Review on Nutrition Mode

Heterotrophs indirectly acquire their energy source by consumption of organic nutrients, for example, consumption of other organisms. Since they acquire carbon in the process they are more appropriately called *chemoheterotrophs*. The process of directly assimilating carbon (for example, atmospheric CO_2) occurs by converting and storing it as simple carbohydrate sugars. This process is known as carbon fixation and is accomplished by all *autotrophs*. If the process utilizes solar energy, the organisms are more precisely known as *photoautotrophs* (e.g., photosynthetic organisms). All autotrophs acquire carbon directly from the environment (**Fig. 7.2**). In contrast, a chemoautotroph is nonphotosynthetic and derives its source of energy and carbon from inorganic matter (hydrogen sulfide, elemental sulfur, ferrous iron, molecular hydrogen, or ammonia). The oxidative processes used by chemoautotrophs to acquire energy occur in extreme environments (e.g., thermal vents) where few other organisms can thrive.

Fig. 7.1

Coevolution is seen in bumble bee and the echinacea flowering plant adaptations that mutually benefit from their interactions.

Source: John Beetham under CC BY 4.0

Photosynthesis

Photosynthesis is essential to maintaining the processes for life to exist on Earth, not only for the photosynthetic plants but for the many heterotrophic animals that depend on it. Phototrophs arose 3 billion years ago (Bya) as microscopic organisms capable of a complex biological process to capture solar energy emanating from the sun and convert it into a usable chemical form of energy. Solar radiation is energy emitted by the sun that is captured during photosynthesis in the first of two phases: the light reactions. It is then converted to chemical compounds with high-energy covalent bonds during the second phase: the dark reactions (a/k/a Calvin Cycle). The chemical energy is initially converted into simple carbohydrate sugars. Species that have evolved mechanisms to perform photosynthesis include multicellular plants, algae, and unicellular bacteria called cyanobacteria (the earliest photosynthesizers on Earth). They are the photoautotrophs, which literally means "self-feeders using light."

a) Polystichum setiferum

b) kelp forest

c) Prochlorococcus

d) Sully hydrothermal vent

e) Beggiatoa chemoautotroph bacteria

Fig. 7.2

(a) Plants, (b) algae, and (c) cyanobacteria. All photosynthesizers produce their own organic compounds. (d) Deep sea vents (e) Chemoautotroph deep sea bacteria which capture energy directly from inorganic compounds.

Source: NASA; NOAA; Undersea Research Center, Public Domain

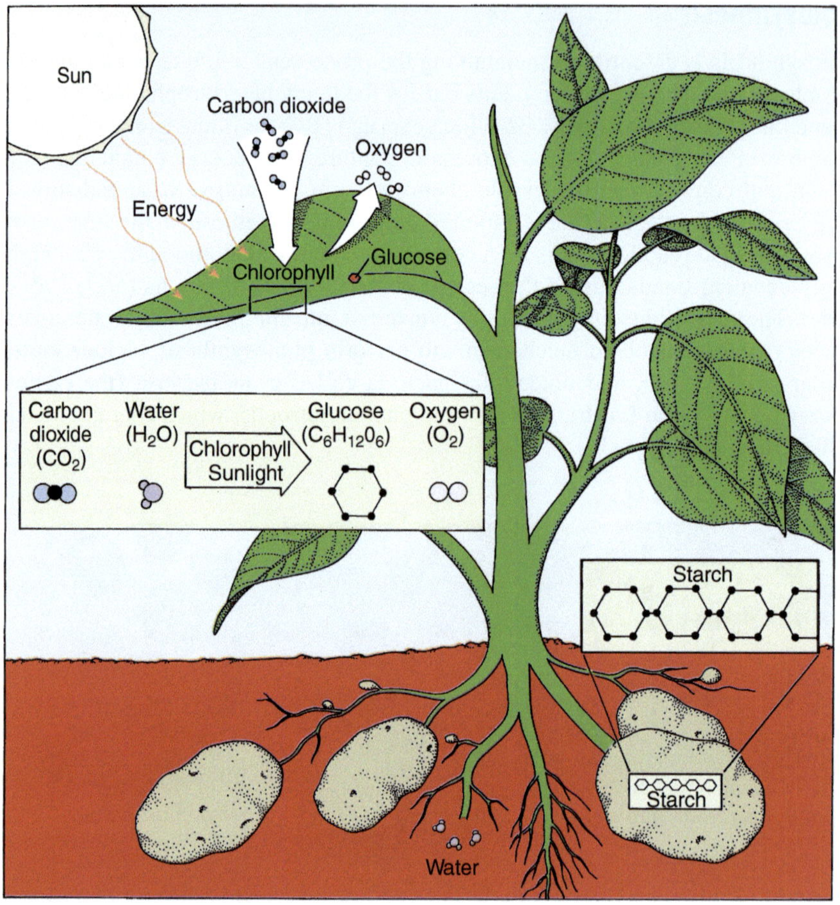

Fig. 7.3

Requirements of photosynthesis. Locations of photosynthesis are the chloroplast organelles of plants and algae, or cellular interior of bacteria.

Photosynthetic processes require sunlight and CO_2 uptake from a plant's aerial parts and H_2O uptake from its roots (**Fig. 7.3**). Both CO_2 and H_2O are chemically viewed as low-energy substrates that will participate in a multistep process to absorb solar energy that converts them into higher-energy molecules. After the process is complete a different gas will be generated in the plant, oxygen. The higher energy 3-carbon compound is a simple sugar (carbohydrate), called **glyceraldehyde-3-phosphate** (G3P). This simple sugar is mostly converted into glucose for metabolism or sucrose, a stored form of energy, dependent upon the needs of the organism.

Although the equation for photosynthesis looks straightforward, there are two separate processes (light and the dark reactions). In most cases photosynthesis occurs in the middle tissue layer of leaves, called the **mesophyll**. Gas exchange allows intake of CO_2 to the mesophyll cells while expelling the metabolic waste product of photosynthesis (oxygen). The gas exchange occurs through tiny

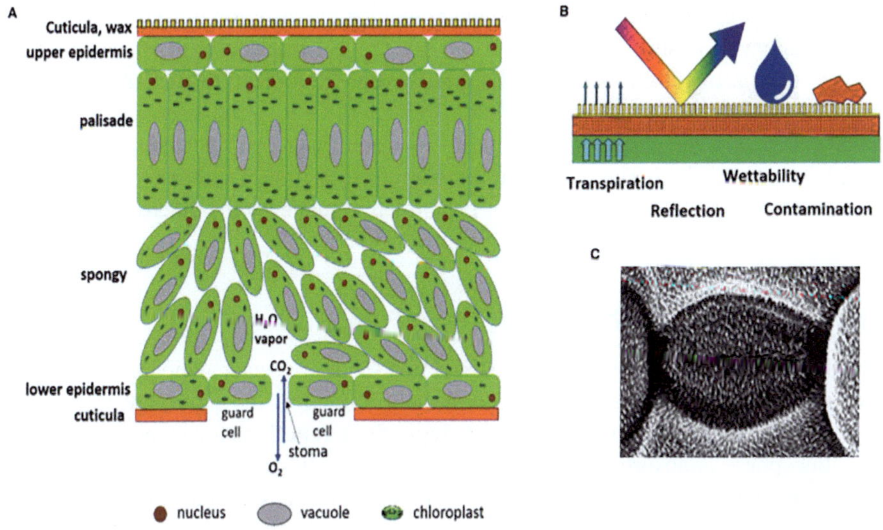

Fig. 7.4

Leaf cross-section with upper and lower epithelia layers, waterproofing cuticle, middle mesophyll layer, and underside openings (stoma).

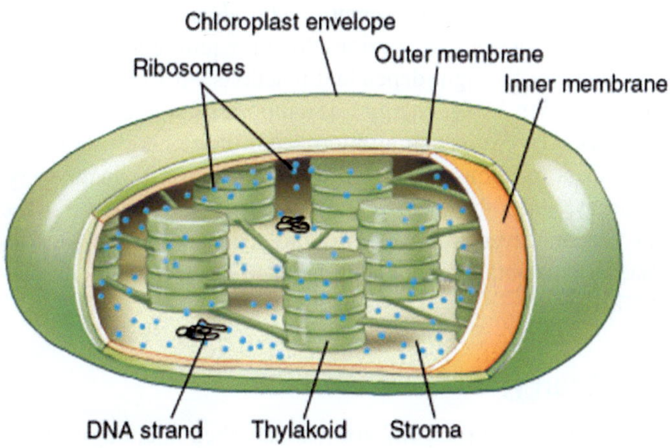

Fig. 7.5

Stacked thylakoid membranes (grana) of chloroplasts.

openings on the underside of leaves, called **stomata**. This underside location helps to minimize water loss through the opened stoma during the heat of the day. Each stoma is flanked by guard cells to open and close the stoma in a regulated fashion. Both the underside location and regulation of the stomata's opening are an adaptation to help conserve water loss.

Inside each leaf's mesophyll tissue are cells that contain specialized organelles called **chloroplast**. The chloroplast has a very unique membrane system, the **thylakoid**, which is optimized for energy capture (**Figs**. **7.4** and **7.5**.). Along these membranes are a multitude of pigment molecules called **chlorophyll**, which are very efficient at absorbing solar energy. But the chlorophyll molecule does not absorb all emitting light waves and some wavelengths are reflected. Among the wavelengths of the light spectrum that is reflected back to the atmosphere is the color green. Leaves appear mostly green to the human eye because their wavelength is not absorbed. The chlorophyll molecule's capture of light energy begins a set of interactions along a complex of proteins that make up the electron transport chain.

The Two Phases of Photosynthesis in Plants

The mechanism of photosynthesis occurs in two sequential stages: the light and dark reactions. These reactions are more commonly known as the light-dependent and light-independent reactions. In the **light-dependent reactions** solar radiation from the atmosphere is absorbed by the chloroplast organelles of photosynthetic tissue. More specifically, the energy gets absorbed by the chlorophyll molecules and immediately converted into a temporary chemical energy. In the next phase, the **light-independent reactions**, this chemical energy is used to drive the biosynthesis. From the carbon atoms extracted from CO_2 new 3-carbon sugars are synthesized. The light-independent reactions that comprise the second phase of photosynthesis are also known as "dark reactions" since they do not require sunlight. Another name, **Calvin cycle**, is used in recognition of the scientist who elucidated its biochemical reactions. While the light-independent reactions do not require light they cannot function without the *products* of the light-dependent reactions. At the end of the first phase (light-dependent reactions) solar energy was temporarily stored using energy carriers or electron carriers. The carriers carefully transfer their energy cargo (electrons) to the light-independent reactions. While doing so they move down an energy gradient (high to low energy). After the energy carrier molecules have delivered their packets of energy, the carriers revert to low-energy molecules and cycle back to the light-dependent reactions to obtain more energy-packed electrons to deliver to the light-independent reactions.

What Is Light Energy?

In Chapter 2 we referred to solar radiation as the phenomenon of rays of sunlight that had a major effect on Earth's climate and temperature. Solar energy is another term used to describe the properties of solar radiation. While the sun continuously emits electromagnetic radiation (solar energy) human vision registers only a fraction of this energy, referred to as "visible light." The fraction of perceived "visible" light varies among animal species. For example, bullfrogs and pit viper snakes detect infrared (IR) light wave energy that is not perceptible to human vision.

Solar energy travels through the atmosphere as energy waves in varying intensities that comprise the entire spectrum of electromagnetic radiation. Scientists can differentiate the types of radiant energy contained in specific wavelengths. The shortest wavelength corresponds to the highest intensity (gamma) and the longest wavelength corresponds to the lowest intensity (radio waves).

It is possible to determine the amount of energy present in **wavelengths** by simply measuring the distance between consecutive crests of a wave of light (**Fig. 7.6**). Visible light perceived by humans consists of a limited range of wavelengths within the spectrum, between 400 nm and 700 nm measured from crest to crest (**Fig. 7.7**).

Within the electromagnetic spectrum there are many undetected forms of radiation (to the human eye at least), including X-rays and UV rays. These high-energy short waves are capable of penetrating tissues of organisms and causing damage to cells and DNA. The 400 to 700 nm visible spectrum is detected by humans because there are pigment receptors in the human eye designed to absorb solar energy at those

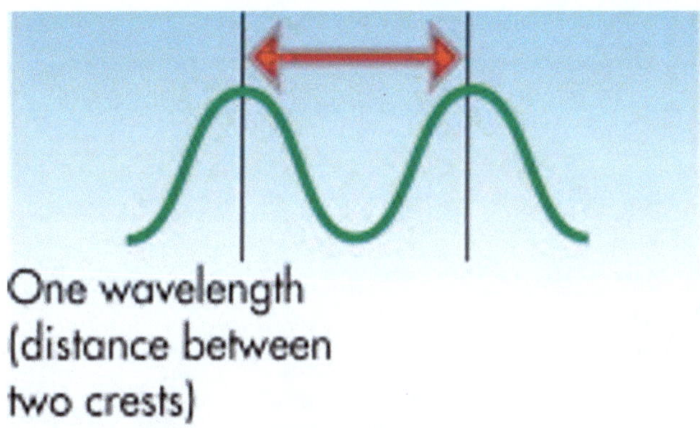

Fig. 7.6

Wavelengths are measured in nanometers (nm).

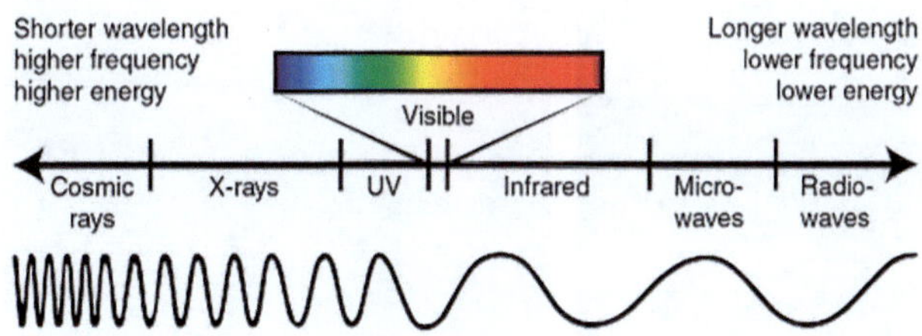

Fig. 7.7

Emission of the sun's electromagnetic radiation. Different wavelengths have their own characteristic radiant energy.

wavelengths. Though the environment surrounding us at all times receives the full electromagnetic spectrum (unless shielded), the human eye will not experience it.

According to NASA's Global Climate Change, almost half of the solar energy that arrives on Earth is the longer wavelengths converted to another energy form, thermal radiation (near IR <4 μm). Thermal energy is outside the range of visible solar energy, with a few exceptions. For example, lightning and natural fires release sufficient energy to be perceived as visible energy to the human eye.

Most warm-blooded animals emit a form of IR thermal radiation to the environment. Cold-blooded snakes don't visualize the world as humans do, lacking our vision receptors. Instead they evolved with receptors capable of perceiving images of their warm-blooded prey. The pit viper snake's viper pit organ (**Fig. 7.8**) detects

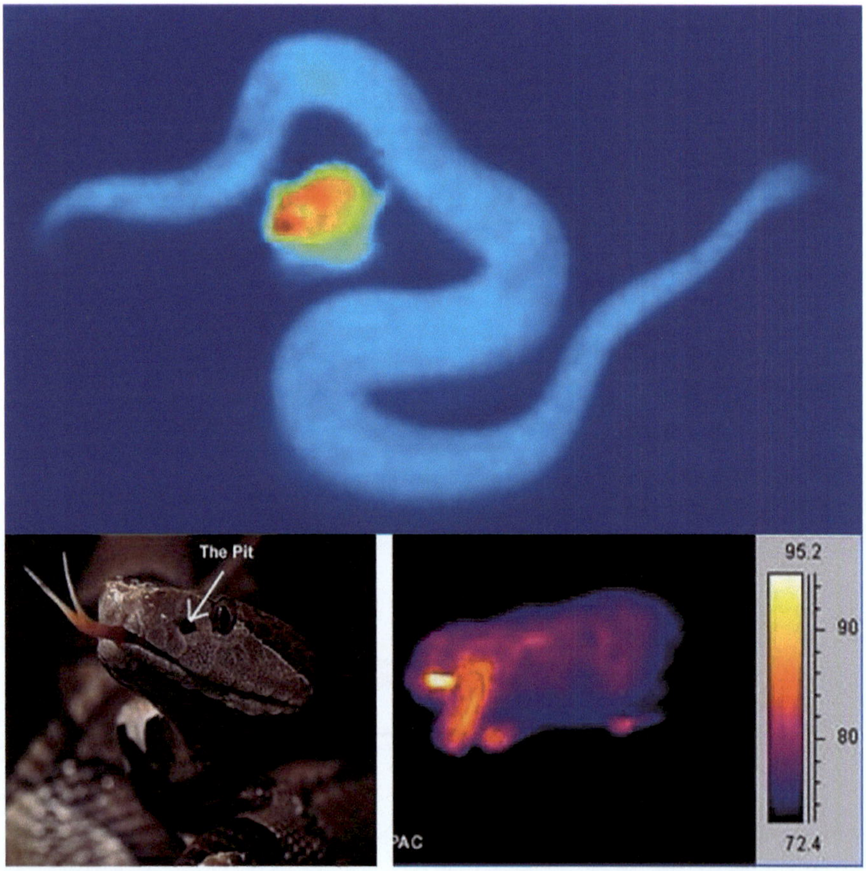

Fig. 7.8

Thermal IR image of warm-blooded animal and pit viper snake.

Source: Modification of Serpent nirvana under CC BY-SA 4.0

IR heat emitted. Plants' adaptations utilize the electromagnetic spectrum for their critical growth and developmental processes. For example, the timing of seasonal changes is detected by the degree of high-intensity UV radiation it receives under different light conditions. During photosynthesis the most important light waves entering a plant leaf are in the visible light spectrum range, 400 to 700 nm. This spectral range is the only solar energy absorbed by the plant's light-capturing pigment molecules and is known as **PAR**.

Absorption of Light

The presence of solar energy initiates the process of photosynthesis. There are several different light-absorbing pigment molecules present in plants. Each will have a range narrower than the full PAR of 400 to 700 nm in which the molecule's electrons interact with incoming light waves. To absorb solar energy across the entire PAR range plants contain many pigment molecules with different absorbance ranges. The same concept applies to the pigments found in the human eye retina (the rods and cones) that absorb visible light.

Outside of the PAR the chloroplast pigments do not absorb solar radiation. Electromagnetic spectrum wavelengths longer than red light (e.g., IR) cannot raise an orbital electron to its excited (quantum) state because they do not contain sufficient energy. Alternatively, wavelengths on the electromagnetic spectrum to the left of blue light (<400 nm) can physically tear the pigment molecules apart, a process known as **photobleaching**. The main pigments found in abundance in higher plant chloroplasts are **chlorophyll a** and **chlorophyll b**. But chlorophyll molecules are not the only pigments found in plants. **Carotenoid** molecules are another class of photosynthetic pigments found in plants and algae that will absorb solar energy. The multiple types of pigment molecules are the reason plant parts have so many different color displays.

Pigment molecules are important to plant physiology and are evolutionary adaptations to available light waves throughout the course of the year. Their activation can trigger growth, senescence, and reproduction, and it acts as a means to attract seed dispersers to their colorful displays. In all cases specific wavelengths of the light spectrum not absorbed by the plant's pigment will be reflected back to the environment, providing the plants distinctive color patterning. During photosynthesis carotenoids function to dispose of excess energy in the violet–blue light range. During a day of full sun exposure a leaf will process an enormous amount of energy. If there are insufficient absorption pigment molecules there is a risk that the intensity of radiation could do significant damage to the plant. The strategic positioning of carotenoids within the thylakoid membrane helps to absorb and safely dissipate that excess energy, in the form of heat, while the chlorophyll molecules proceed in their task of harvesting and converting solar radiation into stored chemical energy (light-dependent reactions). Pigments like carotenoids are retained to capture solar radiation as chlorophyll disappears from the leaf once cooler temperatures prevail with the onset of the fall season. Production of plant pigment variation is an important form of plant adaptation.

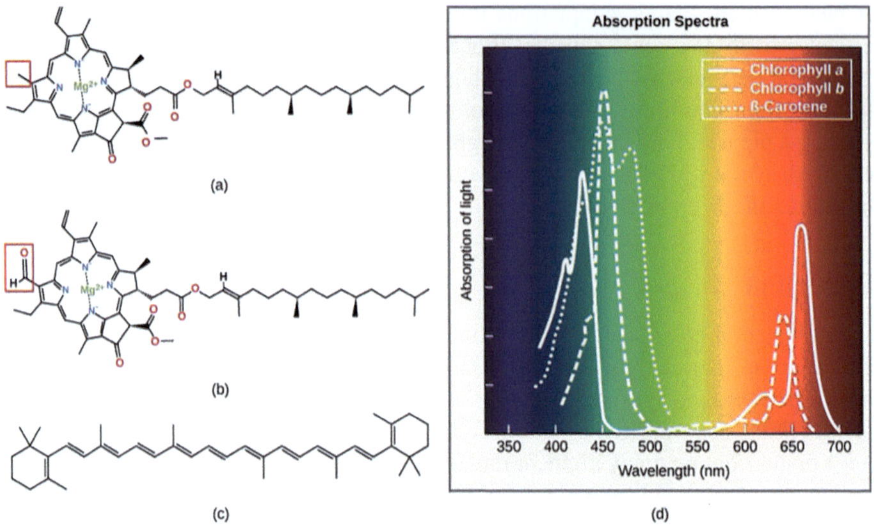

Fig. 7.9

(a) Chlorophyll *a*. (b) Chlorophyll *b*. (c) β-carotene. Shade plants adapt to low levels of light by changing relative concentrations of their chlorophyll pigments, which have unique absorbance spectrum (d).

Source: Jason Hollinger, Boundless under CC BY-SA 4.0

The three key pigments for photosynthesis are chlorophyll *a*, chlorophyll *b*, and β-carotene (a carotenoid). But plants biosynthesize a mixture of pigments covering a wider range of wavelengths, thereby diversifying their productivity. For example, plants on the forest floor will be able to absorb filtered solar radiation that passed through the canopy as scattered light (**Fig. 7.9**). The same holds true wherever organisms grow deep under water as light intensity decreases with increasing depth. Blue pigment molecules like phycocyanin are commonly found in aquatic photosynthesizers (algae and cyanobacteria), perfectly situated to absorb in a wider range. There are even terrestrial plants adapted to growing in shady regions, using far-red light (730 nm) more effectively than plants exposed to the full spectrum.

To differentiate the specific pigments present, an absorption spectrum can be generated using a **spectrophotometer**. Spectrophotometers detect transmitted light to determine the amount of light absorbed by pigments.

Light-dependent reactions convert solar energy into chemical energy in the form of NADPH and ATP (**Fig. 7.10**). This chemical energy will drive the light-independent reactions to complete photosynthesis with the assembly of sugar molecules.

Light-Dependent Reactions

In the light-dependent reactions a pair of multiprotein complexes embedded in the thylakoid membrane work in conjunction with the pigment molecules to produce NADPH

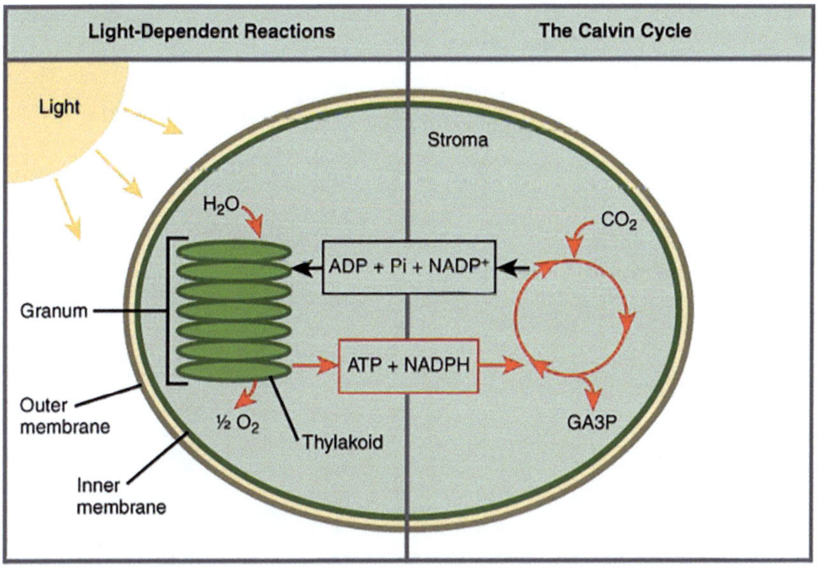

Fig. 7.10

The two phases of photosynthesis.

Source: Boundless Biology under CC BY 4.0

and ATP. The multiprotein complexes are called **photosystems (PS I** and **PS II**). The solar energy is captured using a number of **antennae proteins** bound to chlorophyll pigment molecules surrounding a **reaction center**. The pigments are excited by the solar radiation from ground to excited state. The process begins with absorption of a single light **photon** (a distinct quantity or "packet" of solar radiation) by any of the antennae chlorophyll. The light energy is then captured yet still isn't stored in a useful form. Energy continues to transfer from chlorophyll to chlorophyll within the photosystem's antennae until it is delivered to a pair of chlorophyll in the reaction center (it takes one millionth of a second). During this transfer of energy electrons were excited in their pigment molecules but not removed. This changes once the excitation events reach the **reaction center**, where sufficient energy has now gathered from all the surrounding antennae to *push* an electron away from the reaction center's "special pair" of chlorophyll, breaking its chemical bond and starting its journey along an electron chain on the PS I and PS II multiprotein complexes. **Fig. 7.10** illustrates the processes of photosynthetic light capture at the reaction centers.

In the chloroplast organelle there are many thylakoid membranes surrounded by a liquid stroma (**Fig. 7.11**). Each thylakoid contains numerous reaction centers. The dislodged electron at each reaction center is replenished from a split water molecule. Each electron starts down the electron chain (PS I and PS II), eventually reaching an electron carrier, NADP+, and reduces it to NADPH. This is now in a stored energy form, which can be delivered to phase two of photosynthesis.

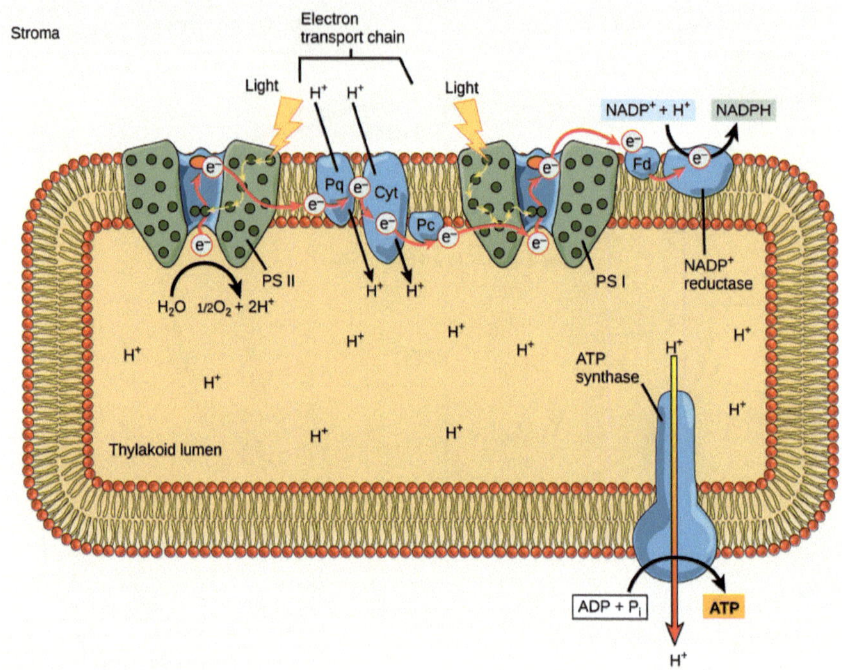

Fig. 7.11

Solar energy is captured in thylakoid reaction centers. The red arrow indicates the path of electrons initially dislodged from H_2O and end up reducing NADP+ to NADPH.

Source: Boundless Biology under CC BY 4.0

Besides solar radiation water is required for photosynthesis to continue. Water molecules will be split during the process, releasing their atoms: two electrons, two hydrogen ions, and one oxygen ($H_2O \rightarrow O + 2\ H$). The oxygen is released through the stomata, the electrons keep replenishing the reaction center electrons, and the hydrogen ions will be used to produce ATP, the other stored energy molecule produced in the light-dependent reactions, powered by its **proton motive force**. We will not cover the details of ATP synthesis in this chapter. In summary, phase one of photosynthesis results in molecules of NADPH and ATP, which are stored forms of energy to be utilized in phase two.

Light-Independent Reactions

Once solar energy is converted into chemical energy in the form of ATP and NADPH molecules the cell begins phase two of photosynthesis. In this process the chloroplast organelles participate in biosynthesis of carbohydrate molecules during the light-independent reactions. Phase two (light-independent reactions) requires the third requirement of photosynthesis: CO_2. *Recall: photosynthesis requires water, carbon, and solar energy.*

Carbon dioxide gas enters the leaf's stomata, where it diffuses through intercellular spaces until it reaches the mesophyll cells, then travels into the stroma of the chloroplast organelles. **Fig. 7.12** illustrates that both light-dependent and light-independent reactions are in close proximity within the mesophyll cell and operate simultaneously.

In the chloroplast stroma CO_2 is utilized in the light-independent reactions by an enzyme, ribulose bisphosphate carboxylase (**Rubisco**), and another substrate molecule, ribulose 1,5 bisphosphate (RuBP). In a continuous cycle Rubisco generates 3-carbon sugars (**G3P**) that can later be converted to 6-carbon glucose, as needed. The light-dependent products, ATP and NADPH, are made available to complete the reactions, delivering needed energy and electrons.

The formula for photosynthesis is:

$$6\ CO_2 + 6\ H_2O + \text{solar energy} \rightarrow C_6H_{12}O_6 + 6\ O_2$$

- The reactants are CO_2, solar energy, and H_2O.
- The products are 6-carbon glucose and oxygen.

During times of hot and dry conditions the reactions of photosynthesis will be at odds with water conservation because gas exchange requires opened stomata. Desert plants must adapt to a constraint during photosynthesis because their environment's high heat and low precipitation can very quickly lead to critical loss of water by the

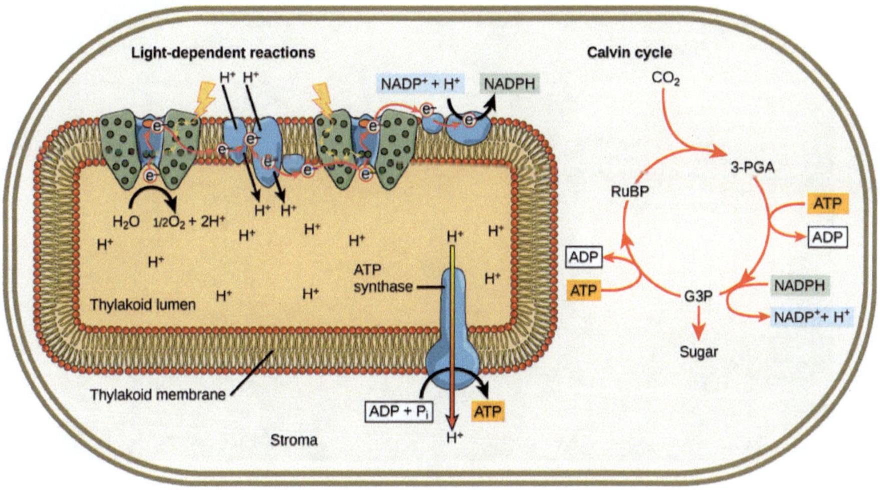

Fig. 7.12

Chloroplast of the mesophyll cell, where light and light-independent reactions occur.
Source: Boundless Biology under CC BY 4.0

process of **transpiration**. The H_2O evaporates through the stomata during uptake of CO_2 in active photosynthesis. Desert plants have evolved a number of adaptations to optimize their conservation of water. Some have evolved mechanisms to separate the two phases of photosynthesis, only performing light-independent reactions during the cooler evenings of the night and closing off the stomata opening during the heat of the day. These are known as **CAM** plants (crassulacean acid metabolism) (**Figs. 7.13a, b**). Another plant adaptation is seen in **C4** plants, optimized to avoid a different complication of photosynthesis: photorespiration. In photorespiration oxygen gas binds to the rubisco enzyme instead of CO_2, leading to futile cycling without producing 3 carbon sugars. Plants that have not developed specialized features to combat desiccation or photorespiration are called **C3** plants.

The activity of photosynthesis powers 99% of Earth's ecosystems and is a measure of an area's productivity. Its sole function is to store solar radiation in high-energy carbon–carbon bonds of carbohydrate molecules. For heterotrophs the consumption of carbohydrates also powers the synthesis of cellular energy (ATP) via cellular respiration (metabolism). The success of photosynthesis impacts all trophic levels, from primary producer through consumers. It is fascinating to think that a top predator is, in essence, at the end of an energy path that began with nuclear reactions on the

Fig. 7.13a

High-intensity heat and low precipitation conditions of deserts led to plant adaptive variations on timing and locations of the light-independent reactions.

Source: Piotr Wojtkowski under CC BY 4.0

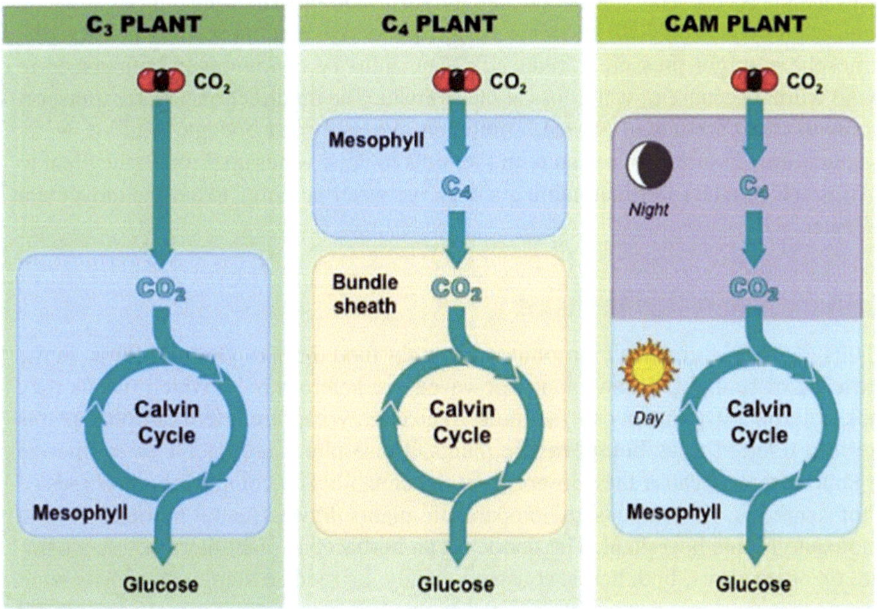

Fig. 7.13b

Carbon fixation pathways.

surface of the sun and followed a pathway to Earth's atmosphere where it became vegetative biomass eventually consumed by primary, secondary, and finally apex consumers.

Water Movement in Plants

During photosynthesis it has been demonstrated that water is required to resupply electrons at the reaction centers to keep energy flowing from solar radiation to chemical energy. Water molecules do not enter plants through the leaves directly. This isn't possible as most plants have protective layers of cuticle wax to prevent desiccation. Plants meet their requirement for water and nutrients through absorption from the soil by microscopic root hairs. The nutrients then travel through the vascular supply lines of plants to all of their living tissues. The vascular supply system is an adaptation that operates under passive processes (no energy input), pulling water upward by the cohesive properties of water's hydrogen bonds. Eventually water molecules arrive at the leafy parts where passage through the opened stomata pores via the process of transpiration ends its journey. A nonwoody plant's ability to remain upright, without a benefit of a skeletal framework, depends upon being fully hydrated. The high-water volume required for its upright habit is stored within large center vacuoles of every plant cell. The volume creates a hydrostatic pressure, pushing against the outer plant's rigid cell wall, known as **turgor pressure**. Whenever water is lost to the atmosphere, turgor pressure is reduced. Water must be continuously replenished to avoid wilting because of water loss at the stomata. The unidirectional water transport (roots to stems) requires: (1) solar radiation energy strong enough to cause water evaporation; (2) sufficient moisture in the roots to draw water up through the plant to its tips; and (3) the roots maintaining a negative water potential to aid the movement of water.

Nutrient Acquisition in Plants

Plants are mostly green, autotrophic, and obtain food through photosynthesis in the presence of sunlight. A few plants, however, are heterotrophic, which means they lack chlorophyll pigments to facilitate light energy capture. Heterotrophic plants are also referred to as **holoparasitic** plants. These plants are unable to synthesize organic carbon because they do not have the capability to complete both phases of photosynthesis. To compensate holoparasitic plants draw essential nutrients from a photoautotrophic host plant. The dodder is an herbaceous plant that lives as a parasite on other plants, both herbaceous and woody. Its reddish-stemmed tendrils wind around the stems of a host plant and feed exclusively on the organic matter produced by its host (**Fig. 7.14**).

Plants often need assistance acquiring some of their nutrients and work in a mutually beneficial partnership with soil microbes. Different species of bacteria and fungi have coevolved with a number of plants to create a symbiotic relationship between the microbe and the plant roots, each contributing to the livelihood of the other. The formation of bacterial nodules in legume plant roots and the fungal/plant interaction of mycorrhization are among the most successful nutritional adaptations that plants have acquired (**Fig. 7.15**). In the case of mycorrhization it is a fungus and provides water and soil nutrients to plants in exchange for fixed carbon from photosynthesis. The mixotrophic plants are parasitic, using mycorrhizal as a conduit to obtain their photosynthetic products. These are comparable to holoparasitic plants, merely functioning in a more indirect way. In the case of bacterial nodules the relationship is also symbiotic similar to the mycorrhizal relationship with plants.

Nitrogen Fixation: Interactions Between Plants and Bacteria

Nitrogen is a **macronutrient** because it is required to synthesize amino acid and nucleic acids. The atmosphere, comprised of approximately 78% N_2, is the largest pool of nitrogen in terrestrial ecosystems. Nonetheless, plants cannot directly acquire N_2 from the atmosphere because they lack enzymes necessary to convert N_2 into a biologically useful form. Nitrogen must first be "fixed" into terrestrial systems by its conversion from gaseous N_2 to NH_3 by either biological, physical, or chemical processes and then made assimilable by plants. In nitrogen fixation N_2 is converted into NH_3 by soil bacteria or aquatic cyanobacteria. NH_3 is preferred by forest trees but most plants uptake nitrogen from the soil in its nitrate

Fig. 7.14

Parasitic dodder (*Cuscuta*) vine taken in Bucks County, United States.

Source: Silvae, Public Domain

form, NO_3^-. The plant family *Leguminosae* (legumes) coevolved with a group of soil bacteria (***Rhizobia***) that convert NH_3 into NO_3^-. Farmers often rotate crops with soybeans (a legume) in alternate seasons to enrich the soil with nitrogen to improve crop productivity. The *Rhizobia* forms specialized structures called nodules that extend from the legume roots as they colonize and fix nitrogen. For mutual symbiosis to occur the bacteria must also receive a benefit from the association. In this example of a coevolutionary interaction the legume plant contributes the carbon source to the bacterium, as well as provides a protected niche in which to thrive.

Mycorrhizae: A Symbiotic Relationship Between Fungi and Root. Soil nutrients can be depleted when there is rapid nutrient uptake, low diffusion rates, low soil moisture, or other means of leaching minerals from the soil. As these conditions may persist in some regions, plants developed adaptations to acquire a steady and adequate nutrient supply. Some plants rely on fungi to facilitate their uptake of minerals from the soil, using symbiotic associations called **mycorrhizae**. The fungi integrate into the living tissue of the plant root and colonize during active plant growth. Through the process of mycorrhization, the plant obtains essential phosphate, zinc, copper, and other minerals from the soil. In return the

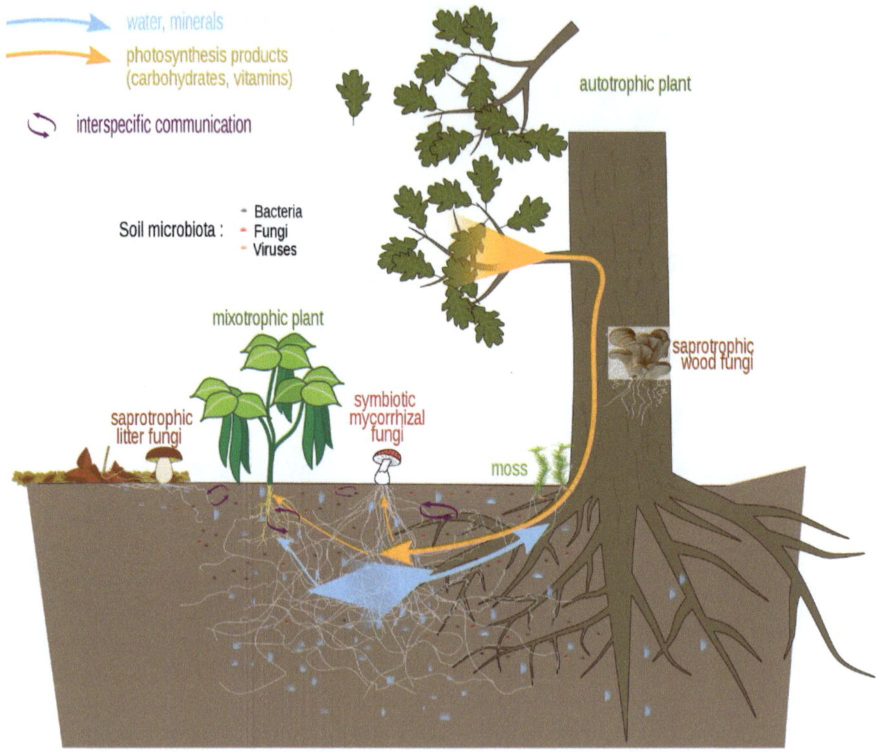

Fig. 7.15

Mixed underground relationships between plants and fungi.

Source: Adapted (translated from) https://commons.wikimedia.org/wiki/File:R%C3%A9seau_mycorhizien.svg under CC BY SA 4.0

fungus obtains carbon compounds produced by the plant's photosynthetic machinery, which was stored in the plant root (**Fig. 7.16**). Mycorrhizae hyphae will spread and ultimately increase the surface area of the plant root system, sometimes beyond the zone of nutrient depletion. Mycorrhizae can function as a physical barrier to pathogen invasions by inducing a generalized host defense mechanism. They have even been known to produce protective antibiotic compounds against pathogens.

Plants Adapt in Environments That Are Deficient in Resources

It is no wonder that immobilized plants have adopted so many partnerships, as they otherwise must rely on their roots, stems, and leaves to reach essential sunlight, water, soil nutrients, and oxygen resources. In less-than-ideal habitats plants developed a number of diverse adaptations to enhance their survival when

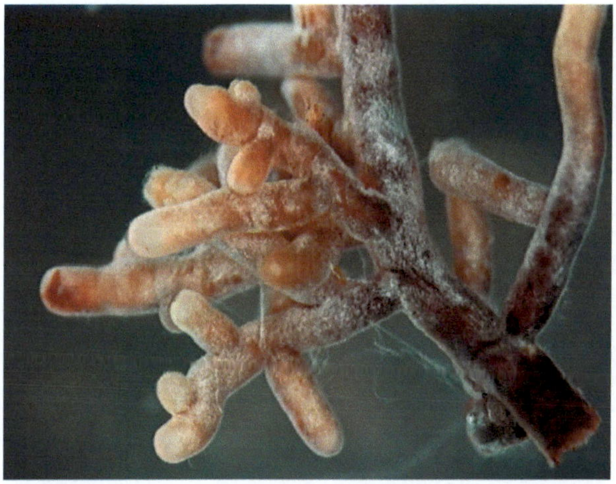

Fig. 7.16

Root tips infected with mycorrhizal fungus appear as off-white fuzz.
Source: https://doi.org/10.1186/1471-2105-6-178 under CC BY 2.51360-1385

resources are in short supply. In tropical rainforests access to sunlight is scarce below the dense canopy in which the tallest trees and plants grow close together and effectively block sunlight from reaching the forest beneath. One adaptation commonly seen in tropical plant species is the presence of broad leaves to maximize capture of any scattered sunlight.

Another adaptation is the **epiphyte** lifestyle: low-growing plants that gain access to sunlight by growing on top of other plants. Epiphytes, also called air plants, are able to live higher up in the canopy, atop the branches of other trees, where sunlight is more plentiful, and they can acquire resources by feeding on rain, minerals, and composting debris collecting in the branches and leaves of the supporting tree. Bromeliads, ferns, and orchids are all examples of tropical epiphytes (**Fig. 7.17**). Epiphytes are not parasitic because they take no nourishment away from their supporting tree. Some developed specialized tissues to efficiently capture and store water.

Aquatic plants developed specialized adaptations to allow them to "breathe" under water. While aquatic plants produce their own oxygen during photosynthesis oxygen consumption occurs 24/7 in all of their living tissues to perform needed metabolic processes. Most aquatic plants store oxygen produced during the day in plant **lacunae**, which can account for as much as 70% of the plant tissue. Similar to terrestrial plant vacuoles storing water for turgor pressure, the aquatic plant lacunae accumulate oxygen gas for their use when photosynthesis is not available.

Oxygen uptake has also evolved in swamp-residing plants where plant roots are submerged under water at least some of the time, while their aerial parts remain above the water. In typical terrestrial plants oxygen is part of the nutrient uptake by their

Fig. 7.17

Spanish moss (*Tillandsia usneoides*), an epiphyte in the tropical flowering bromeliad plant family.

Source: Gh5046, Public Domain

roots through the soil. But waterlogged roots have limited oxygen access. Mangrove trees (*Rhizophora sp.*), which grow in coastal waters, produce aboveground roots adapted to help support the tree in the soft sediment soils (**Fig. 7.18**) and to facilitate their oxygen exchange. The specialized aerial roots have open passages called **pneumatophores** to uptake atmospheric oxygen. Wild rice is another aquatic plant with pneumatophores, as its air-filled root tissue functions like a snorkel to pocket oxygen while excluding water.

Fig. 7.18

Mangrove trees develop aerial roots from their branches, which descend into the ground to help anchor the trees.

Source: Image courtesy of Paul Marek, Mangrove.at

Carbon Uptake in Aquatic Plants

Aquatic plants do not have stomata for gas exchange. Instead CO_2 diffuses to and from the atmosphere into its watery environment where it will be converted to an assimilable form. Aqueous CO_2 enters the carbonic acid cycle, forming HCO_3^- in equilibrium-dependent reactions, according to temperature and other factors present (**Fig. 7.19**). Marine water contains approximately 90% inorganic carbon in the form of HCO_3^-.

It is *bicarbonate ions* that will be taken up freely by submerged aquatic plants to supply their carbon needs. If the aquatic plant has aerial parts, however, it will preferentially uptake CO_2 directly using stomata. This is because CO_2 diffuses directly into tissues and cell membranes whereas HCO_3^- has a negative charge and cannot freely diffuse throughout the plant tissues and cells. Bicarbonate ions require an enzyme (carbonic anhydrase) to convert them back into CO_2, or the presence of specialized membrane transporters.

Atmospheric Temperature Influences Plant Adaptations

During times of temperature increases (above freezing) photosynthesis rates increase proportionally. Photosynthesis declines under extreme temperatures

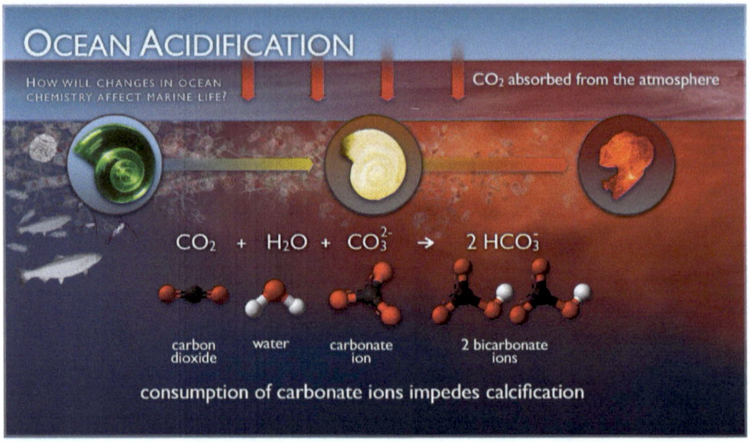

Fig. 7.19

CO_2 enters aquatic systems through the carbonic acid cycle.

Source: Image courtesy of author

outside the optimal range of 20°C to 40°C. At higher temperatures the energy absorbed becomes intense and will potentially disrupt photosynthetic processes. As long as the plant has an adequate water supply evaporation is one means of releasing rising internal leaf temperatures. Because water holds a high specific heat capacity loss of water results in lowering photosynthetic tissue temperature. But as mentioned previously, organisms need to maintain their water balance. Thus a balance must be achieved between lowering temps for integrity of photosynthetic processes and avoiding desiccation.

Photosynthesis is the reverse reaction of **cellular respiration**. Cellular respiration occurs within all cells by a series of reactions to break down chemical energy as needed to fuel metabolic processes. One involves carbon uptake and the other releases carbon back into the environment. To survive, grow, and reproduce, plants must maintain a positive carbon balance to synthesize their glucose molecules. This requires adequate solar radiation, CO_2, water, and mineral nutrients. The plant must also be able to tolerate the prevalent environmental factors, including temperature range, pH, and salinity levels.

Acclimation vs Adaptation in Plants

Adaptation and acclimation describe strategies taken by plants and animals when their natural habitat changes. Adaptation describes how a species has adjusted to its habitat. This is typically done over time through genetic changes under natural selection and affects the entire species. Plants that evolved to thrive under extremely low water or extremely high temperatures are examples of adaptation at work. For example, the acacia tree that grows in Hawaii, the southwestern United States, and Mexico thrives in arid conditions by its deep tap roots that reach groundwater sources. The tropical rainforest presents a very different challenge. Here the leaves of trees have adapted by producing thick, waxy leaves with "drip tips" to shed rain quickly and avoid an overgrowth of fungus and bacteria that flourish in the warm, wet tropical rainforest. These structural changes are the result at the molecular level of genetic changes, which is the definition of evolutionary processes that affect the entire population.

Acclimation is a temporary adaptation to gradual changes in the natural habitat, occurring during the lifespan of the organism, but is not an evolutionary process. Acclimation does not affect the entire species. The adjustment is made by modifying physical or behavioral reactions (to environmental changes) within the range that an individual's genotype permits. An example would be a freshwater plant being placed in an aquarium. The new habitat could mimic the freshwater environment with respect to temperature and dissolved solutes but there will be new challenges. If it can, the plant will acclimate to its new surroundings by developing new responses and behaviors. Should the plant be placed back in its natural habitat, it would likely return to its previous physical responses to its environment. No genetic change occurred. The capacity to change how it responds is written within its genetic code.

Acclimation in animals occurs during forced migration, for example, when tigers move into higher altitude terrains due to human population encroachment in their native habitat. The mountain air will have less oxygen, and at first it is more difficult for the tiger to breathe. Within a few months the tiger's red blood cell production will increase to help offset the constraint of gas exchange in higher altitudes. The increased blood cell count did not happen by a genetic change nor does it reflect a trade-off. Rather, the capacity to alter blood cell count was already present within the tiger's genome so it could quickly acclimate its physiological processes. It is important to recognize that acclimation may or may not occur in a habitat.

7.2 Animal Adaptations

In this section students should be able to:

- Explain why the concept of animal size is a fundamental constraint.
- Describe strategies animals use to acquire energy and nutrients.
- Distinguish between conformer and regulator modes of responding to the environment.
- Understand the importance of water balance in animals.
- List various means of thermoregulation.
- Describe animal habitat as a reflection of constraints, adaptations, and trade-offs.

Size Is a Fundamental Constraint

Animals are described and categorized by size within taxonomic groups. Each animal has corresponding morphological and physiological constraints based on its overall size. Organisms across all taxonomic groups range in body weight from 2 μg (unicellular microorganisms) to 100,00 kg (blue whale).

Each individual cell is metabolically active, whether comprising a unicellular organism or representing one cell of many in a multicellular organism. A critical constraint on evolution occurs at this cellular level due to the relationship between surface area of a cell and its internal volume. Cells are limited to the maximum size of 0.1 μm to 5.0 μm for prokaryotic cells and 10 μm to 100 μm for eukaryotic cells. The physiological and biochemical processes that must occur to maintain each cell are massive, particularly considering the cell's small size. Every living cell must perform respiration and there must be a large diffusive intake of oxygen. It is the constant need to transport products across its membrane (both uptake and release of products) that specifies how large the cell can become. The larger the cell size the more constrained it will be to move sufficient quantities across its cell surface.

Many of the morphological adaptations at organismal, or cellular level, are dedicated to maintaining exchange with the environment for essential nutrients, gases, water balance, and thermal energy. A cell adaptation evolved for body regions with higher demands for transport, for example, the gastrointestinal microvilli cells. These absorptive cells have a surface membrane that is convoluted or wrinkled to increase nutrient intake. Whenever the surface area along an inner body tube is increased size constraints will be reduced in the species. This is possible because more nutrients can now be assimilated and converted into energy reserves.

As animals become larger and more complex they must process greater quantities of food reserves to meet their energy needs. Therefore additional transport and absorption strategies are needed, including the presence of lungs, digestive systems, and body cavities.

Strategies Animals Use to Acquire Energy and Nutrients

The strategies to acquire adequate nutrients for an organism's energy needs are a major selective pressure on biodiversity, and there are many examples throughout the animal kingdom. Nutrient acquisition can be said to have always driven animal adaptations. Physiological ecologists study how animals have adapted physiologically to their environmental constraints. A first step in obtaining essential nutrients is to procure food by a series of behaviors and adaptations. After consumption the nutrients must be digested and absorbed by the body to be utilized by cellular processes. Specialized body systems and organs are needed for the entire process, including the appropriate ingesting and chewing structures like mandibular jaws. Animals that consume fibrous foods that resist enzymatic digestion often establish symbiotic relationships with microbes inhabiting their gastrointestinal tracts, which are capable of breaking down the tough fibers for them. For example, the wood-eating termite does not chemically break down those woody fibers it is notorious for destroying. Termites developed symbiotic relationships with quite a few microorganisms, including bacteria and protozoans, which reside in their digestive systems and are responsible for digesting the complex sugars in wood (cellulose and lignin). In theory each animal's digestive tract contains enzymes it produces and the **symbionts** it harbors, which correlates to the chemical nature of their diet.

Animals can be classified according to which resources they exploit for their survival:

- **Herbivores**—feed exclusively on plant tissues
- **Carnivores**—feed exclusively on animal tissues
- **Omnivores**—feed on both plant and animal tissues
- **Detritivores**—feed only on dead, decaying plant and animal matter

For the **herbivore** the challenge is consuming adequate fat and protein, which is much lower in plant tissue. Plants produce varying amounts and types of carbohydrates and lignans. Herbivores may select to consume only specific plant material in their diet. For example, **grazers** only consume leafy material like grasses, **browsers** feed on woody material, and **granivores** feed on seeds. There are also **nectivores** and **frugivores** (nectar and fruit feeders, respectively). The mostly indigestible cellulose and lignan layers of plant tissue are unavoidably consumed by both grazers and browsers, requiring assistance to break down their tough fibers. This is because animals do not contain the correct enzyme to break down the type of carbohydrate bond present in dense fibers. Animals rely on the microbial fermenters that inhabit their digestive tracts and can enzymatically break down the fibers. The symbionts are benefitted in this exchange by receiving access to the nutrients of the digestive tract. Animals that graze all day long are known as **ruminants**. Their compartments are significantly expanded for extensive microbial action. The rumen compartments are analogous to a minifermentation industry (**Fig. 7.20**).

Microbial "fermenters" in the guts of all mammals, including humans, break down the tough fibers for digestion and synthesize fatty acids, amino acids, proteins, and vitamins for their host. According to Anna R. Schwyter and Karen L. Vaughan, University of Wyoming, plants have a calculated ratio of carbon to nitrogen of 50:1. In animals the ratio of carbon to nitrogen is reduced to only 10:1. A protein diet rich in nitrogen is optimal for animal herbivores because nitrogen is essential for their growth (needed for proteins and DNA synthesis). Some herbivores preferentially feed on the higher nitrogen ratio found in new growth, located in the tips, leaves, and buds of plants. Another adaptation is to feed during seasonal growth periods. The very short life span of insects, for example, can be completed entirely during spring and summer months before leaf fall. Animal birthing cycles during early spring are even an adaptation, developed to derive the greatest advantages for their new offspring's growth when higher-quality food sources are available.

Carnivores consume food from tissue structures quite similar to their own and don't encounter the digestion problems associated with cellulose and lignin fibers. The carnivore's greatest feeding challenge is access to sufficient food sources to meet

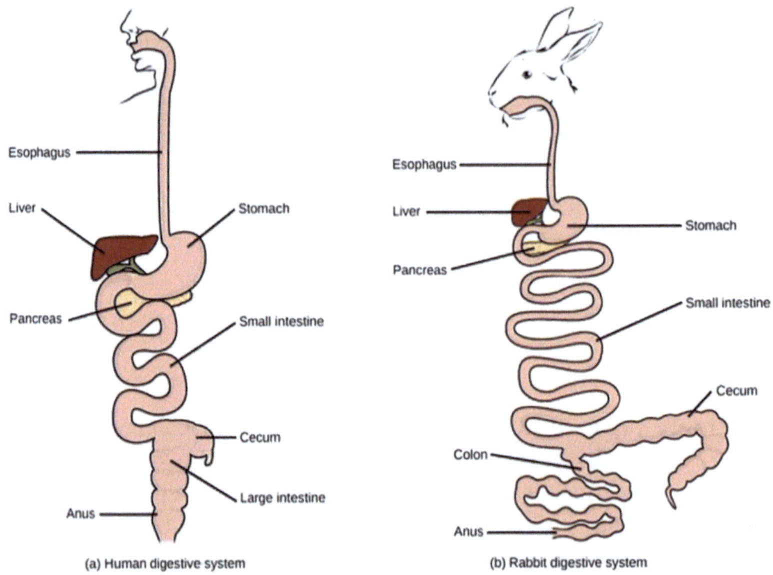

Fig. 7.20

Ruminant digestive system of the rabbit reflects a food acquisition lifestyle high in herbivory.

Source: Voer under CC BY 3.0

its energy needs. This constraint led to a multitude of adaptive strategies between predator–prey, competition, and habitat selection, topics discussed in greater detail in later chapters (population dynamics and community interactions).

Omnivores often adjust their feeding habits to the season and their life-cycle stages—for example, acquisition of available seeds and berries when in season and switching to a carnivore feeding cycle after vegetative sources become limited.

Conformer and Regulator Modes of Response to the Environment

Most ecosystems experience fluctuation in environmental conditions with respect to their most influential abiotic factors, light availability, and temperature. Animal diversity varies from the local effects of changing winds and moisture levels. When animals adapt by conforming their internal environment to the existing external conditions they are called **conformers**. When animals maintain a defined internal range that is distinctly different than what is happening in the environment surrounding them they are called **regulators**. Conformers maintain their internal conditions by energy-consuming regulatory mechanisms. The various mechanisms work by adjusting internal temperatures, water content, pH, salinity, and tissue oxygen levels. The purpose is to maintain an internal homeostatic balance. Homeostatic maintenance offers many advantages to the animal but requires an energy cost investment, which reflects a trade-off. Regulators devote an enormous amount of their energy to homeostatic maintenance of internal conditions. Regulation is energy consumptive under widely varying climatic conditions. Since conformers exist by allowing their internal environment to adjust to the external environment much less energy will be required for their lifestyle. The conformer's physiological and biochemical parameters function adequately even as food availability becomes limiting. Conformers are capable of slowing down their metabolism "to ride out the storm," an adaptive strategy that includes torpor in the amphibian species.

The physical environment will still impart a constraint on the conformer lifestyle as there are limits on suitable habitats. Regulators, on the other hand, can maintain their internal environment by adjusting physiological, biochemical, and behavioral activities over a broader range of external environmental conditions. But in extreme temperatures a regulator's maintenance of body temperature to a defined internal range can account for as much as 90% of its total energy budget. The higher energetic costs require consuming more food resources. Some regulator animals are continually feeding to replenish spent energy supplies to survive in their environment.

Very different costs and benefits are apparent between regulator and conformer lifestyles. The conformer's lower energetic expenditure means it will have lower activity levels (think of the laid-back tortoise). Reproduction and growth are reduced in conformers as their altered internal state may not be conducive to these energy-requiring life processes. In contrast the regulator moves quickly. When internal structures are kept at optimal ranges performance is enhanced to acquire resources and more robust reproductive processes and growth rates are possible during their respective life stages.

An important physiological function in regulator animals, not present in conformers, is the need to maintain fluid balance. Regulators accomplish this using a process known as **osmoregulation**. Fluid levels are maintained by balancing electrolyte salts (primarily sodium and chloride ions). Because of the movement of water by osmosis solutes are maintained at internal concentrations optimal to water balance inside cells and body tissues. Regulator animals in both aquatic and terrestrial environments maintain the necessary concentration of solutes in body fluids through various mechanisms occurring with ingestion and excretion.

Internal Conditions Are Maintained by Homeostasis

Homeostasis is the process by which regulator animals maintain internal conditions in a stable, steady state. The optimal range is maintained despite changing external climate, diet, or activity levels. Variables that require a set range include body pH, temperature, and solutes including blood glucose. These set ranges are necessary to optimize cellular enzymes, transport, and the organism's structural components at high-efficiency rates. Homeostatic regulation operates as a system and includes (1) a receptor, which senses a change in the environment; (2) a control center, which determines the appropriate response to a perceived change (e.g., central nervous system); and (3) an effector, which is the means of bringing about a needed change to restore internal conditions back to their narrow, optimal range. A **negative feedback mechanism** is then initiated, which informs the receptor to stop further signaling to the control center to avoid over adjusting in the opposite direction. It is called a *negative* feedback loop because it continues to swing from out of range back to normal, negating any risk of imbalance and potential risk to the organism. An example of a negative feedback loop in humans occurs in maintaining blood sugar levels. The hormones, insulin, and glycogen are the effectors regulating blood sugar levels in the bloodstream.

The Importance of Water Balance

Plants and animals consist of approximately 75% to 95% H_2O. Water is essential for biochemical reactions and serves as the medium for regulatory processes. Terrestrial and aquatic animals have widely different mechanisms to maintain their internal water balance. In terrestrial animals water is acquired through food and water intake and as a byproduct of cellular respiration. They lose H_2O through excretory processes including urination, feces, and evaporation through the epidermal layer (skin) that are part of its osmoregulation. Reabsorption of H_2O during normal excretory processes is tightly regulated to maintain a proper balance between H_2O intake and H_2O loss. Animals accomplish this through kidneys, salt glands, or cloaca that modify salt concentrations for H_2O balance and osmotic regulation. For animals in arid climates additional mechanisms are needed, including behavioral adaptations to avoid the hottest point of the day and structural features like thicker outer cuticle layers that prevent desiccation. During the dry, hot periods of the desert some animals enter a summer version of hibernation, called **estivation**. Semiaquatic and terrestrial animals may burrow under the soil and insects may undergo **diapause** (e.g., arrested development).

Aquatic animals resolve water balance constraints in their environment quite differently. Although bathed at all times in H_2O, H_2O balance is still necessary to maintain healthy tissues and membrane structures. In marine environments salts are higher in the external environment, which by osmotic pressure (H_2O follows salts) can lead to animal desiccation, even while bathed in H_2O! In freshwater environments it is the internal tissues that have higher concentrations of salts compared to the surrounding H_2O. Kidneys or other mechanisms function in freshwater aquatic animals to regulate salt intake and excretion of excess H_2O as needed to maintain their H_2O balance. Some invertebrate aquatic animals are **isosmotic**, with body fluids taking on the same osmotic conditions as their external environment. This includes jellyfish, mollusks, and sea anemones. These animals are known as **osmoconformers** with respect to H_2O balance. However, if osmoconformers have other body systems under internal regulation, they will still be classified as regulators.

Thermoregulation in Animals

Another animal adaptation is **thermoregulation**. Adjusting to temperature variation in the environment is a problem that terrestrial animals tackle in a variety of ways to conserve energy reserves. For example, an outer insulating layer aids heat exchange efficiency between the organism's core and the external conditions. Changing the internal metabolic rate is another means of adjusting to the external climate. Faster metabolisms emit more heat as a byproduct of cellular respiration and cold climates are more habitable if the organism has a higher metabolic rate. The characteristics of an organism's outer coverings interact with its environment by utilizing one of several heat exchange processes: **conduction**, **convection**, **radiation**, and **evaporation** (**Fig. 7.21**). **Thermoconformers** invest less into thermoregulation and experience less thermal stress over a broader range.

In aquatic settings thermal stress is less of an issue, requiring less regulation, as the environment is energetically more stable with fewer temperature fluctuations. This is because H_2O is better at absorbing solar radiation than air. Overall most regulator organisms modify body temperature through behavior, physiology, and morphology.

The terms thermoregulation and thermoconformer may not adequately describe the interplay of heat exchange occurring between organisms and their environment. New terminology was applied to cover the broader distinctions of heat exchange. Organisms that maintain body temperatures through an exchange of thermal energy with the surrounding environment are known as **ectotherm**. They can be classified as thermoconformers, but *only* if they do not actively engage in changing their body temperature by other means such as altering body temperature through behavioral modifications. Aquatic species are generally ectotherms but many do not fit in the classification thermoconformers. Thermoregulator organisms are heavily invested in maintaining body temperatures through internally generated heat production or

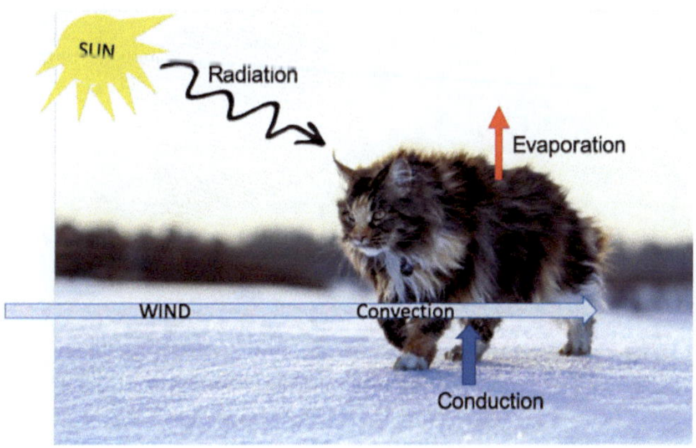

Fig. 7.21

Heat exchange in a cat.

Source: Image by author

behavioral modifications and are known as **endotherms**. Mammals and birds are endotherms, and dinosaurs are believed to have been endotherms.

In reality all animals generate some internal heat through their own metabolism and all animals will be affected to some extent by external sources of thermal energy. The differences are in the degree of regulation. For example, amphibians and reptiles are ectotherms but they can also be classified as thermoregulators. The terms thermoregulators or thermoconformers, as well as the terms cold-blooded and warm-blooded, are somewhat outdated and inadequate to distinguish animal heat exchange types.

Animals that primarily modify their body temperature through behavior have long been classified as **poikilotherms**. Many reptiles fall in this category (ectotherm but not thermoconformer). The desert lizard can maintain its precise temperature, within a degree of two of its lethal temperature ranges, through behaviors such as undergoing torpor and seeking shady shelters during the heat of the day (**Fig. 7.22**). While activity is severely restricted the desert lizard does not expend a great deal of energy to maintain its internal temperature. Overall there is a wide range of adaptations employed by species to modify internal temperatures to safe ranges.

Animals that hibernate are behaviorally endothermic while active but resemble ectotherms when they hibernate in winter. For example, larger fish are capable of generating and conserving sufficient heat to raise their body temperature above the surrounding body of water in some instances but they are not classified as true endotherms. This is because they don't maintain body temperatures within a specific range. Fish with endothermic characteristics include swordfish and some sharks. Both have the ability to maintain their central nervous systems above ambient temperatures to

Fig. 7.22

Collared lizard (*Crotaphytus collaris*), Black Canyon, Gunnison National Park.

Source: Daniel Schwen, under CC BY 4.0

increase their ability to detect and react to prey. And tuna fish maintain the internal temperatures of their muscles, a useful adaptation to improve stamina for swimming at high speeds. To differentiate between endotherms and ectotherms remember that endotherms will always maintain higher metabolic rates than the environment and burn calories (food) to maintain those internal temperatures.

The internal furnace that generates body heat is cellular respiration. In these processes chemical energy in the form of glucose is oxidized, releasing its energy to the internal environment for capture and conversion to a more usable form of chemical energy, ATP. The process isn't very efficient (much less so than your home's furnace, which probably runs at 80%–90% efficiency). Your body's respiratory furnace is only 40% to 55% efficient. The body's excess energy is released as heat. During winter months animals rely on the inefficiency of metabolism to survive. In the summer months slowing down internal metabolism by reducing activity avoids excess heat energy build-up in the body. Other adaptive characteristics that contribute to body temperature regulation include insulating layers of fat to conserve heat and sweat glands or pores to release excess heat. Since water has a very high specific heat capacity sweating, panting, and evaporative cooling all effectively release moist heat to the atmosphere. However, this solution is viable only if the water balance isn't disrupted and leads to desiccated tissues. If you ever wondered how your domestic pet's thick fur coat doesn't cause them to die of heat stroke, it is because the design of the mammalian coat includes an air layer, helping to shield the animal from the full effect of hot weather. This modification also means that dogs don't have sweat glands on their skin, except for the paw area. Instead they are prone to excessive panting when overheated, which releases heat in the form of warmed air from the lungs. Establishing a balance between releasing heat but not disturbing the critical H_2O balance represents an important trade-off in animals, as well as plants, in warm climates.

In colder climates animals may conserve heat through an added high-density insulating layer called brown fat (regular fat cell layers are white). Brown fat is more metabolically active than other body tissues. Involuntary muscle shivering is another regulatory mechanism that helps to generate heat by increasing the organism's metabolism.

In summary, for animals to adapt and acclimate to their habitats various trade-offs are navigated to obtain needed advantages while mitigating any disadvantages to their survival. Strategies that allow an organism to excel under certain conditions may cause a deficit under different conditions. Habitats that favor one strategy may prove to be uninhabitable if the climate changed abruptly or severely. For example, the freedom from environmental variation that endotherms enjoy, by way of managing their own internal regulation, requires constant replenishment of their energy reserves. Foraging is a major daily preoccupation to offset metabolic consumption. If food sources become scarce, for example, as temperatures rise dramatically, the strain on endotherms would understandably become excessive. In contrast an ectotherm may not experience the same degree of stress due to changing climate [b].

Camouflage or Biological Mimicry

Prey species have evolved many adaptations in body coloration, chemical production, and behaviors that provide them with a survival advantage against predators. Camouflage is the first line of defense, particularly for insects, lizards, snakes, and frogs, which develop colors and patterns that match leaves or twigs they inhabit. Sometimes even the morphology will match the shape of twigs or leaves. This strategy is more effective, of course, when the animal isn't moving. The production of toxic chemicals can make a prey poisonous to eat, a learned behavior that leads to predator avoidance of certain species. Experiments have shown that it often takes just one encounter with a toxic prey species for the predator to avoid it. Bright coloration patterns can even accomplish a warning signal for predators to stay away from certain species, known as **aposematic coloration**.

Mimicry takes advantage of camouflage strategies without the added energy to produce toxic chemicals or evolving structures for its safe storage or dispersal. Animals instead mimic pigment colorations of other, poisonous animals that predators have already learned to avoid. This is known as **Batesian mimicry**, after the British naturalist who first identified it. A similar approach, **Müllerian comimicry**, occurs when two prey species share a similar look and are both toxic. This strategy takes a lesser toll on both species by predators who are still learning to avoid that particular appearance. A well-known example is the Viceroy and Monarch butterflies. The concept of mimicry or camouflage will be further explored in the species–species interactions chapter.

A rare example of camouflage combined with an adaptation for energy acquisition was discovered in the leaf sheep, or leaf slug (*Costasiella kuroshimae*). *C. Kuroshimae* is a tiny marine caterpillar in the mollusk family that lives in the Indo–Pacific Ocean. Acquisition of energy is through a chemical process called "kleptoplasty." As the name suggests this tiny animal is *stealing* energy. As it feeds on algae it retains its intact chloroplast organelles, which then perform photosynthesis directly for the leaf sheep. The chloroplast, filled with chlorophyll pigment, provides green cover for the animal. Camouflage occurs because the now-green *C. kuroshimae* is able to blend into the algal background to avoid predation by species higher up in the food chain.

Habitats Reflect Trade-Offs

The habitat defines the diverse range of relationships between an organism and its environment. Each environmental niche around the globe has a unique distribution of species that coexist. These species occupy a unique geographic area in which to grow and reproduce. A fundamental constraint on species distribution is its ability to obtain essential resources and inhabit environments conducive to its continued life and reproductive success. This is a fundamental requirement under Darwin's "survival of the fittest." Various abiotic factors, including solar radiation, temperature, pH, and salinity, have direct influences on the physiological processes of life. Every organism has its own limited range of conditions upon which it can thrive. While the environment must provide the essential resources for life, including prey, the habitat reflects a variety of adaptations that organisms adopt to take up residency. For animals that are mobile specific locations to inhabit are less selective to their suitability than they would be for immobile plants. Nonetheless **habitat selection** is a widely studied area of ecology. According to ecologist Dr. Jones, 2001, "Habitat selection refers to a hierarchical process of behavioral responses that may result in the disproportionate use of *habitats* to influence survival and fitness of individuals." That is, a plant species may be important to an animal's habitat selection, particularly if it provides a preferred high-nitrogen food source in its seeds and fruits. Species life histories and species–species interactions will be covered in upcoming chapters in more detail.

Glossary

cellular respiration
proton motive force
rubisco
ectotherm
endotherm
plant vacuole
lacunae
pneumatophores
photobleaching
Batesian mimicry
Müllerian co-mimicry
aposematic coloration
mesophyll
glyceraldehyde-3-phosphate (G3P)
CAM plants
C4 plants
thylakoid
wavelengths
visible light
chlorophyll a
isosmotic
chlorophyll b
carotenoid
ß-carotene
spectrophotometer
Habitat selection
negative feedback mechanism
poikilotherms
turgor pressure
holoparasitic plants
conduction
convection
radiation
evaporation
osmoregulation
osmoconformers
thermoconformers
symbionts
ruminants
grazers
granivores
nectivore
frugivores

Citations

1. Patricia Sanchez-Baracaldo, Cyanobacteria and biogeochemical cycles through Earth history, Trends Microbiol. Review 30 (2) (2021) 143–157. https://doi.org/10.1016/j.tim.2021.05.008
2. P. Meyer, B., Van de Poel, B. De Coninck, UV-B light and its application potential to reduce disease and pest incidence in crops, Hortic Res 8 (2021) 194. https://doi.org/10.1038/s41438-021-00629-5
3. Iram Shafiq, Crop photosynthetic response to light quality and light intensity, J Integr Agric 20 (1) (2021) 4–23. https://doi.org/10.1016/S2095-3119(20)63227-0
4. R.F. Carvalho, M. Takaki, R.A. Azevedo, Plant pigments: the many faces of light perception, Acta Physiol Plant 33 (2011) 241–248. https://doi.org/10.1007/s11738-010-0533-7
5. H. Hashimoto, Carotenoids and photosynthesis, Subcell Biochem 79 (2016) 111–139. https://doi.org/10.1007/978-3-319-39126-7_4

6. R. Fernandes, et al., Exploring the benefits of phycocyanin: from spirulina cultivation to its widespread applications, Pharmaceuticals (Basel) 16 (4) (2023) 592. https://doi.org/10.3390/ph16040592.
7. A. Casadesús, S. Munné-Bosch, Holoparasitic plant-host interactions and their impact on Mediterranean ecosystems, Plant Physiol 185 (4) (2021) 1325–1338. https://doi.org/10.1093/plphys/kiab030
8. B.P. Meena, "Legume green manuring: an option for soil sustainability,", in: R. Meena, A. Das, G. Yadav, R. Lal., (Eds.), Legumes for Soil Health and Sustainable Management, Springer, Singapore, 2018, pp. 387–408. 10.1007/978-981-13-0253-4_12
9. M. Fujita, M. Kusajima, M. Fukagawa, et al., Response of tomatoes primed by mycorrhizal colonization to virulent and avirulent bacterial pathogens, Sci Rep 12 (2022) 4686. https://doi.org/10.1038/s41598-022-08395-7
10. Sandhya Srikanth, Mangrove root: adaptations and ecological importance, Trees 30 (2) (2015). https://doi.org/10.1007/s00468-015-1233-0
11. M. Ejiri, Y. Sawazaki, K. Shiono, Some accessions of Amazonian wild rice (Oryza glumaepatula) constitutively form a barrier to radial oxygen loss along adventitious roots under aerated conditions, Plants (Basel) 9 (7) (2020) 880. https://doi.org/10.3390/plants9070880
12. Emma Steggles, The potential for deep groundwater use by Acacia papyrocarpa (Western myall) in a water-limited environment: Deep groundwater use by trees in water-limited environments, Ecohydrology (2016). https://doi.org/10.1002/eco.1791
13. W. Palm, C. Thompson, Nutrient acquisition strategies of mammalian cells, Nature 546 (2017) 234–242. https://doi.org/10.1038/nature22379
14. S. Boudsoca, Evolution of nutrient acquisition: when adaptation fills the gap between contrasting ecological theories, Proc Royal Society P. 278 (2011) 1704. https://doi.org/10.1098/rspb.2010.1167
15. J.A. Venter, M.M. Vermeulen, C.F. Brooke, Feeding ecology of large browsing and grazing herbivores, in: I. Gordon, H. Prins, (Eds.), The Ecology of Browsing and Grazing II. *Ecological Studies*. 239, Springer, Cham, 2019. https://doi.org/10.1007/978-3-030-25865-8_5
16. M. Yamasaki, K. Kikuzawa, Temporal and spatial variations in leaf herbivory within a canopy of *Fagus crenata*, Oecologia. 137 (2) (2003) 226–232. http://www.jstor.org/stable/4223753
17. Richard T. Corlett, The ecology of animals: foods and feeding. The Ecology of Tropical East Asia, 2nd ed (online edition, Oxford Academic, 2014. 20 Nov. 2014). accessed 22 Feb. 2024 https://doi.org/10.1093/acprof:oso/9780199681341.003.0005
18. S. Giroud, The torpid state: recent advances in metabolic adaptations and protective mechanisms, Front Physiol (2021). https://doi.org/10.3389/fphys.2020.623665
19. D. Rozen-Rechels, When water interacts with temperature: ecological and evolutionary implications of thermo-hydroregulation in terrestrial ectotherms, Ecol Evol 9 (17) (2019) 10029–10043. https://doi.org/10.1002/ece3.5440
20. P. Withers, C.E. Cooper, Dormancy: Brian Fath, *Encyclopedia of Ecology*, 2nd ed, Elsevier, 2019.309314 ISBN 9780444641304 https://doi.org/10.1016/B978-0-12-409548-9.11167-4
21. G.J. Ragland, P.A. Armbruster, M.E. Meuti, Evolutionary and functional genetics of insect diapause: a call for greater integration, Curr Opin Insect Sci 36 (2019) 74–81. https://doi.org/10.1016/j.cois.2019.08.003

22. J. Wiemann, I. Menéndez, J.M. Crawford, et al., Fossil biomolecules reveal an avian metabolism in the ancestral dinosaur, Nature 606 (2022) 522–526. https://doi.org/10.1038/s41586-022-04770-6
23. D. Moore, A. Stow, M.R. Kearney, Under the weather? The direct effects of climate warming on a threatened desert lizard are mediated by their activity phase and burrow system, J Anim Ecol 87 (3) (2018) 660–671. https://doi.org/10.1111/1365-2656.12812
24. K.A. Dickson, J.B. Graham, Evolution and consequences of endothermy in fishes, Physiol Biochem Zool 77 (2004) 6.
25. D. Mota-Rojas, Efficacy and function of feathers, hair, and glabrous skin in the thermoregulation strategies of domestic animals, Animals (Basel) 11 (12) (2021) 3472. https://doi.org/10.3390/ani11123472
26. S. Clusella-Trullas, T.M. Blackburn, S.L. Chown, Climatic predictors of temperature performance curve parameters in ectotherms imply complex responses to climate change, Am Nat 177 (2011) 738–751 https://doi.org/10.1086/660021 10.1086/660021
27. K.E. Smith, C.G. Halpin, C. Rowe, The benefits of being toxic to deter predators depends on prey body size, Behav Ecol 27 (6) (2016) 1650–1655 https://doi.org/10.1093/beheco/arw086 Epub 2016 Jun 21.
28. T.N. Sherratt, The evolution of Müllerian mimicry, Naturwissenschaften. 95 (8) (2008) 681–695 https://doi.org/10.1007/s00114-008-0403-y Epub 2008 Jun 10.

CHAPTER

Ecosystem Energetics

8

Chapter Outline

8.1 Energy Is Governed by Physical Laws of Nature

By the end of this section students should be able to:

- Apply laws of thermodynamics to energy capture and movement within ecological systems.
- Understand why a loss of energy occurs with each energy transfer.
- Relate the terms "order" and "disorder" to entropy state.
- Give an example of how entropy increases or decreases in ecosystems.

The sun is the ultimate source of energy that drives the dynamics of the planet. Solar energy enters Earth's atmosphere as electromagnetic radiation (see Shortwaves, Chapter 2). The unique phenomenon of solar radiation is that it is both a wave of light and a particle of energy known as a photon. It is the energy from solar radiation that gets harnessed within plant chloroplasts until it is high enough in intensity to eject electrons into the electron transport chain during photosynthesis (see Light Reactions, Chapter 7). Electromagnetic radiation possesses radiant energy, which is converted to another energy form, *heat*, as it warms the air, land, and oceans after entering our atmosphere. Heat energy is explicitly connected to the hydrologic water cycle, affecting currents and winds. From the two phases of photosynthesis solar

Ecology for the 21st Century. DOI: https://doi.org/10.1016/B978-0-443-24031-7.00008-8

energy is ultimately converted into chemical energy, captured in new carbon bonds. It is the transformation of solar energy into chemical energy that will now drive food webs and ecosystems. Since ecological processes are directly tied to energy transfer, a review of how energy is obtained and converted and its utilization efficiencies are essential to fully describing ecosystems.

Thermodynamics is integral to the discipline of physics, as it is the study of energy and its transformations. All matter in a specified environment, including ecosystems, and all its energy transfers or conversions are classified as a **system**. Everything that exists outside of the defined system is called the **surroundings**. For example, using a stove to heat a pot of water constitutes a system in which energy in the form of heat can be quantitated between stove to pot to water. The energy transfers between each component are still part of the system and are measurable. But once some energy is transferred outside of the system, for example, as steam, it is considered energy loss to the surroundings. As the laws of thermodynamics state, you'll still have the same amount of total energy but some remains in the system and some is now part of the surroundings.

There are several different forms of energy, such as:

- Sound or vibrational energy
- Chemical energy, a type of potential energy that is stored in chemical bonds
- Radiant energy from the sun
- Electrical energy, the movement of electrons through matter
- Mechanical and kinetic energy of moving objects.

Of these energy forms chemical and radiant energy are most applicable to our review of bioenergetics. There are two types of biological systems to consider: open vs closed. In an open system energy can be transferred between the system and its surroundings (like when heating water and some heat escapes as steam). Any system that is prevented from transferring energy to its surroundings is considered a closed system. The study of a biological organism is inherently an open system because energy *must* be exchanged with its surroundings. The organisms will need to consume energy-storing molecules and in turn release energy back to the surroundings, for example, through heat exchanges and their waste products (**Fig. 8.1**). The study of ecology mostly applies to open system bioenergetics because the ecologist is looking at interactions between species and their environment. A mostly closed system exists in aquatic tank ecology because it is closed off from most exchanges with its surrounding environment. In contrast lake ecology is always an open system, freely exchanging energy with its environment. The B2 at the University of Arizona represented an attempt to create a closed ecosystem (though under scrutiny it was shown that researcher participants failed to completely avoid interactions outside the domes).

Fig. 8.1

A high-energy hummingbird hovers in air as it feeds on nectar, a specialized food source high in fructose to help it maintain prolonged periods of flight. Unlike other birds flight is powered using upward and downward strokes, beating 60–80/sec. Because they expend so much energy, hummingbirds feed every 10 min, consuming an equivalent of two-thirds of their body weight daily.

Source: Modification of work by Cory Zanker, CC BY 4.0

There are two states in which energy can exist: potential or kinetic. Potential and kinetic energy represent, respectively, energy stored in the system and energy that is being used to do work. **Potential energy** and **kinetic energy** states are governed under the laws of thermodynamics and can be quantitatively measured.

First Law of Thermodynamics—Energy Is a Constant

The **first law of thermodynamics (also referred to as the Law of Conservation of Energy)** states that the total amount of energy available in the universe is (constant) unchanging, neither created nor destroyed. Concentration of energy pockets is possible, and energy can be relocated. It can also convert between its many forms (mechanical, radiant, sound, etc.). The energy of the universe is the sum of the system and the surroundings.

The SI Unit of energy is the Joule with units of energy also being expressed in watts, or calories. For example, it is possible to track how light bulbs transform electrical energy into light energy (watts or Joules) or how a gas stove transforms the chemical energy in natural gas (potential state) into heat energy (kinetic state). In summary, the total energy is unchanging but can change into many different forms and will always be found in one of its two energy states (potential or kinetic).

The photosynthetic processes that occur in plants are essentially energy transformations performed on a large scale, converting the energy emissions from solar

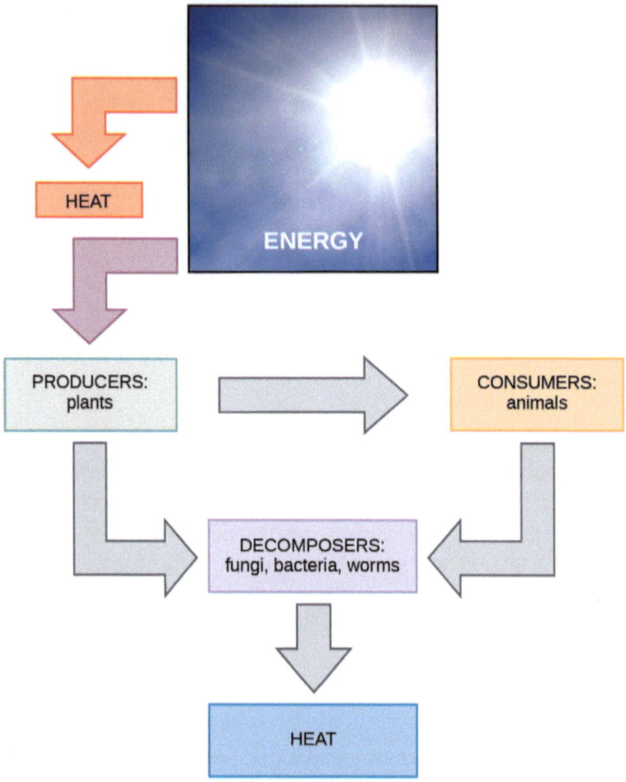

Fig. 8.2

Energy transformations begin with solar radiation trapped by the system (Earth) through producers, which in turn becomes a food source when its biomass is consumed by other Earth dwellers. Bioenergetics traces the flow of energy through transformations and eventual breakdown (decomposition) and its release as heat back into the surroundings.

Source: OpenStax College under CC BY 4.0

radiation into a form of potential chemical energy stored in organic molecules. In the study of bioenergetics ecologists might trace the flow of energy in a primary producer plant, through its energy conversions to complete its metabolic processes, and upon its consumption by heterotrophs to power the consumers life cycles. The bioenergetics of energy transformations are shown in **Fig. 8.2**.

Since only producers are capable of direct conversion of light energy all other organisms must obtain energy from their surroundings indirectly, as organic molecules. During cellular respiration another series of chemical reactions transforms the potential energy of these organic molecules into ATP, its energy now easily accessible to do the body's work.

For living organisms some proportion of their acquired energy will be transformed into heat energy during cellular metabolism. This is because energy transfer from one form to another is never completely efficient. During metabolism an organism loses up to 60% of its acquired energy in the form of heat, released back into its surroundings. According to the first law of thermodynamics the total amount of energy remains constant—either contained within the system or released to its surroundings.

Second Law of Thermodynamics—Random or Not?

The second law of thermodynamics provides an explanation of why energy transfers and transformations within a system are never completely efficient. Some amount of energy will be "lost" to the universe and unusable by the system. Typically, the loss of energy is by thermal transfer, or heat energy. When measuring work being done by the system the transfers of energy are not occurring through heat loss. For example, in a vehicle's gasoline-powered mechanical system 25% of the chemical energy from gasoline will be converted into usable energy that keeps the engine running. To also move the vehicle, it is necessary to transfer some gasoline energy again, this time into motion as wheel axles turn. But now only 20% of the chemical energy stored in gasoline is available to move the vehicle. A sizable proportion of energy (55%) was lost to the surroundings, transformed into thermal energy through exhaust fumes, the car's coolant and lubricant systems, or as friction between its moving parts.

An important property of the physical world is the measure of order vs disorder (randomness). This property is explained by the Second Law of Thermodynamics, which informs us that disorder will always increase (energy lost to an unusable form, as shown in the vehicle). When a system's energy is transformed and escapes as heat to the surroundings there is now less order and more randomness. Note it is necessary to account for both the system and the surroundings to determine whether order or disorder prevails. The degree to which disorder or randomness changes within a system under study is known as its **entropy**. In any discussion of bioenergetics a higher value in a system's entropy equates to a greater disorder and, by association, less energy available to do work (**Fig. 8.3**).

To relate the changes of entropy within the system, envision the degree of order in a child's bedroom. Without the addition of energy to do work the room would quickly become disorderly. This is a system with high entropy (disorder)! Once the child expends the energy to put everything away by performing physical work the room is elevated to a state of cleanliness and order. The room would now be a low entropy system. But maintaining this state requires a constant input of energy. The analogy applies just as well to a vehicle, house, or forest ecosystem. Left alone entropy gradually increases in each through rust and degradation in the case of the vehicle or house … but in what ways does forest ecology increase in entropy?

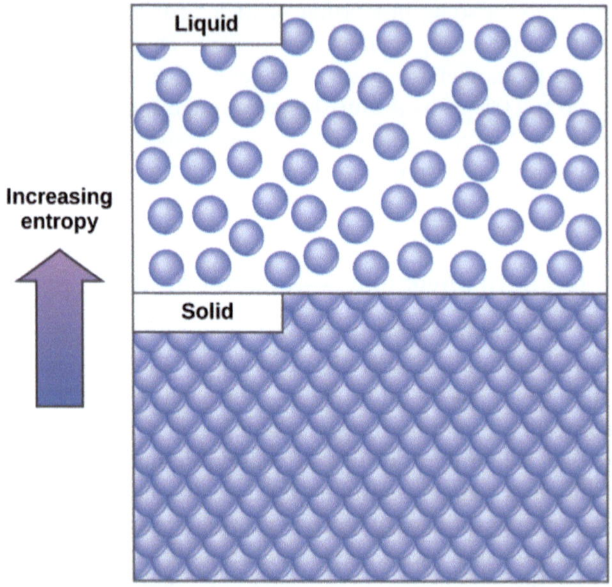

Fig. 8.3

Entropy measures randomness or disorder in a system. Molecules in their gas phase possess higher entropy than the liquid phase, which has higher entropy than the solid phase.

Source: Image by author

Forest fires are an example of a system changing from order into devastating losses with resulting high entropy as energy is released along with CO_2 and water vapor to the surroundings. The Second Law of Thermodynamics can be observed at all levels within the ecosystem, as simply as water crashes over a waterfall, ice melting, or the natural cycling of old growth forest die off. And while the evolution of species appears to contradict increasing entropy in ecosystems, with more efficient and orderly life forms selected for, this is a faulty assumption because it doesn't account for the surroundings. Organisms exist by virtue of increasing entropy within the environment surrounding them. They do so, for example, when irradiating excess heat to the atmosphere and by consumption and breaking down of larger complex molecules (in tissues) into small molecular weight waste products. In contrast low entropy ecosystems result in an energy cost investment, for example, when soil reclamation, pollution control, or other ecological management systems are applied. An analogy would be energy cost required to restore or clean up the child's room.

The second law of thermodynamics states that with every energy transfer there will be an increase in the entropy of the universe. This is aptly relevant to ecological food webs, with energy moving sequentially up the food chain with less total energy present at each level higher.

To exist an organism requires a highly ordered system in a state of low entropy. In the process it is essentially fighting against the universe, where the total entropy is constantly increasing. Retaining order is a continuous uphill battle, only possible with the continuous effort of producers to capture solar energy to cycle again through the food webs. In conclusion, a common theme of energy flow in the natural world is that:

- Organisms are highly ordered (low entropy), requiring constant energy input from their surroundings, whereas,
- Only autotrophs are capable of directly converting solar radiation into chemical energy, whereas,
- All living systems consume chemical energy (high-energy molecules) that will be transformed through biochemical reactions to meet their energy needs, while,
- Some quantifiable amount of usable energy will be returned to the surroundings in the form of heat during its energy transformations.

These guiding principles in living organisms are tightly connected by the physical laws of nature under which all ecological systems must obey.

8.2 Energy Is Stored Temporarily Within Organisms as ATP

By the end of this section students should be able to:

- Describe how redox reactions move electrons in energy transfer.
- Identify trophic levels and food pyramids as energy conversions.
- Discuss net primary productivity (NPP) efficiencies according to light and nutrient availability.

Bioenergetics is the term biologists have adopted to discuss concepts of energy flow through living systems and individuals. Living organisms require energy to perform virtually every task. Animals in the wild often use most of their energy acquiring food and avoiding predators. A source of energy is needed for optimal activity, including movement, growth, and reproduction. And the lack of an energy supply may restrict lifestyles, leaving an animal vulnerable or weakened. Cells are continuously metabolizing, consuming, and regenerating their own energy source from food intake. Ingested nutrients and minerals are digested to smaller molecules, like glucose, amino acids, and fatty acids. Energy can then be extracted from the small molecules through a multistep process during cellular respiration, in which molecular bonds are broken and molecular energy is transferred into a usable cellular energy form.

This energy form must be accessible to cells to synthesize their macromolecules, proteins, steroids, and complex carbohydrates, as needed for everyday functions. The cells (and tissues and organs of the multicellular organisms) will need to perform many specialized tasks, including the transport of material, body movement, heat generation, and other important processes of life.

Energy Transfers and Electrons

Energy flows through the transfer of electrons between molecules, as chemical bonds are made or broken, ordered or less ordered, respectively. In biochemical reactions molecular bonds are broken and formed through movement of electrons, changing the potential energy between the two molecules. Basically, bond breaking and bond making involve the loss and gain of energy, and the net difference between the two is what releases or consumes energy. A more structured molecule will have more bonds, which will also mean it is lower entropy. When a molecule is less structured it has fewer bonds and higher entropy. It is appropriate to say that energy moves through biological systems with the movement of electrons. Your chemistry instructor may have alluded to this fact by using the phrase, "Follow the electrons" to explain chemical reactions. When an electron is transferred from an element within an atom in a molecular compound, it is considered *oxidized*, with a corresponding decrease in its potential energy. The electron does not remain unbonded for long, however, as it is shifted to a new compound. The element within a compound receiving the new electron is now considered *reduced*, with a corresponding increase in its potential energy. Simultaneous, or the coupling of oxidation and reduction reactions, are called **redox reactions** and energy exchanges result.

During the shift of an electron from compound A to compound B some potential energy is transferred with it. The energy stored in organisms and used to perform biological functions is held in high-energy electrons. Whether we are examining the bioenergetics of a food web or an entire ecological system, the species present are continually seeking a means of energy capture. Heterotrophs do so through extraction of chemical energy from foods with high-energy bonds (e.g., carbohydrates). The flow of energy is traced through the flow of electrons, which moves from high potential transfer energy molecules to lower potential transfer energy molecules.

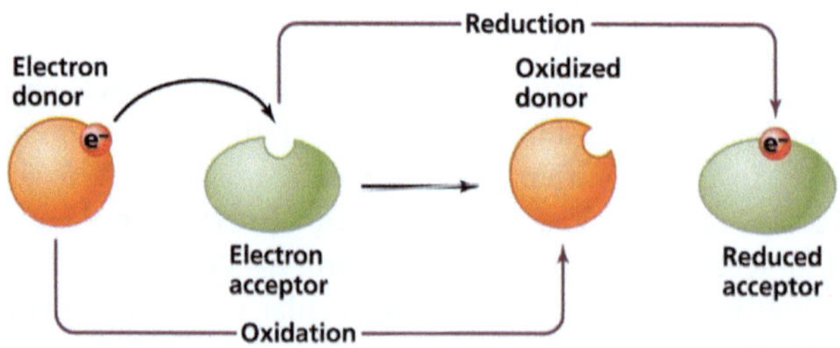

Fig. 8.4a

Redox reactions involve flow of electrons from a donor to an acceptor. The electron donor is oxidized as the electron acceptor is reduced. In NAD+ to NADH/H+ two electrons are added during its reduction.

Source: Images by author

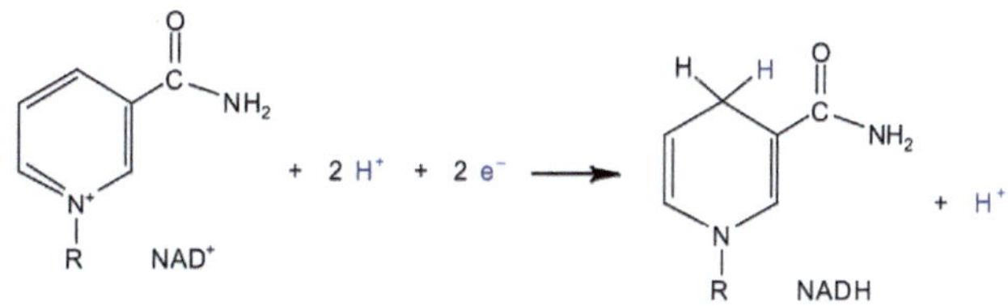

Fig. 8.4b

Electron carrier (NAD+) is oxidized on the left. Reduced NADH/H+ is shown on the right with two more hydrogen ions and two more electrons than NAD+.

Source: Images by author

Electron carriers. Within biological systems electrons are moved around by compounds known as electron shuttles, or electron carriers. Their purpose is to temporarily bind and carry high-energy electrons from one compound in a metabolic pathway to another. An example of an important electron carrier in humans is **n**icotinamide **a**denine **d**inucleotide (NAD, chemically is written as NADH/H+, or NAD+). NAD is a derivative of niacin, also known as vitamin B_3 (**Figs. 8.4a, b**). Electron carriers like NAD function by first accepting elections (a reduction reaction), then transferring electrons to another molecule, which is, in turn, reduced by the NAD carrier. NAD, in its oxidized form, is notated as NAD+, having already donated electrons. NAD in its reduced state is noted as NAD**H/H+**, having just gained electrons. By using notations for reduced or oxidized molecular states in chemical equations it is easy to determine whether the electron carrier is delivering electrons or has already done so. Another important electron carrier, flavin adenine dinucleotide (FAD), is derived from riboflavin, also known as vitamin B_2. Its reduced form is FAD$\mathbf{H_2}$, and its oxidized form is FAD+. Both electron carriers, NAD and FAD, are extensively used by heterotrophs in energy extraction from sugars, temporarily raised in potential energy until they deliver their cargo (electrons to form new bonds). In plants a similar electron carrier, nicotinamide adenine dinucleotide phosphate (NADP), plays an important role in metabolic reactions, including processes of the photosynthetic reactions. NADP has two energy states: NADP+ and NADP**H**, oxidized and reduced, respectively.

Energy and Metabolism

Sources of energy and minerals are essential to ecosystem structure, function, and composition. The cellular processes occurring in organisms maintain their biomass, including building and breaking down of complex molecules needed for their physiological processes. Some chemical reactions are spontaneous and release energy, whereas others require energy to proceed. All chemical reactions occurring within cells, both those that use energy and those that release energy, make up the organism's entire metabolism. What's left over from its energy acquisition is available to be utilized by organisms up the food chain. But the energy transferred

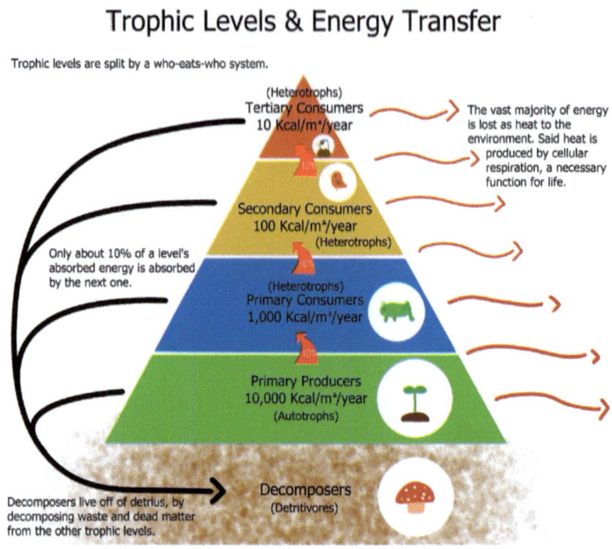

Fig. 8.5

Energy losses in trophic levels of the food pyramid.

Source: WDiment under CC BY 4.0

represents only a fraction of the energy that had been available in the level below (**Fig. 8.5**). This fact provides an explanation as to why there are far more plants, microbes, and insects lower in the food chain than there are large mammals higher up the food chain.

For most organisms energy derived from foods, and released through oxidizing its chemical bonds, is captured as another intermediate molecule. This versatile molecule is called **ATP** and represents an energy currency that is available for immediate expenditure. ATP's energy currency is spent during the breaking of its high-energy terminal phosphate bond. When the terminal phosphate group is released the remaining molecule is called adenosine diphosphate, or ADP. The released inorganic phosphate (p_i) can be reattached to ADP to replenish the cell's supply of ATP.

During the various processes of life ATP is continuously broken down safely into ADP and, like a rechargeable battery, ADP will be continuously regenerated back into ATP. Where does the energy come from to keep regenerating ATP? This goes back to that original six-carbon glucose, which photosynthesis made possible; it is the main fuel source of nearly all organisms. As glucose is metabolized during cellular respiration it releases the energy needed to make more ATP.

ATP is a small molecule, composed of a nucleic acid, adenine (also found in DNA), a ribose sugar, and three oxygenated phosphate groups (**Fig. 8.6**). ATP is an energy powerhouse because both of its two outer phosphate bonds are "high energy."

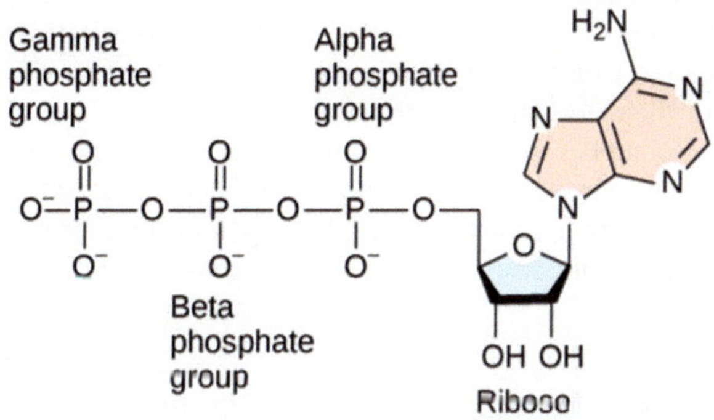

Fig. 8.6

ATP is the primary energy currency of the cell. Note three phosphates with extensive negative charges. By release of the gamma, or terminal group, ATP is converted to ADP.

Source: Image by author

The release of its terminal phosphate group as ATP is hydrolyzed into ADP and occurs in a favorable reaction because it releases energy.

$$\mathbf{ATP + H_2O \rightarrow ADP + p_i + free\ energy}$$

The hydrolysis of ATP into ADP is reversible but going from ADP + pi, regenerating ATP, requires an input of energy to remake those high-energy bonds.

$$\mathbf{ADP + p_i + free\ energy \rightarrow ATP + H_2O}$$

The energy released by ATP hydrolysis is used to perform the cell's work, often occurring during a reaction strategy known as **energy coupling**. Cells couple the energy release of an **exergonic** reaction of ATP hydrolysis simultaneously with an **endergonic** reaction, which requires an absorption of energy. Coupling reactions in this manner allow the body's energetically consuming biochemical reactions to go forward. According to the laws of thermodynamics biochemical reactions are no more efficient than other energy transfers, with approximately 60% of their energy transformed into heat and lost to the surroundings. This fact explains why the more metabolically active an organism is, the more its body heats up. And why the involuntary activity of muscle shivering is an adaptive response for many endothermic organisms during extreme cold temperatures.

A continual source of energy must be extracted from the surrounding environment throughout an organism's lifespan. Within the richly biodiverse world we inhabit many strategies exist to obtain energy. The most efficient is to draw energy directly from solar radiation emitted by the sun, which, of course, is exactly the strategy of primary producers at the start of the food web.

To recap, energy enters an ecological system as light energy, gets transformed into chemical energy in the form of organic molecules (carbohydrates), and then into energy currencies (ATP) during cellular respiration. That energy will be available to do the work of the cell. Ultimately the energy will be converted to heat energy, according to the second law of thermodynamics. Heat energy dissipated back to the surroundings is in an energy form that cannot be recycled, in accord with increasing entropy. This means there is a continued need for solar energy, otherwise biological systems would quickly shut down. Earth is considered an ***open system*** with respect to its energy source.

Net Energy Intake Is Modified in the Hummingbird

The hummingbirds' use of hovering flight, which allows them to move quickly and easily access energy-rich nectar in flowers, is made possible because of a marvelous modification to their metabolism. The hummingbird feeding style is nectivorous, reliant exclusively on a mix of sucrose, glucose, and fructose found in nectar. Nectivores, like nectar bats and hummingbirds, are able to directly metabolize exogenous fructose and glucose from nectar in a more energy-efficient process than how prolonged flight

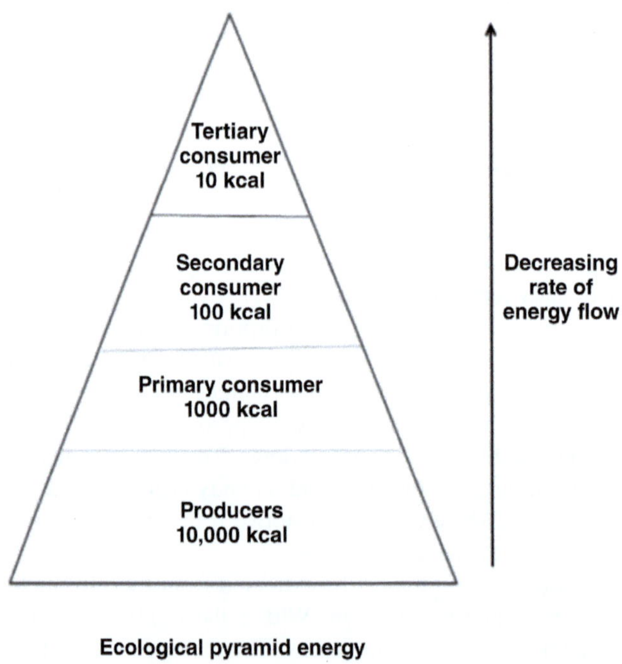

Fig. 8.7

Ecological pyramids.

is typically powered. The trade-off for enhanced maneuverability of **hovering** is that it is energetically expensive, with both upward and downward strokes as fast as 80 flaps/sec. Only the hummingbird and nectivorous bats are known to use hovering flight, with nectar as their primary food source being the key distinction with other avian taxa.

Hummingbirds enhance their net energy intake for hovering capacity through a combination of adaptations in diet, behavior, and an extremely high **mass-specific metabolic rate**. Most birds feed on pollen and insects to obtain adequate protein and lipids for their energetic needs. And while other migrating birds oxidize endogenous fatty acids and amino acids to fuel flight, nectivores have acquired the trait of powering flight muscles directly through oxidation of recently ingested simple sugars (a much faster and energy-efficient process). The implications of the specialized dietary ecology of hummingbirds, without the need to increase protein consumption or store dense glycogen for prolonged flight, is that these tiny animals have evolved to be lightweight with acute control over their net energy gain.

The tiny ruby-throated hummingbird (*Archilochus colubris*) has been shown to fly nonstop for 2200 km, the equivalent of flying half-way across the United States! This is possible because hovering flight allows quick feeding off nectar-bearing flowers without perching.

8.3 Productivity

By the end of this section students should be able to:

- Explain why primary productivity marks the entry point of energy into ecosystems.
- Know the main biochemical processes by their chemical equations.
- Calculate biomass and other metrics used to describe energy transfer in trophic levels.
- Apply the ecological footprint as a life choice.
- Discuss abiotic factors that determine primary productivity.

Cells use ATP as energy currency to perform immediate work. The supply of ATP is increased whenever glucose or other molecules are oxidized during cellular respiration. If an organism has met its energy needs rather than oxidizing more glucose into ATP it can be modified and stored "in the bank" as starch in plant roots or as glycogen in animal muscles. These large energy-storing polymers are readily broken down and mobilized into glucose, which can enter any cell to be credited back into ATP currency whenever a cell's energy status runs low.

Primary productivity. In the solar energy conversion process (e.g., photosynthesis) CO_2 and water (H_2O) are taken up from the surrounding environment and used to fix carbon into chemical energy (primary productivity). This is the first form of stored energy in ecosystems.

Two major biochemical processes, photosynthesis and respiration, are reverse reactions—one in which solar energy transforms into potential energy (glucose, starch) and the other in which potential energy is released to perform work (i.e., growth, movement, and reproduction).

In chemical equations substrates are on the left of the arrow and reaction products are on the right of the arrow. The two equations are as follows:

1. $6\ CO_2 + 6\ H_2O + \text{solar energy} \rightarrow C_6H_{12}O_6 + 6\ O_2$
2. $C_6H_{12}O_6 + 6\ O_2 \rightarrow 6\ CO_2 + 6\ H_2O + \text{chemical energy (ATP)}$

Even if you were not familiar with the equations, could you determine which one is cellular respiration and which one is photosynthesis? In the two equations observe that one requires an input of energy (that's photosynthesis) and one shows a release of energy in its product (that's cellular respiration).

In these metabolic processes fixed carbon is shown as glucose ($C_6H_{12}O_6$). Because glucose is the product of photosynthesis it is an indirect measure of productivity.

Productivity in Trophic Levels

Coral reef ecosystems along the continental shelf are classified as high-productivity ecosystems. Because deeper ocean waters do not have access to adequate sunlight, they are considered among the lowest productivity. Ecosystem productivity is defined as the percentage of energy entering the ecosystem and incorporated into biomass along a particular trophic level. Biomass is defined as the total mass per unit area (g/m^2) of living organisms. It is possible to measure the amount of biomass present in each trophic level.

Ecosystems have characteristic amounts of energy and corresponding biomass at each trophic level, which decreases as energy flows up the food chain (**Fig. 8.7**). Productivity is a reasonable measure of the primary producers present and able to photosynthesize. The rate at which a primary producer can incorporate solar energy into biomass is called its GPP. **Fig. 8.8** illustrates GPP and energy flow in the Silver Springs, Florida, aquatic ecosystem. GPP by Silver Springs primary producers is 20,810 kcal/m^2 per year, calculated as usable energy (kcal) per given area (m^2) per time period.

Primary producers will need to consume some of the captured energy to maintain its biological functions and some energy will escape as metabolic heat loss. Scientists generally prefer to use the NPP of an ecosystem. NPP is the available energy in primary producers, after accounting for the organisms' metabolic and growth needs and losses back to the environment.

Most solar radiation enters Earth in wavelengths unsuitable for photosynthesis. This is because 98% to 99% is reflected by the leaves, absorbed by other surfaces, or otherwise dissipated. From the solar radiation that is available for primary production (estimated 1.5% of total solar radiation), approximately 25% of captured solar energy must be subtracted. This is to account for energy expended by the primary producer for its own life processes. Ultimately just over 1% of the sunlight reaching Earth's surface at any given moment in time is

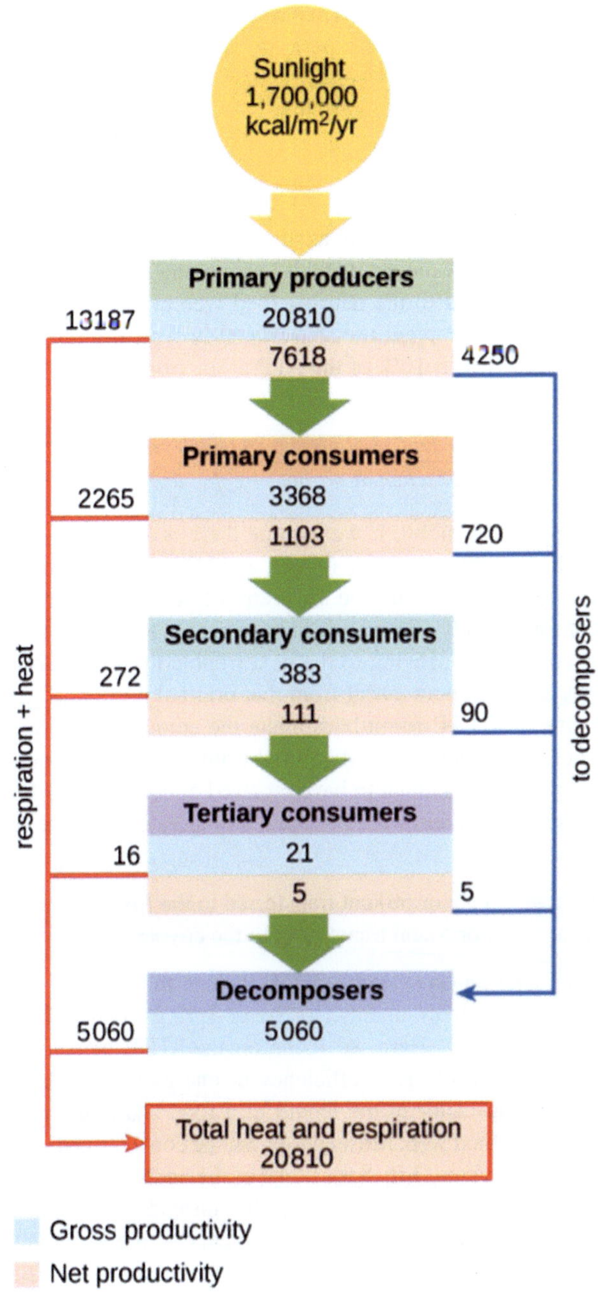

Fig. 8.8

Energy flow in a spring ecosystem in Silver Springs, Florida.

Source: Lumen Biology under CC BY 4.0

converted into NPP. Furthermore, each ecosystem's NPP capacity varies according to the amount of precipitation, temperature, and nutrient requirements of the producers. Thus the difference between GPP and NPP is the excess energy acquisition not used up by primary producers that is represented as biomass available for consumption. NPP reflects what is available to primary consumers at the *second* trophic level.

A useful marker of ecological health is its efficiency of energy transfer between two successive trophic levels, known as the **trophic level transfer efficiency (TLTE)**, which describes how much energy is lost moving up the grazing food chain. Energy losses can be tracked to the detritus food web (wastes, dead organic matter) or lost as metabolic R. Percent transfer efficiency typically ranges from 5% to 20%. As a general rule NPP is 10% of the NPP at the trophic level below it, sometimes called the "Ten percent law."

$$TLTE = \frac{production\ at\ present\ trophic\ level}{production\ at\ previous\ trophic\ level} \times 100$$

Using the example of a rabbit, if we assume there is 1000 kcal of energy to be transferred from the primary producer (grass) and 90% is lost during transfer we can project that only 100 kcal would be transferred to the rabbit. From this trophic level (primary consumer) only 10 kcal would be available to a fox in the next trophic level (secondary consumer). Moving to the next trophic level (tertiary consumer) a coyote would now acquire about 1 kcal from that original portion of grass fed upon by the rabbit. Using the TLTE formula estimate the amount of energy available to a fox or coyote consuming a rabbit if the initial primary producer source contained 4500 kcal and the rabbit was known to have received 500 kcal during its first energy transfer. To solve plug in the known values and extrapolate using the rule of thumb (10% transfer):

- 500 kcal/4500 kcal = 11% or 50 kcal transferred to the fox
- 50 kcal/500 kcal = 1% or 5 kcal transferred to the coyote

Factors that can alter an ecosystem's TLTE include the types of species present and the complexity of the environment. The Silver Springs food web (**Fig. 8.8**) was first introduced in Chapter 1. Here we focus on the TLTE between the first two trophic levels, which is 14.8%. Low efficiency of energy transfer between trophic levels is a major limiting factor in the length of a food chain and total number of levels observed within a food web. Most energy will be concentrated in the producers at the bottom of the food chain (**Fig. 8.9**). Some of the energy, however, doesn't pass directly into the plant-to-carnivore food chain but instead enters the **detritus food chain**. Detritus feeders (bacteria, fungi, and scavengers) are eventually consumed by other organisms. Typically, after four to six energy transfers, there isn't enough energy left to support another trophic level. The Silver Spring food chain began with GPP 20,810 kcal/m^2/yr, with an estimated 13,187 kcal/m^2/yr consumed, leaving an

NPP value of 7632 kcal/m²/yr for transfer into the primary consumer trophic level (14.8%).

Assimilation refers to the biomass acquired from food intake of the present trophic level after accounting for losses from its own metabolic respiration, heat loss, and *any indigestible parts of the food.* Incomplete ingestion accounts for instances where consumers eat only a portion of their food. For example, many predators will eat everything except for the hide and bones of their prey. Since not all energy transfers are equal, another measure is used to help determine which species are more efficient at incorporating their food source into biomass, which species consume far more than they contribute, and even which ecosystems have high nonpredatory deaths and don't contribute to the next trophic level. This is known as **Net Production Efficiency (NPE)** or exploitation efficiency. When there is incomplete ingestion of biomass, NPE is reduced, which affects the total NPP at the next level. And since carnivores like the lion discard energy-rich bone marrow and other parts of their prey (e.g., antelope), they don't fully consume all the energy or calories the prey species accounted for in its trophic level below. Given that metabolic rates vary among species, species assimilate their food sources differently, and causes of nonpredatory death can be random, NPE values vary among species and ecosystems across a wide range. One reliable measure of how efficiently an organism assimilates its food source is performed by using ^{14}C radioisotopes and measuring the change in radioactivity.

Ectotherms have low metabolic rates and maintain a lifestyle requiring far less energy than endotherms. The extra heat generated by the endotherm's faster

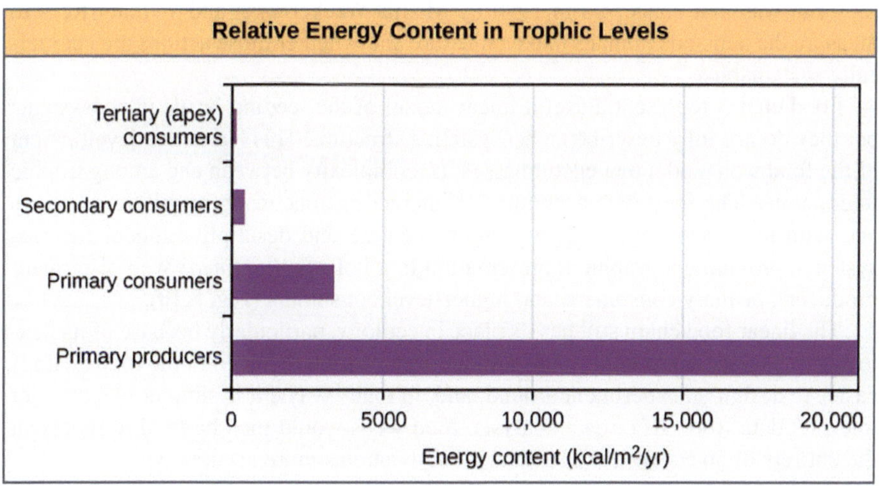

Fig. 8.9

Relative ecosystem energy levels in Silver Springs, Florida.
Source: Great Lakes Environmental Research Laboratory (GLERL), NOAA, Public Domain

metabolic rates is useful to maintain its mobility in colder environments but is offset by higher energy consumption. The higher activities in endotherms lead to lower energy transfer efficiencies. We could predict a much lower NPE for endotherms compared to ectotherms in the same habitat. Endotherms will have less energy to contribute to the next trophic level. In general the NPE for ectotherms is 10 times greater than endotherms. For example, the ectotherm caterpillar feeding on leaves has a calculated NPE of 18 %, whereas the endotherm squirrel eating acorns has a calculated NPE of 1.6 %. The squirrel is an order of magnitude less efficient.

NPE and Human Food Supplies

Understanding energy transfer efficiencies (NPE) is relevant in human food supplies, particularly in areas where population overgrowth coincides with limited land availability for agricultural and animal husbandry use. The disadvantages of raising endotherm animals (i.e., cattle, pigs) as the world's food supply is directly related to the inefficiency of their energy transfer efficiency. The meat industry requires copious amounts of crops to feed its livestock, yet much of the energy consumed in animal feed will not be transferred to the next trophic level (i.e., humans). The difference in costs between raising crops vs meat stock is an astounding twenty-fold. Also, cattle require an inordinate allocation of land. It is estimated that 1 acre of corn can produce 12.3 million calories of food, compared to cattle (raised on 1 acre) producing only 1.1 million calories of food (USDA). Growing awareness of the limits of Earth's resources has inspired a worldwide movement in reorganizing the food supply chain and promoting consumption of more nonmeat and nondairy foods. The mission is to create a more sustainable future with less energy waste from feeding animals for the meat industry, thereby lowering the **ecological footprint**. The term ecological footprint was first conceived in 1990 by Mathis Wackernagel and William Rees to measure the demands humans make on nature and whether those actions are ecologically sustainable.

Food chains represent a useful linear model of the feeding levels in ecosystems but they do not fully describe an ecosystem's structure. This led to the development of the food web model that encompasses the complexity between and among trophic interactions. The food web accounts for interactions occurring in different species and with their environment. It is a more accurate and descriptive model for ecosystems, providing a graphic representation in a holistic, *nonlinear* web of primary producers, primary consumers, and higher-level consumers (**Fig. 8.10**).

The linear food chain still has its place in ecology, particularly because of its flexibility in analytical modeling. The simpler organization depicted in the food chain is easier to design an experiment around, and, in many ways, it is simpler to extract or interpret data from. After data analysis, food webs would then be used to represent the entirety of an ecosystem's structure and dynamics more accurately.

NPP Is Determined by Light and Nutrient Availability

NPP can vary according to several environmental conditions (including human activities like urbanization) that impact the rate of photosynthesis. The key

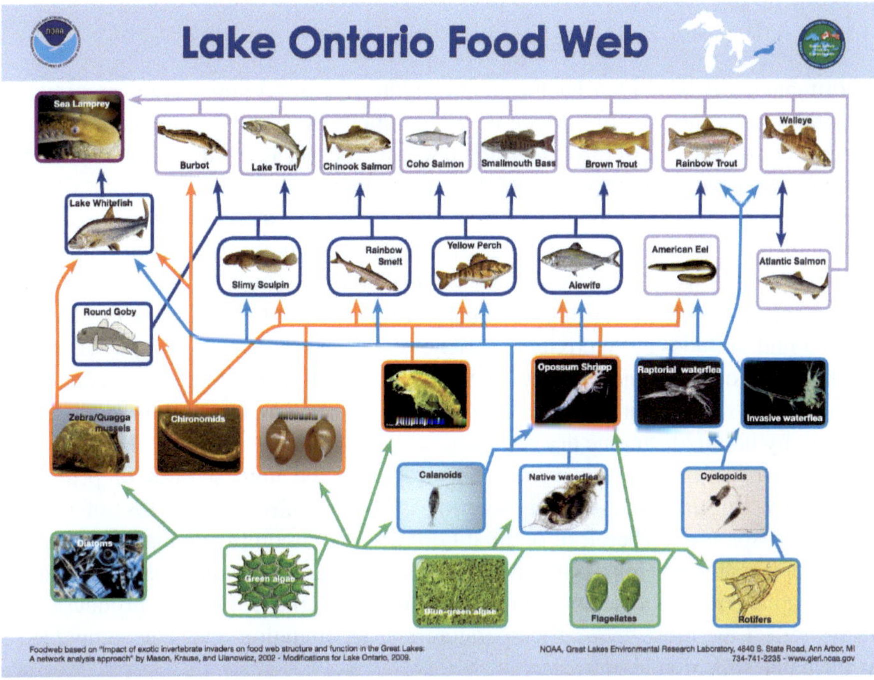

Fig. 8.10

Food web in Lake Ontario ecosystem illustrating interactions between organisms across trophic levels. Primary producers (green); primary consumers (orange); secondary consumers (blue); and tertiary or apex consumers (purple). Arrows point away from an organism that gets consumed and toward the organism that consumed it.

Source: Earth Observatory, NASA, Public Domain

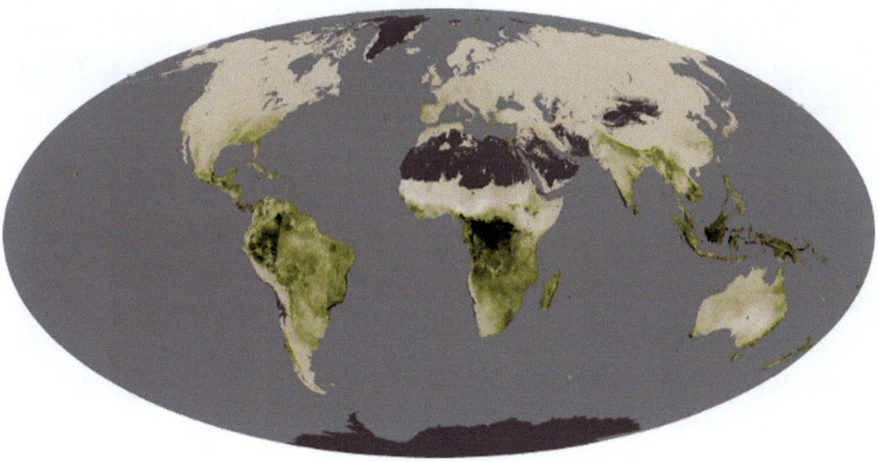

Fig. 8.11

Earth's terrestrial NPP (February 2000 to December 2013).

Source: Global Climate Change, NASA, Public Domain

determinants will be driven by the major abiotic factors: temperature, precipitation, and nutrient availability. Mean annual temperature directly reflects length of growing season (period of days in which photosynthesis can occur). Globally, NPP increases whenever a combination of the mean annual temperature and rainfall are optimal. In **Fig. 8.11** the global map from February 2000 shows areas in shades of green indicating seasonally high productivity is located in the equatorial regions.

Data from NASA's Terra satellite. Value range: 0 to 6.5 g/m^2/day. Negative values occur where more carbon is released back into the atmosphere than the producers fixed. At mid-latitudes productivity experiences seasonal change, peaking in summer months. Year-round, tropical forests in South America, Africa, Southeast Asia, and Indonesia will retain their high productivity. The Amazon basin exhibits its highest productivity in fall, during its dry season.

Access to ample sunlight is necessary to reach the highest rates of primary productivity. But unrelenting solar radiation is accompanied by high loss of water through transpiration. Therefore optimal GPP and NPP require adequate precipitation to replenish water lost to the atmosphere. The plant roots replenish lost water through uptake from the surrounding soil. Global patterns of primary productivity always reflect the most optimal temperature and precipitation levels, accompanied by amenable soils rich in nutrients.

Any study on climate patterns will reveal high year-round NPP across the equatorial regions. There is a general decline in productivity moving inward from coastlines, in conjunction with climate patterns of decreasing precipitation and temperature. Aquatic NPP is highest in proximity to coastlines, including estuaries, coral reefs, and algal beds. The highest terrestrial NPPs are found in tropical rainforests,

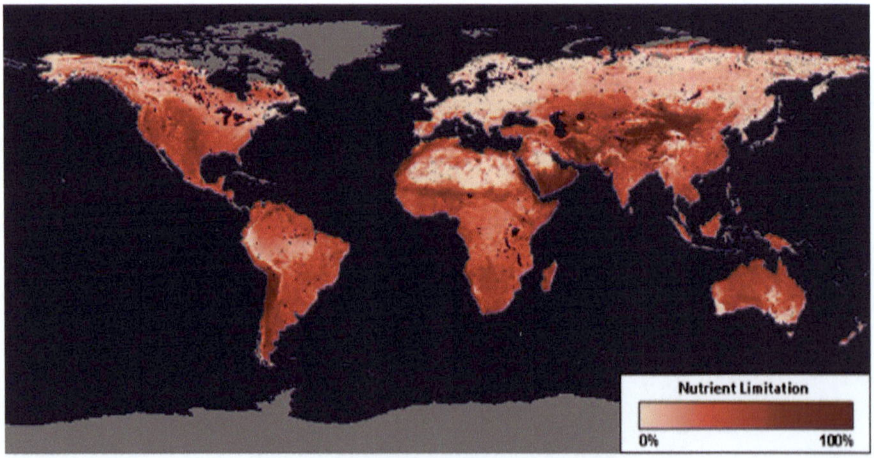

Fig. 8.12

Global percentage vegetative growth limited by soil nutrient availability.
Source: SeaWiFS Project, NASA/Goddard Space Flight Center, Public Domain

swamps, and marshlands. In contrast the lowest NPP occurs in open oceans and deserts. To protect worldwide productivity ecologists and conservationists understandably focus on maintenance of critical high-productivity areas.

Nutrient Acquisition and NPP

The major essential elements in plants are carbon, hydrogen, oxygen, and nitrogen (covered in more detail in Chapter 9), with acquisition of the first three directly converted into plant biomass during photosynthesis. The fourth, nitrogen, is vital to synthesize proteins and nucleic acid and must be present in the soil for plant uptake in adequate quantities. Without nitrogen plants would not be able to synthesize the two photosystems that harness energy or any of the enzymes that fix and process carbon during photosynthesis. Agricultural studies on nutrient-rich soils demonstrate that nitrogen is a key component in maintaining adequate NPP levels (**Fig. 8.12**). Essential nutrients for plant growth and productivity (including phosphorus, potassium, magnesium, etc.) enter the plant through its roots, which tells us that good fertile soil is another requirement for high NPP.

Nutrient availability and light availability are important criterion for optimal NPP in photosynthetic aquatic organisms as well. Recall that photosynthetically active radiation (PAR) is strongest at the surface of bodies of water and declines on an exponential gradient through the aquatic zones. Meanwhile the nutrients that accumulate in the lower depths need to be made available to **phytoplankton** (and **zooplankton**).

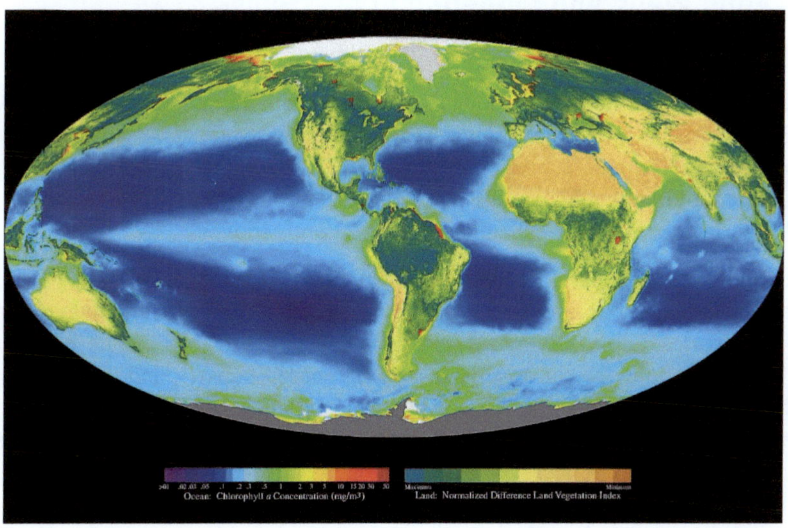

Fig. 8.13

Global oceanic and terrestrial photoautotroph abundance, September 1997 to August 2000, estimating primary producer biomass.

Source: SeaWiFS Project, Public Domain

Phytoplankton are comparable to terrestrial photosynthesizers, and zooplankton are comparable to the small animals and microbes at the bottom of the food chain. The most limiting essential nutrients for oceanic primary producer growth are nitrogen, phosphorus, and iron. Locations where aquatic NPP is the highest is inevitably hugging the shorelines because that is where ease of nutrient transport from the bottom sediment is greatest due to wave and tidal action (**Fig. 8.13**).

Along rivers and streams the degree of foliage overgrowth or tree canopies impacts NPP through limiting light availability to producers beneath their canopies. Nutrient quantities along rivers and streams are mostly influenced by moving waters sweeping across a stream bed or riverbank. And nutrients will be leached from land into rivers and streams. The concentration of phosphorus is dependent on leaching from local rocks and soil, whereas nitrogen concentrations are dependent on the nitrogen cycle (e.g., nitrogen fixation by soil bacteria) or by natural runoff, sewage, and agriculture seepage. Primary producers' ability to uptake available phosphorus and nitrogen vs its displacement further downstream is a process known as **nutrient spiraling**. Typically NPP is lower in rivers and streams compared to standing water, which is evident in stagnant swamps and marshlands having the highest terrestrial NPP. In lake systems nutrient turnover is enriched seasonally during upwelling and downwelling events of lake turnover. Lakes generally have access to greater light availability than rivers and streams, especially if the moving waters are in wooded areas with impeding foliage-blocking light availability.

8.4 Food Webs and Modeling Ecosystems

By the end of this section students should be able to:

- Describe how ecological pyramids function as a modeling tool in energy flow.
- Provide an example of how biomass and productivity reflect very different trends in ecosystems.
- Discuss environmental consequences of biomagnification.
- Appreciate the impact of chemical pollutants on food webs.

The structure of ecosystems and its dynamic food web are graphically visualized by the ecological pyramid, which was first described in the 1920s by the pioneering work of Charles Elton. Ecological pyramids are designed to illustrate parameters like the type and number of organisms present, energy transfer, and biomass across trophic levels (**Fig. 8.7**). Pyramids can be displayed either upright, inverted, or even diamond-shaped, depending on the ecosystem, to depict data collected on organisms or biomass. They can reflect biomass or productivity levels and provide a basis for comparison between ecosystems.

Using the metrics described previously to calculate biomass (NPP and NPE), pyramids visually describe the amount of energy converted into living tissue at the different trophic levels. In Section 8.3 we established that only a fraction of the

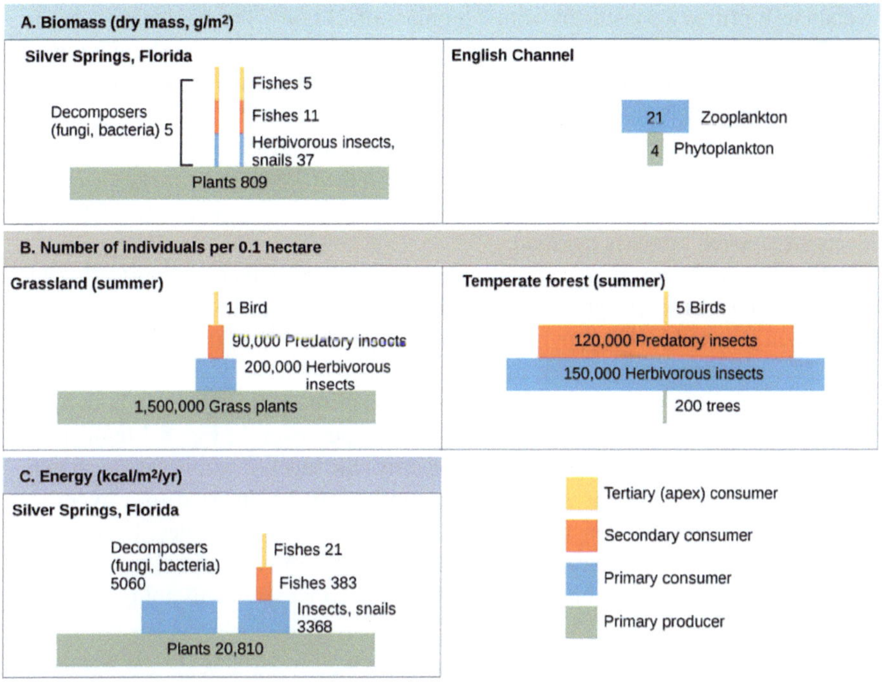

Figs. 8.14a–c

Ecological pyramids illustrating different parameters. Biomass pyramids are typically upright following the pattern of flow of energy (productivity) across trophic levels but can be inverted, as shown in the English Channel and temperate forests.

Source: Van Hoof, GLERL, NOAA, Public Domain

energy available at one trophic level is ever transferred to the next trophic level, following the 'ten percent law'. With biomass transfer there is a wide range of values. Generally, we could predict a bottom-heavy ecosystem structure. But biomass does not necessarily have to decrease in a linear fashion across trophic levels. Food webs remind us that feeding does not strictly occur from the next trophic level below.

The pyramid data in **Fig. 8.14b** illustrates a typical grassland during the summer with a large base of diverse plants (e.g., grasses), decreasing in numbers of organisms at each rising trophic level. In contrast, during the summer months in a temperate forest the base of the pyramid is quite small, with far fewer primary producers (e.g., trees) compared with primary consumers in the next level (a very large insect population). One explanation for the inversion of this particular ecological pyramid is that the physical height of trees in temperate forest ecosystems dominates any other plants.

Fig. 8.14a describes aquatic biomass in the English Channel, a widely studied aquatic ecosystem. Its primary producers have a biomass of just 4 g/m^2 in contrast to the channel's primary consumers with a biomass of 21 g/m^2. The explanation has to do with the characteristics of its stable primary consumer population (high biomass crustaceans, mollusks, and fish), compared to high turnover rate of its phytoplankton primary producers (algae and bacteria). But phytoplankton reproduce quickly. While the high turnover rate results in low biomass at any given time, phytoplankton still contributes considerable productivity to sustain the Channel's primary consumer energy needs. Like the seasonal forest ecosystem biomass in the English Channel reflects an inverted biomass pyramid.

In **Figs. 8.14a, b** two aquatic systems are compared: the Silver Springs freshwater ecosystem shown as an upright biomass pyramid and the English Channel ecosystem shown as an inverted biomass pyramid. The difference between the two ecosystems is that Silver Springs has a large primary producer biomass whereas the English Channel's marine ecosystem has rapidly consumed primary producers. In both pyramids biomass is measured as stored in living tissue per level. In **Fig. 8.14c** the rates by which biomass is incorporated into the next trophic level (kcal/m^2/yr) are shown.

From open oceans to forests, a food pyramid may be inverted to a hyperboloid (hourglass) shape when massive top predators (sharks, whales, bears) feed across many trophic levels (20). In these cases the generally accepted size basis of trophic levels decreasing in biomass as predator size increases will not prove to be accurate.

In summary, biomass is more concentrated in the upper trophic levels whenever secondary consumers are generalists or the primary producers are consumed rapidly.

Consequences of Food Webs: Biomagnification

An ecosystem dynamic with important implications is a phenomenon called biomagnification. **Biomagnification** occurs when the ingestion of toxic substances that persist in the food web begin to accumulate in organisms, specifically in higher trophic levels in increasing concentrations. One well-known substance that biomagnifies in toxic dosages as it moves up the food chain is the banned (in most countries) chemical DDT (**d**ichloro**d**iphenyl**t**richloroethane). Public awareness of the serious ecological consequences from use of DDT was first published in the 1960s bestseller "Silent Spring." The author, **Rachel Carson**, was a biologist who raised the alarm that DDT, the most common pesticide in the United States at the time, led to grave health concerns in animals and humans. She became disturbed by the widespread and indiscriminate application of pesticides known to affect wildlife. As an ardent environmentalist she traced the mysterious increase in bird deaths at a nearby farm to heavy application of DDT.

In trophic cascades organisms consume many organisms from lower trophic levels. If the ingested biomass is contaminated with a toxic compound it begins to accumulate. The effects can be particularly detrimental if the chemical persists in body tissue. Studies conclusively showed that birds (apex consumers) fed a diet of DDT-contaminated fish accumulated sufficient amounts of the DDT toxin to cause adverse effects. Further, high levels of DDT have been linked to fragility in bird eggshells.

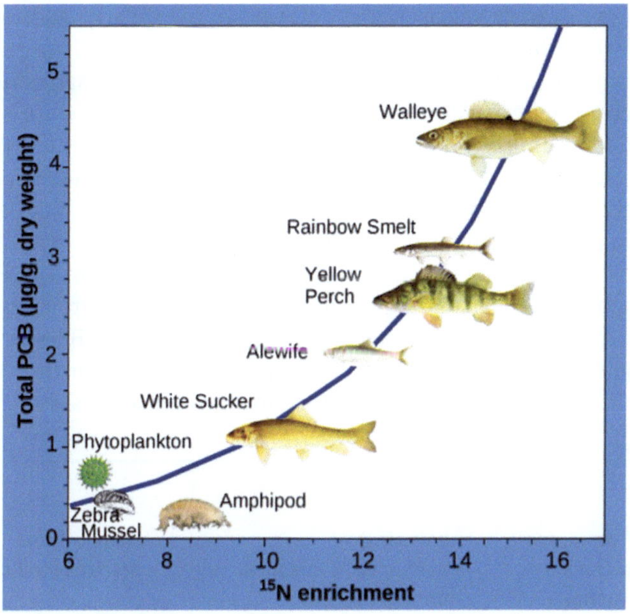

Fig. 8.15

Saginaw Bay Ecosystem, Lake Huron. Chart of PCB concentrations in the trophic levels. X-axis reflects enrichment with heavy isotopes of nitrogen (^{15}N), the marker used to trace bioaccumulation.

Source: OpenStax modified under CC BY 3.0

Female birds fed high DTT-contaminated diets produced fragile eggs that were more prone to egg breakage during nesting. This led to fewer live births, which in turn threatened the survival rates of the entire bird population.

Carson and the Audubon Naturalist Society, which opposed US spraying programs, together sought to publicize the misinformation spread by chemical companies. Use of DDT was banned in the United States by the 1970s. At first the media ignored the message in Rachel Carson's book and downplayed any negative impact of DDT. But once the president of the United States at the time, John F. Kennedy, read "Silent Spring" he called for a science advisory council to investigate the claims made by Carson. This led to swift and decisive action to limit the environmental damage and health risks related to mass application of a synthetic pesticide.

Today a much larger list of chemicals has been shown to biomagnify when released into the environment. These are known as **PBT** (persistent, bioaccumulative, and toxic chemicals). PBTs include polychlorinated biphenyls (**PCBs**), once used in coolant liquids in the United States (banned in 1979), heavy metals (mercury, lead, and cadmium), anthracene, and asbestos.

The pathway of toxic substances in ecosystems is regularly studied in aquatic ecosystems, where fish species accumulate toxic substances. Chemical pollutants reach waterways by a number of ways, often as industrial or agricultural run off reaching streams and groundwater, but sometimes airborne as industrial discharge (atmospheric deposition by rain and snow). Several studies performed in the Saginaw Bay of Lake Huron (**Fig. 8.15**) demonstrated that PCBs were found in trace amounts in phytoplankton, with increasing concentrations through the trophic levels of the lake's aquatic food web. For example, Walleye fish (an apex consumer in this ecosystem) was shown to have more than four times the concentration of PCB pollutants compared to the phytoplankton trophic level. Based on results from this and other studies, birds that feed solely on contaminated fish could have PCB levels at least one order of magnitude higher than lake fish. Humans consuming a high-fish diet laced with any of the listed PBTs would be exposed to similar health risks.

Glossary

system v. surroundings
exergonic reactions
endergonic reactions
hovering
mass specific metabolic rate
phytoplankton
zooplankton
detritus food chain

Citations

1. B.D.V. Marino, H.T. Odum, (Eds.), Biosphere 2: Research Past and Present, Elsevier Science (1999), https://en.wikipedia.org/wiki/Special:BookSources/0-08-043208-5.
2. Y. Demirel, S.I. Sandler, "Thermodynamics and bioenergetics.", Biophys Chem 97 (2–3) (June 2002) 87–111. https://doi.org/10.1016/S0301-4622(02)00069-8
3. E.V. Osilla, J.,L., Marsidi, K.R. Shumway, et al., Physiology, Temperature Regulation: StatPearls [Internet], StatPearls Publishing, Treasure Island (FL), 2024. https://www.ncbi.nlm.nih.gov/books/NBK507838/
4. J. Thomas, Drive cycle powertrain efficiencies and trends derived from EPA vehicle dynamometer results", SAE Int J Passeng Cars Mech Syst 7 (4) (2014) 1374–1384. https://doi.org/10.4271/2014-01-2562
5. M. Baglione, M. Duty & G. Baglione, Vehicle system energy analysis methodology and tool for determining vehicle subsystem energy supply and demand. SAE Technical Paper 2007-01-0398, 2007 SAE World Congress, Detroit, Michigan, 2007.
6. S.A. Cushman, Entropy, ecology and evolution: toward a unified philosophy of biology, Entropy (Basel) 25 (3) (2023) 405. https://doi.org/10.3390/e25030405
7. S.N. Nielsen, F. Müller, J.C. Marques, S. Bastianoni, S.E. Jørgensen, Thermodynamics in ecology—an introductory review, Entropy (Basel) 22 (8) (2020) 820. https://doi.org/10.3390/e22080820
8. R.K. Suarez, K.C. Welch, Sugar metabolism in hummingbirds and nectar bats, Nutrients 9 (7) (2017) 743. https://doi.org/10.3390/nu9070743

9. E.R. Eberts, M.F. Dick, K.C. Welch, Jr., Metabolic fates of evening crop-stored sugar in ruby-throated hummingbirds *(Archilochus colubris)*, Diversity 11 (1) (2019). https://doi.org/10.3390/d11010009.
10. T.J. Zenzal, Jr., Frank Moore, Stopover biology of ruby-throated hummingbirds (Archilochus colubris) during autumn migration, The Auk 133 (2016) 237–250. https://doi.org/10.1642/AUK-15-160.1
11. W.M. Kemp, W.R. Boynton, Productivity, trophic structure, and energy flow in the steady-state ecosystems of Silver Springs, Florida, Ecol Model 178 (1–2) (2004) 43–49. https://doi.org/10.1016/j.ecolmodel.2003.12.020
12. D. Lawlor, Photosynthesis, Third Edition, Bios Scientific Publishers Ltd, Oxford, 2001.
13. R.L. Lindeman, The trophic-dynamic aspect of ecology, Ecology 23 (4) (1942) 399–418. https://doi.org/10.2307/1930126
14. A. Souloumiac, et al., Short-term accumulation and elimination of carbon-14 in the common carp Cyprinus carpio under laboratory conditions, J Environ Radioac (2021) 233. https://doi.org/10.1016/j.jenvrad.2021.106585
15. https://www.footprintnetwork.org/our-work/ecological-footprint/
16. A.V. Rocha, M.L. Goulden, Why is marsh productivity so high? New insights from eddy covariance and biomass measurements in a Typha marsh, Agric For Meteorol 149 (1) (2009) 159–168. https://doi.org/10.1016/j.agrformet.2008.07.010
17. J.H. Brown, J.F. Gillooly, A.P. Allen, V.M. Savage, G.B. West, Toward a metabolic theory of ecology, Ecology 85 (2004) 1771–1789.
18. C. Woodson, J.R. Schramski, S.B. Joye, A unifying theory for top-heavy ecosystem structure in the ocean, Nat Commun 9 (2018) 23. https://doi.org/10.1038/s41467-017-02450-y
19. R.L. Carson, Silent spring, Penguin, London, 2002.
20. G.J. Kolaja, D.E. Hinton, Effects of DDT on eggshell quality and calcium adenosine triphosphatase, J Toxicol Environ Health 3 (4) (1977) 699–704. https://doi.org/10.1080/15287397709529604
21. L. Lear, Rachel Carson: Witness for Nature, Henry Holt and Company, New York, 1997. ISBN 978-0-8050-3428-8
22. J. Blais, Biogeochemistry of persistent bioaccumulative toxicants: processes affecting the transport of contaminants to remote areas, Can J Fish Aquat Sci 62 (2005) 236–243.
23. V.J. Bierman, D.M. Dolan, Modeling of phytoplankton-nutrient dynamics in Saginaw Bay, Lake Huron, J Great Lake Res 7 (4) (1981) 409–439.

CHAPTER

Nutrient and Biogeochemical Cycling

9

Chapter Outline

9.1 The Biogeochemical Cycle

By the end of this section students should be able to:

- Explain how nutrients enter ecological systems.
- List the major elements that make up living organisms.
- Provide an example of biogeochemical cycling.
- Distinguish between the energy and biogeochemical systems.

Matter is any substance that occupies space and has mass, including living and nonliving structures. The fundamental unit of matter is the atom. There are different types of atoms—a total of 118 different elements—that are the unique atoms and that form matter (**Fig. 9.1**). Each element has its own specific chemical and physical properties. The four most common elements found in all living organisms are oxygen (O), carbon (C), hydrogen (H), and nitrogen (N) (**Table 9.1**). These elements are also abundant to varying degrees in the atmosphere, soil, and water.

The flow of energy through an ecosystem can be told by following the movement of the element carbon, starting with photosynthesis and the fixation of carbon from atmospheric carbon dioxide (CO_2). The energetics of primary producers through trophic levels follows its carbon atoms (i.e., electrons), first as carbon–carbon bonds, mostly in the form of carbohydrates, then through cellular respiration as it is oxidized to extract energy held in its bonds (electron flow). As introduced in Chapter 8 the full story of primary productivity depends on the uptake of several essential nutrients

Ecology for the 21st Century. DOI: https://doi.org/10.1016/B978-0-443-24031-7.00009-X

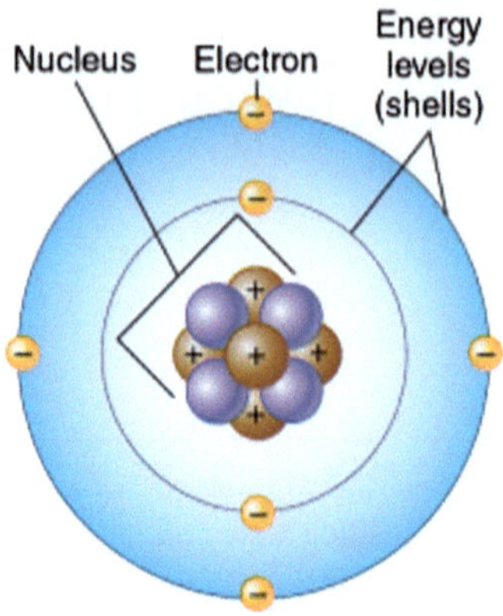

Fig. 9.1

Atoms are the building blocks for molecules throughout the biosphere, including in the air, soil, water, rocks, and within cells of all living organisms.

Table 9.1 Approximate Percent of Elements in Living Organisms (Humans) Compared to the Nonliving World

Element	Life (Humans)	Atmosphere	Earth's Crust
Oxygen (O)	65%	21%	46%
Carbon (C)	18%	trace	trace
Hydrogen (H)	10%	trace	0.1%
Nitrogen (N)	3%	78%	trace

in addition to carbon, which will be needed to make up living tissues and perform physiological processes.

The main elements enter living organisms by different pathways. Plants obtain inorganic carbon from the surrounding atmosphere and its other macronutrients (hydrogen, oxygen, and nitrogen) through water or soils. Animals mostly obtain elements directly from consuming other organisms. Metabolism then transforms the ingested organic macromolecules into smaller compounds through digestive and cellular processes. Eventually those nutrients will return to their inorganic form via excretion and decomposition. For example, atoms of carbon will no longer be bound as molecules of organic matter but will break down into one of its inorganic forms, CO_2, bicarbonate ion (HCO_3), and carbonate (CO_3^{2-}), respectively. Phosphorus in organic matter decomposes to dissolved inorganic phosphates and nitrogen generally decomposes to soil nitrites and nitrates. Often microscopic organisms complete the breakdown of organic molecules in a process called decomposition or mineralization.

The origin and movement of an element through an ecosystem is known as its **biogeochemical cycle**. The initial source or reservoir of elements is typically the atmosphere, rocks, or dispersed in soil or water. Elements whose supply tends to be limiting on biological activity are called nutrients. The cycle begins as primary producers take up essential nutrients and from there the nutrients move through the food chain. When an organism dies, the nutrients found in **DOM** are returned to Earth and enter the decomposer (detrital) food chain. For the most part the nutrients will remain within the ecosystem.

Because decomposition of nutrients and minerals is neither destroyed nor lost, Earth is considered a ***closed system*** with respect to the elements. Recall that Earth is an ***open system*** with respect to its energy flow. Elements are cycled continuously between their biotic and abiotic states within ecosystems. Once the microbial decomposers break down DOM into its mineral form it is available for plants to uptake and incorporate into new living and growing tissues.

9.2 Terrestrial and Aquatic Decomposition

By the end of this section students should be able to:

- Define the term catabolism with respect to energy flow.
- Explain how leaf litter bags are used to analyze decomposition.
- List factors that affect rates of decomposition in the environment.
- Describe decomposition in terrestrial and aquatic environments.

Decomposition is a complex process and involves several intermediate steps and activities by microorganisms, including fungi, protozoans, and bacteria. In this process the chemical bonds that hold together organic molecules are broken in a chemical process called catabolism. Catabolism also occurs in cellular respiration when bonds are broken to release energy. The release of energy during organic decomposition fuels microorganisms in the same manner that ingested carbohydrates fuel cellular metabolism. Decomposition also describes break down and release of inorganic minerals from rocks and soils. Whether by leaching, fragmentation, ingestion, changing of chemical structures, or excretion of waste products from living organisms, the processes involved are all decomposition reactions. With respect to the nutrient cycle humans and other heterotrophs are decomposers. This is because in the process of ingesting and breaking down nutrients to altered chemical forms for energy extraction there is a partial release of excreted waste products. The largest class of decomposers is found in microscopic soil organisms

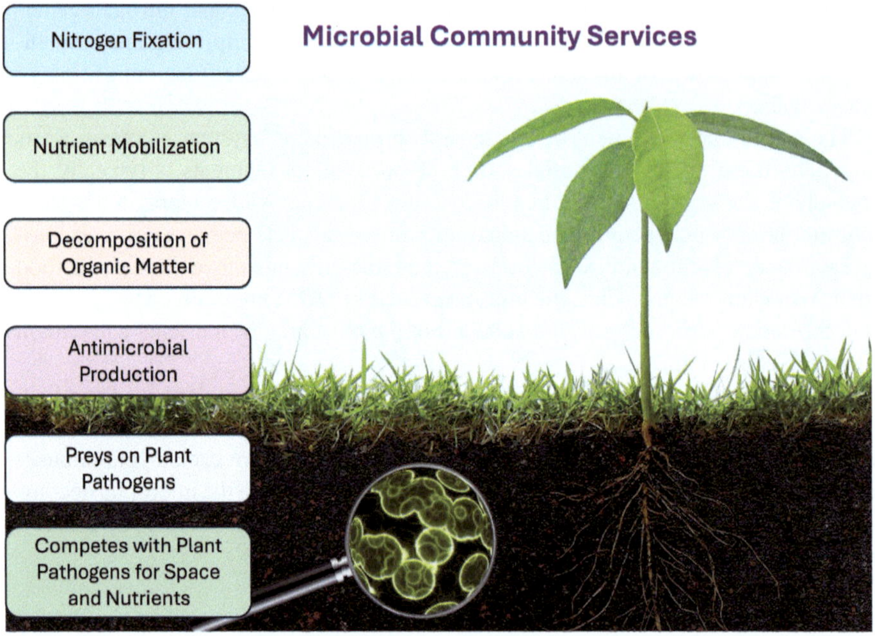

Fig. 9.2

Soil microbial communities play an important role in plant health and soil quality. Beneficial microbiomes improve plant productivity, health, and stress tolerance.
Source: Hallrob3 BY CC BY-SA 4.0

that feed on decomposing plant and animal parts, feces, or detritus. The feeding lifestyle of a decomposer is detritivore (**Fig. 9.2**).

Decomposers are found everywhere on Earth, in all crevices and niches where organic matter can be found. This includes areas with high oxygen levels or the complete absence of oxygen (anaerobic). Microorganisms are so ubiquitous they can be found in all places within Earth's biosphere, including deep within its crust or spread into the far reaches of its atmosphere. Bacteria are the most dominant decomposers of dead animal forms, and fungi are the most dominant decomposers of plant forms.

Fungi are interesting detritivores, excreting enzymes to any dead organic matter in proximity of their environment. The fungal enzymes will then decompose the DOM into an assimilable form for uptake from the soil. The process is somewhat analogous to enzymatic digestion occurring in the mammal's intestine. Pancreatic enzymes are secreted into the intestinal tract to break down whatever is present, typically large food particles, decomposing it into smaller molecules that can be easily transported across the intestinal lining. Some fungi are also capable of invading dead tissues by growth of thin hyphal extensions. And it has been shown that plants gain better access to nutrients in soil in the presence of fungal hypha, which loosens and breaks up soil particles. Some bacteria even secrete enzymes during decomposition processes to make nutrients available for absorption by nearby fungi and bacteria. These interactions are symbiotic, or at least commensal, in nature.

Invertebrates like nematodes, mites, millipedes, and earthworms aid the process of decomposition by fragmentation of any larger detritus. In aquatic environments fragmentation is accomplished by snails, annelid worms (small crustaceans), mollusks, and crabs (primary consumers). In summary, the many families and genus of microbes, with differing abilities to break down and absorb chemical nutrients, successively decompose organic matter in the soil until it is reduced to its inorganic elemental form and assimilable for plant use through its synergistic processes.

By following the fate of DOM ecologists can answer many questions about an environment under study, such as:

- Does decomposition vary in different places or according to microorganism size?
- What environmental factors (moisture, temperature, nutrients) most significantly affect decomposition?
- Do different leaves (maple vs oak) or fallen trees decompose at different rates?

The study of **leaf litter** is sometimes thought of as the "poor man's rainforest" because the decomposing matter exposes a wealth of data from local biodiversity. Most inhabitants are soil arthropods but there will also be present a diverse population of protists, bacteria, fungi, and worms. Without traveling to exotic places it's possible to collect information on environmental conditions that affect rates of decomposition.

Leaf litter experiments begin by developing a collection protocol with preset time periods and collection sites. A known mass of litter (e.g., forest leaves) will be placed into a permeable bag (holes must be just large enough for microbes to enter and leave freely) and the litter bag will be left on a field site for the specified time period. At the end of the period (typically several weeks) leaf litter bags are collected, and whatever is retained in the bags will be reweighed. The amount of remaining biomass is a function of its decomposition rate. During the process decomposers extract some of its energy in the form of carbon. Leaf litter studies are also used to study decomposition in stream ecology. In fact, leaf litter forms a considerable amount of DOM in aquatic ecosystems, adding energy and nutrients for the food chain. During leaf litter studies in aquatic environments the litter is placed in mesh bags (or leaf packs) and anchored within a stream of flowing water. By varying pore size it is even possible to restrict access (or egress) to evaluate the contribution of different-sized decomposers to the process of energy extraction and biodegradation (**Fig. 9.3**).

Litter leaf studies have helped to reveal the factors that have the greatest influence on rates of decomposition. The specific types of carbon compounds present (typically 45%–60% total dry weight of plant tissue) can affect rates of decomposition. For example, the complex molecular structures like lignin and cellulose fibers in plant cell walls are less readily decomposed. Lignin is a major component in wood and is the most difficult to decompose. Some species of single-celled protozoans have adapted to breaking down lignans into digestible simpler sugars, particularly

Fig. 9.3

A Tapajos National Forest, Brazil, decomposition study with litterfall in leaf litter bags (20 × 20 cm, 1 mm^2 pore size) collected weekly from marupa leaves and dried at 50°C (122 °F) for 72 hours.

Source: McGroddy, Oak Ridge National Laboratory, NASA Earth Data, Public Domain

those that inhabit the hindgut of termites. However, most decomposition of wood is accomplished in nature by a large subdivision of fungi called the basidiomycetes. Over time the leaf litter bags will have an increasing percentage of lignin as the residual biomass since other carbon sources have faster rates of decomposition. Lignin content in leaf litter provides a good indicator of the quality of DOM available for decomposers.

Ecological field studies have shown that the physical environment has a role in decomposition rates. According to Shizing Zhou, same-source leaf litter bags placed into different latitudes and biome types will decompose at different rates, depending on the environmental conditions and microbial communities present. Lower temperatures and drier conditions tend to be more limiting as a habitat for microorganisms. Decomposition rates occur slower in these dry, cool habitats. In contrast moist, warm habitats are conducive to decomposition because the habitats are optimal for diverse microorganisms to thrive in and multiply.

> In aquatic systems phytoplankton are an easily decomposed sources of carbon. But grasses and reeds, ubiquitous in estuaries, will be high in lignans, which decompose slowly, making them poor sources of carbon. Rates of decomposition will vary dependent on oxygen levels. Anaerobic bacteria will be actively decomposing in the thick, muddy sediments of an aquatic habitat. But fungi that are needed to break down any lignin content are aerobic species, making decomposition rates extremely slow in anaerobic areas where grasses and reeds are abundant.

The largest carbon sink is in the lithosphere, stored as carbonate. This includes sedimentary rocks (limestone), hydrocarbons (fossil fuels), and marine sediments.

As organic sources of carbon are catabolized by decomposers they are converted to either CO_2 released to the atmosphere or retained by the soil as carbonates and hydrocarbons. In waterways carbon is released as dissolved carbonic acid or stored as calcium carbonate and limestone sediments.

During DOM consumption decomposers transform other nutrients into inorganic compounds in the process called **mineralization**. Mineralization of nitrogen proceeds through the same pathway of nitrogen gas (N_2) into soils—by ammonification and then nitrification to nitrates (NO_3^-). Nitrogen is a macronutrient needed for protein and nucleic acids, and decomposers will consume some of the nitrogen for their own use in another process called **immobilization**. The remaining nitrogen in leaf litter bags is expressed as a percentage of the original content.

As decomposition progresses the mineralization of DOM results in soil rich in organic matter. In leaf litter bags this is observed as dark brown or black humus. In open aquatic environments decomposition can happen rapidly (except in anaerobic marshy habitats). The organic matter is broken into particulate organic matter (POM), which drifts toward the benthic bottom where it is made available for anaerobic bacteria to feed on and complete its breakdown. In flowing streams larger twigs, leaves, needles, and other matter rapidly lose biomass (up to 30% in first 24 hours) by leaching, which allows colonization of the litter by microbes that begin to change its structure. Decomposition then proceeds by the feeding action of shredders (detritivorous macroinvertebrate species) fragmenting the litter to POM, which becomes a food source for filter feeders.

In summary, decomposition rates will be affected by carbon type and the physical environment, particularly the content of oxygen, moisture, and temperature. Nutrient cycling represents the direct link between NPP and decomposition.

9.3 The Major Nutrient Cycles

By the end of this section students should be able to:

- Describe the five major biogeochemical cycles.
- Distinguish between renewable and nonrenewable resources.
- Show how Chesapeake Bay Area highlights the need for legislative cooperativity to address ecological issues.
- Explain how excess nutrients can lead to eutrophication.
- Discuss the factors leading to dead zones and their global impact.

Energy flows within and through ecosystems, entering as solar radiation and leaving as heat, throughout its many energy transfers during metabolic processes and as energy moves through the trophic cascades. Geologic processes are important to recycle inorganic elements trapped in soil, rocks, the atmosphere, and even in Earth's crust. Natural geologic processes involved in cycling of inorganic matter are weathering, erosion, water drainage, and subduction of the continental plates. The cycling of organic and inorganic matter between living organisms and the environment is called the biogeochemical cycle because geology and chemistry are major parts of the recycling process for the molecules of life. Each individual element of the biogeochemical cycle has its own distinct cycling patterns and interactions in the environment. Ecologists and environmentalists track biogeochemical cycles because disruption of nutrient cycles leads to degradation of soil and aquatic dead zones. The depletion of nutrients from soils has been responsible for the collapse of many civilizations, including the Sumerians of Mesopotamia and the Mayans. Here are the most important of the nutrient cycles:

- **Hydrologic cycle**: aqueous solvent (50%–60%) of living things, made up of hydrogen and oxygen atoms, essential to all living processes; continual water movement and storage occurs in the hydrosphere between soil, oceans, and atmosphere
- **Carbon cycle**: organic macromolecules are made of carbon, with reservoirs in mineralized rocks, sediments, and shells and small amounts in the atmosphere (0.04%) but its major reservoir in the ocean's limestone
- **Nitrogen cycle**: found in nucleic acids and protein macromolecules of living things, in high concentrations in the atmosphere (78%), in water and soils rich humus and organic matter
- **Phosphorus cycle**: found in nucleic acids of living things, in water, rocks, and soils as sediment and organic matter but without a major gaseous form is not found in the atmosphere
- **Sulfur cycle**: found in disulfide bonds in living things, in meteorites, reactive gasses in the atmosphere, and as dioxides in rocks and Earth's crust

Each nutrient cycle follows its unique pathway but they all interconnect across ecosystems as they move between the biotic and abiotic worlds. The movement of water through terrestrial environments also plays a key role in nutrient cycling, for example, leaching nitrogen and phosphate into waterways.

The Hydrologic Cycle

Water has a major influence on climate and the productivity of all ecosystems. It is essential for all living processes. Two-thirds of Earth's surface is covered with water. The human body is more than 50% water and human cells are more than 70% water. Living things are dependent on a constant supply of fresh surface water; a lack thereof has detrimental effects on ecosystem dynamics. The water, or hydrologic cycle, includes many phases and locations of water.

The hydrosphere encompasses all rivers, lakes, oceans, groundwater, polar ice caps, glaciers, and even the water vapor in the atmosphere. According to the USGS, of the stores of water on Earth, 97.5% is **nonpotable** salt water (**Fig. 9.4**). Of the remaining 2.5% that is potable water, most of it (99%) is locked underground as water or as ice. There is less than 1% freshwater that is easily accessible, mostly located in lakes and rivers.

Although the pursuit of potable water has been ongoing throughout human history, the supply of freshwater has newly risen as a major issue in modern times.

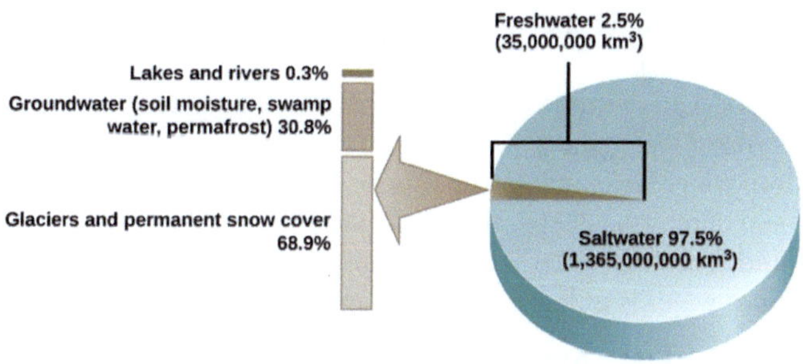

Fig. 9.4

Only 2.5 % of water on Earth is freshwater and less than 1% is easily accessible.

Source: Voer under CC BY 3.0

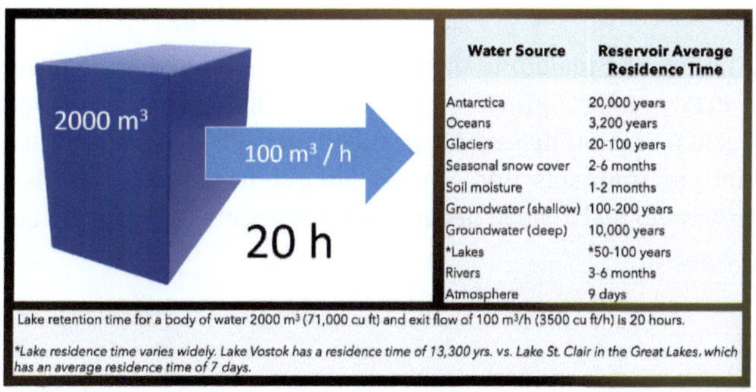

Fig. 9.5

Average residence time for water molecules in Earth's water reservoirs.

Source: Modification of Tournesol – under CC BY SA 4.0

According to the US National Climate Report 2023, droughts and heatwaves in the US Midwest states have impacted crop yields and herd grazing, costing billions of dollars. In the west reservoirs and aquifers have dropped to record low levels from irrigation and urban sprawl into arid landscapes. These trends are exasperated by climate change and expose the fragility of human reliance on freshwater systems, even as freshwater is becoming less abundant due to misuse of water reserves.

Residence time is a measure of the average time an individual water molecule stays in a particular reservoir (**Fig. 9.5**). Since a large amount of Earth's water is locked in place in reservoirs as ice, beneath the ground, or in the ocean it is unavailable for short-term cycling. This is because only surface water interfacing with the atmosphere can evaporate. The versatility of water cycling in its various forms includes:

- **Evaporation**: water evaporates into the atmosphere from oceans and other bodies of water.
- **Transpiration**: surface water loss from plants accounts for 50% of land-based evaporation.

- **Condensation**: water vapor converts to a liquid phase (e.g., dew drops, formation of clouds).
- **Precipitation**: water returns to Earth as rain, sleet, or snow; run-off into lakes, rivers, or ground water sinks.
- **Infiltration**: water sinks into Earth's surface, filtered and purified according to soil type and depth (the deeper the water, the cleaner it becomes).
- **Melting and freezing**: water is "locked up" in glaciers and ice sheets.
- **Sublimation**: water is unlocked from ice and returns to water vapor (gas phase).

The water cycle (**Fig. 9.6**) is driven by the sun's energy as it warms the surface water of oceans and lakes. As water is heated it evaporates or sublimates (to water vapor from water or ice, respectively). As water vapor saturates the atmosphere it condenses to precipitation and snow, cycling back to Earth's surface. Terrestrial water permeates into groundwater, lakes, and streams. Below the surface it will be stored for much longer periods of time, which also removes it from the water cycling process.

Chemical pollutants, including those that can bioaccumulate in tissues like mercury, enter waterways through rain and surface runoff. This is the same pathway that most nutrients—including carbon, nitrogen, phosphorus, and sulfur—are cycled from terrestrial environments into aquatic environments. High levels of nutrients entering waterways can also have negative environmental effects, including contributing to dead zones.

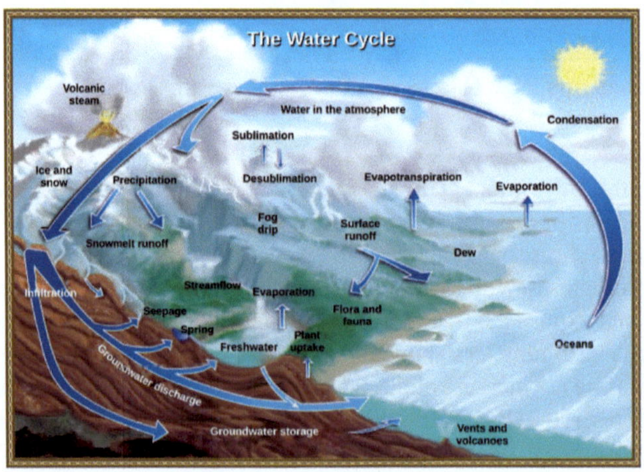

Fig. 9.6

Water from land and oceans enters the atmosphere by evaporation or sublimation, where it condenses into clouds and falls as rain or snow. Precipitated water enters freshwater bodies or infiltrates the soil, which completes the cycle.

Source: John M. Evans and Howard Perlman, USGS, Public Domain

The Carbon Cycle

Carbon is present in high concentrations in all organic molecules and serves as the backbone structure of macromolecules. Because of the unique bonding properties of carbon atoms and the presence of extensive carbon–carbon bonds it is high in energy, which is released during catabolic processes of cellular respiration. In the environment carbon is stored as carbonates but also exists in high-energy forms of fossilized organisms that have been the main source of fuel energy since the Industrial Revolution of the 1800s. The global demand for this limited resource has risen drastically ever since. As fossil fuels are burned they release energy as carbon enters the atmosphere as carbon dioxides. As of April 2020 climate experts at the Global Carbon Project (GCP) estimate that the burning of fossil fuels produces nearly 35 billion metric tons of atmospheric CO_2 per year. The rising atmospheric carbon dioxide levels are known to contribute to climate change and other disturbances of Earth's ecosystems, leading to atmospheric CO_2 being a major environmental concern worldwide. A **carbon footprint** is used to measure environmental impacts due to gas emissions and constitutes approximately 60% of humanity's ecological footprint.

The carbon cycle has two interconnected cycles: rapid carbon exchange among living organisms and long-term cycling through geologic processes (**Fig. 9.7**). The interconnected cycles are also known as the Biological Carbon Cycle and the Biogeochemical Carbon Cycle.

The Biological Carbon Cycle. Living organisms are connected between ecosystems through the carbon cycle. Terrestrial autotrophs mostly obtain their carbon directly from the atmosphere. Marine autotrophs acquire their carbon to be dissolved in water as carbonic acid, $H_2CO_3^-$. Aquatic and terrestrial photosynthetic organisms

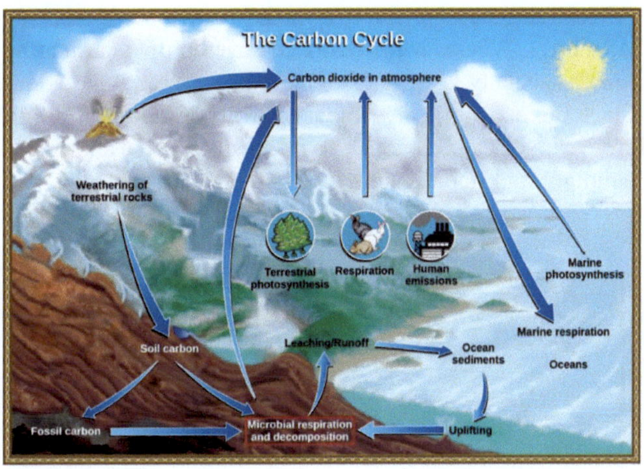

Fig. 9.7

The carbon cycle.

Source: John M. Evans and Howard Perlman, USGS, Public Domain

are essential to cycle atmospheric carbon (and energy) into the primary producer trophic level biomass. Rapid cycling gas exchange through the atmosphere and water is an important way in which carbon cycling connects ecosystems and all living organisms on Earth.

The Biogeochemical Carbon Cycle. The movement of carbon through the biosphere is complex. Carbon reservoirs store inorganic carbon for long periods, mostly in the oceans as sediment in limestone but also in soil, land sediments (e.g., as fossil fuels), and, to a smaller extent, in the atmosphere. In the atmosphere CO_2 is the major reservoir. The concentration of CO_2 in the atmosphere (typically around 0.04% of the air) is continually influenced by the inorganic carbon reservoirs, for example, during burning of fossil fuels (**Fig. 9.8**). As CO_2 dissolves into water it is quickly converted into carbonic acid then ionized into carbonate and bicarbonate ions based on chemical equilibrium constants.

Typically 90% of oceanic carbon is in the form of bicarbonate ions. Free bicarbonate ions combine with the mineral calcium, forming $CaCO_3$, which is a major component of marine organism shells. The calcium is continuously supplied to waterways by leaching from rocks, soils, and organic matter.

Eventually formed shell structures of dead organisms will form the detrital sediments on the ocean floor. And over geologic time $CaCO_3$ forms limestone in the benthic floor. In some ways this is analogous to the extensive formation of coral reefs. Limestone on the ocean floor represents the *largest carbon reservoir* on Earth. Any regions that were once underwater for a long period of time will also be rich in limestone. For example, during the early Carboniferous period (*Latin*: coal-bearing) large portions of the continental United States were a shallow tropical sea. The waters were teeming with biodiversity and dense masses of algae, snails, and bryozoans that created a layer of thick limestone. The North American land uplifted after its collision with Gondwana, causing the rise of the Rocky Mountains and even the Appalachian Mountains. Wherever the sea once covered the land marine limestone deposits are found, especially after glaciation events lowered the sea level exposing more terrain. Since the economic development of the United States, limestone has been quarried from canyons and caverns as a plentiful resource and is found in many of its historic buildings and statutes across the country.

Step 1: CO_2 (atmospheric) $\rightleftharpoons$ CO_2 (dissolved)

Step 2: CO_2 (dissolved) + $H_2O \rightleftharpoons H_2CO_3$ (carbonic acid)

Step 3: $H_2CO_3 \rightleftharpoons + H^+ + HCO_3^-$ (biocarbonate ion)

Step 4: $HCO_3^- \rightleftharpoons H^+ + CO_3^{2-}$ (carbonate ion)

Fig. 9.8

CO_2 reacts with water to form bicarbonate and carbonate ions.

Source: Image by author

Carbon deposition on land follows a similar long-term process. Carbon enters the soil as DOM or gets released with the weathering of rocks. Carbon leaches into water reservoirs by dissolving in water and as surface runoff. The formation of fossil fuels is another reservoir of carbon. The process occurs over millions of years from the decomposed remains of plants. Rich coal and oil deposits are found deep underground anywhere that was once home to highly prolific flora (e.g., Carboniferous 300–400 Mya). Fossil fuels are a **nonrenewable resource** because their use far exceeds their rate of formation. Nonrenewable resources are regenerated very slowly (e.g., aquifers) or not at all, in contrast to renewable resources.

Carbon is released to the atmosphere from Earth's crust during volcanic eruptions. There are also underwater volcanic hydrothermal vents. Finally, inorganic carbon from ocean sediments is brought to the surface by the actions of subduction and geologic lift.

When scientists model and predict the impact of atmospheric carbon emissions they also consider the natural sources of carbon release. The farming practices needed to raise heterotrophic, endothermic animals as a food supply result in unavoidable methane production. This becomes a significant source of atmospheric CO_2 contributing to global warming because of the large numbers of land animals being raised to feed Earth's growing population.

The Nitrogen Cycle

The movement of atmospheric nitrogen into the living world is a difficult transition, regardless of the fact that nitrogen gas comprises 78% of the air we breathe. Animals, plants, and phytoplankton are unable to directly assimilate inorganic nitrogen from the atmosphere. N_2 exists as a tightly bonded, triple-covalent molecular structure that is stable and not reactive. Most of the biological nitrogen found in the living world enters through free-living and symbiotic bacteria that possess a nitrogen fixation pathway. Cyanobacteria, which live in most aquatic ecosystems wherever sunlight is present, play a key role in nitrogen fixation. In terrestrial environments *Rhizobium* bacteria live symbiotically in the root nodules of legumes (such as peas, beans, and peanuts) and have the capacity to fix nitrogen, providing plants with organic nitrogen in its most assimilable form (ammonia, NH_3). Free-living bacteria, such as *Azotobacter*, and some fungi are also important nitrogen fixers.

Because of inorganic nitrogen's limited availability organic nitrogen is especially important to biogeochemical cycling and ecosystem dynamics. Primary growth is dependent on nitrogen to synthesize proteins and nucleic acids needed for growth and development. As shown in **Fig. 9.9** the nitrogen that enters living systems by nitrogen fixation into organic nitrogen will eventually be converted back to nitrogen gas at the end of an organism's life cycle by the action of decomposers and detritivores. The decomposition process occurs in three steps in terrestrial systems:

- **Ammonification:** conversion of nitrogenous waste from living animals or dead animals into ammonium (NH_4^+) by bacteria and fungi

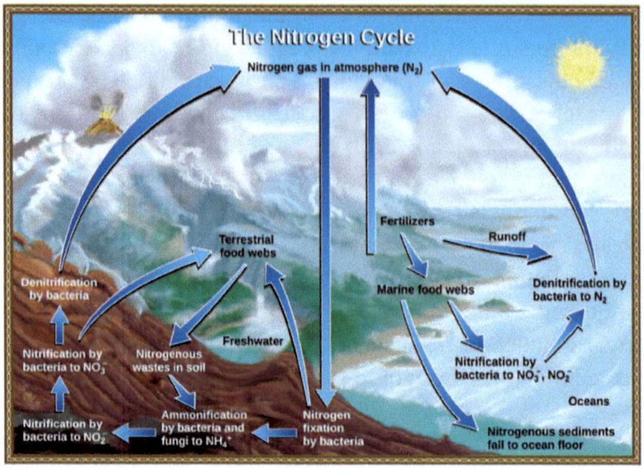

Fig. 9.9

Nitrogen enters the living world from the atmosphere via nitrogen-fixing bacteria.

Source: John M. Evans and Howard Perlman, USGS, Public Domain

- **Nitrification:** conversion of NH_4^+ to nitrites (NO_2^-), which in turn is converted to NO_3^- by nitrifying bacteria
- **Denitrification:** conversion of NO_3^- back to nitrogen gas to reenter the atmosphere, by bacteria

Human activity releases atmospheric nitrogen into the environment by two primary means: combustion of fossil fuels as nitrogen oxides and artificial fertilizers used in agricultural practices entering lakes, streams, and rivers as surface runoff. Atmospheric nitrogen is associated with several detrimental effects on Earth's ecosystems, including the production of acid rain (as nitric acid, HNO_3) and its role as a greenhouse gas (nitrous oxide, N_2O). A serious imbalance occurs in freshwater ecology whenever excess agricultural fertilizer runoff enters waterways, which results in **eutrophication**. In this process the overabundance of fertilizer (nitrogen and phosphorus) causes overgrowth of aquatic microorganisms, which in turn depletes the dissolved oxygen levels that results in suffocating the ecosystem's fauna. In freshwater environments it is typically phosphorus excess and in marine environments it is typically nitrogen excess that leads to dysfunctional productivity levels.

In the marine nitrogen cycle ammonification, nitrification, and denitrification processes are performed by marine bacteria, with some nitrogen falling to the ocean floor as sediment. Over geologic time the nitrogen trapped in ocean sediment moves onto land during events that uplift Earth's surface, and it is incorporated into terrestrial rock. The movement of nitrogen from rock directly into living systems has been traditionally thought to be insignificant compared to fixation of atmospheric nitrogen. However, recent studies suggest nitrogen movement into rock is more significant than originally thought and needs to be included in the global nitrogen cycle.

The Phosphorus Cycle

Phosphorus is an essential nutrient for living processes as it is a major component of nucleic acid, phospholipids, and as calcium phosphate in the supportive components of skeletal bones. Phosphorus is often a limiting nutrient in aquatic ecosystems but in excess is responsible for eutrophication (**Fig. 9.10**).

In nature phosphorus occurs in the form of phosphate ion (PO_3^-). In addition to phosphate in agricultural nutrient runoff, natural surface runoff occurs when it is leached from phosphate-containing rocks through natural weathering processes. Surface runoff, groundwater, and river flow release phosphate into oceans, where it cycles into marine food webs and eventually falls to the ocean floor as sediment. Like nitrogen cycling phosphate atoms in sediment move to the surface over geologic time during periods of Earth's uplifting and mass movement. Volcanic ash, combustion aerosols, and mineral dust contribute a significant source of the total atmospheric phosphate. But phosphorus is not exchanged to a significant degree with the atmosphere since there are few stable forms of atmospheric phosphorus.

Excess phosphorus (and nitrogen) entering ecosystems from agricultural fertilizer runoff or sewage overflow leads to excessive growth of microorganisms. This in turn leads to depletion of aquatic dissolved oxygen. Low oxygen, or hypoxic conditions, eventually results in death of fauna, including sensitive shellfish and finfish. Eutrophication is responsible for dead zones in coastal areas, lakes, and at the mouths of many major rivers (**Fig. 9.11**).

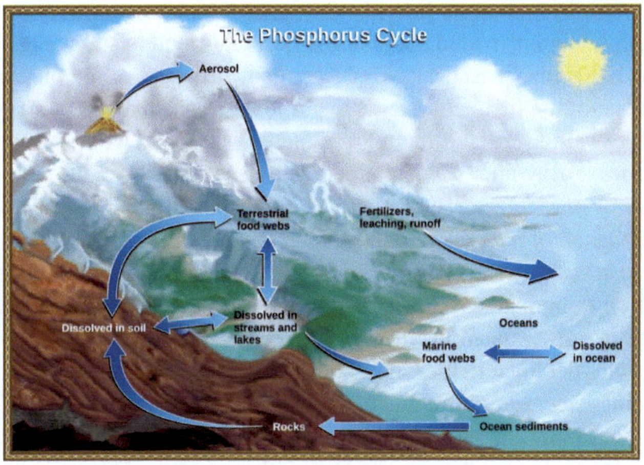

Fig. 9.10

Weathering of rocks and volcanic activity releases phosphate into the soil and water.
Source: John M. Evans and Howard Perlman, USGS, Public Domain

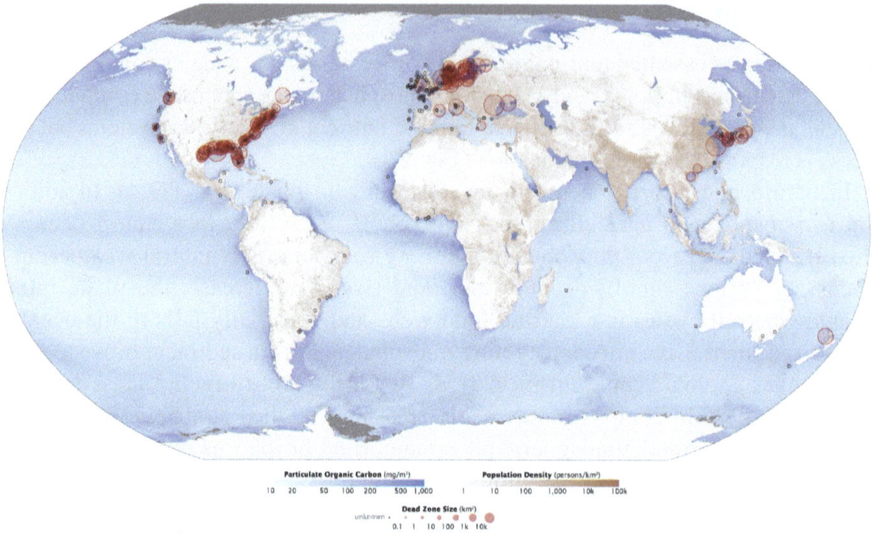

Fig. 9.11

Large dead zones are found worldwide along the coastlines of high population density.
Source: NASA Earth Observatory, Public Domain

Dead Zone: Chesapeake Bay Area

The Chesapeake Bay is a **watershed** estuary in the Mid-Atlantic United States covering the six states of Delaware, Maryland, New York, Pennsylvania, Virginia, and West Virginia. Until the mid-20th century the bay area was valued as one of the most scenic and bountiful regions. Its humid subtropical climate was home to more than 300 aquatic species and provided some of the best crabs and oysters (**Figs. 9.12a, b**).

During the 1970s the Chesapeake Bay was renowned for a different reason. It was recognized as one of the first ecosystems identified as a dead zone, with loss of many fish and bottom-dwelling species, including clams, oysters, and worms. Approximately 40% of Chesapeake Bay area was classified as a dead zone. A **dead zone** is defined as a freshwater or marine ecosystem that has experienced massive losses of its flora and fauna due to lack of oxygen and eutrophication. Dead zones have been known to develop from excessive trauma to aquatic ecology, including oil spills, toxic chemical dumping, and other human activities. Dead zones and eutrophication are topics that will be covered in further detail in Chapter 14, under Causes of Biodiversity Loss.

Several species declined in the Chesapeake Bay as a direct result of surface water runoff containing excess nutrients from artificial fertilizer. But sources of fertilizers (always high in nitrogen and phosphate) are not limited to local agricultural practices. Urbanization surrounds the Chesapeake Bay with more than 150 rivers and streams emptying into the bay, carrying fertilizer runoff from urban lawns and gardens. The decline of the Chesapeake Bay is a complex issue, and its restoration efforts require the cooperation of industry, agriculture, and everyday homeowners.

(a) (b)

Fig. 9.12

(a) Satellite image of the Chesapeake Bay, an ecosystem facing eutrophication.
(b) Member of Army Corps of Engineers working on the oyster restoration efforts.
Source: NASA/MODIS; U.S. Army, Public Domain

The Chesapeake Bay area is one of the best studied estuaries in the world. Of keen interest to conservationists is its oyster population. It is estimated that more than 200,000 acres of oyster reefs once thrived in the bay (era 1700 s) but declined to only 36,000 acres within just 200 years. Oyster harvesting was a major economic boon for the Chesapeake Bay but saw a decline of 88% between 1982 and 2007. The steep decline is due to the combination of fertilizer runoff, dead zones, and another human-derived activity: overharvesting. Studies revealed the direct cause of the oyster's sharp decline in the bay ecosystem—oyster reproduction requires a minimum population density with oysters in close proximity. Overharvesting prevents oysters from reaching population densities conducive to their reproduction.

While oysters are desirable to eat they are most important as a critical component to help clean up the bay. Sessile mollusks, like oysters and clams, are filter feeders that filter water and absorb some of the chemicals and pathogens present in their environment. It is estimated that in colonial America less than a week was sufficient for the Chesapeake Bay's oyster population to filter the entire volume of water in the bay. Today the much-reduced oyster population would require nearly a year to completely filter the same volume of water.

Restoration efforts in cleaning up the bay, ongoing for several years by nonprofit organizations like the Chesapeake Bay Foundation, have reported mixed results. The restoration goal is to find a way to increase population density so the oysters can reproduce more efficiently. Many disease-resistant varieties of oysters (developed at the Virginia Institute of Marine Science for the College of William and Mary) have been used in the construction of experimental oyster reefs, with promising results. Current efforts to clean and restore the bay area require cooperation between the states that border the bay. But commitment to restore Chesapeake Bay efforts by the states of Virginia and Delaware have been hampered, partly because much of the pollution entering the bay originates upstream in states not in their jurisdiction. The sociopolitical implications of ecological projects can be complex, requiring consensus from opposing priorities across legislative bodies. Conservationists and environmentalists emphasize the need for a coordinated ongoing interstate or national effort for any restoration project to succeed.

Dead zone areas are on the rise worldwide; there are well over 500 dead zones. The Gulf of Mexico of the United States, where fertilizer runoff from the Mississippi River basin and repeated oil spills wreak havoc, is one the world's largest dead zones, a massive 8463 square miles. Fertilizer runoff is one of the major factors behind all dead zones in the United States, including several lakes and bay ecosystems.

The Sulfur Cycle

Sulfur is an essential mineral nutrient in the human body, required for the biosynthesis of proteins. Sulfur is abundant in nature, particularly near volcanoes, and found in deposits in its pure form or as its sulfide or sulfate mineral form. As shown in **Fig. 9.13**, sulfur cycles openly between waterways, land, and the atmosphere.

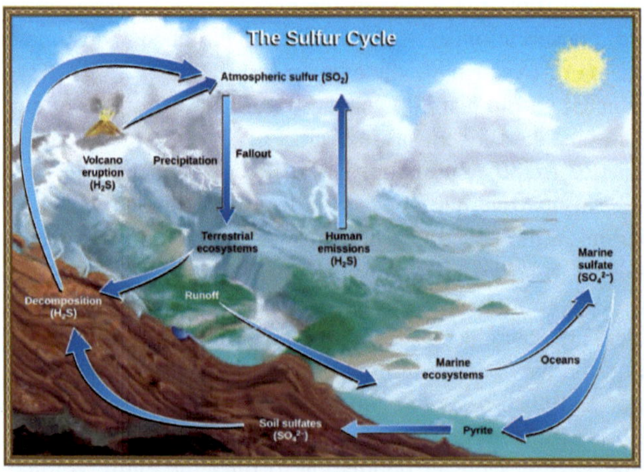

Fig. 9.13

Atmospheric sulfur dissolves in precipitation as weak sulfuric acid (high concentrations in acid rain).

Source: John M. Evans and Howard Perlman, USGS, Public Domain

Atmospheric sulfur dioxide (SO_2) is released during volcanic activity, geothermal vents, and with fossil fuel burning. Precipitation of atmospheric sulfur dioxide enters land dissolved in rainfall as weak sulfuric acid (H_2SO_4). Sulfur also enters land (or water) directly from the atmosphere in a process called fallout. In terrestrial ecosystems, sulfur is processed in the soil to SO^{2-4} for assimilation by organisms.

Rock weathering and the decomposition of organic matter recycle sulfur back into the environment where it now enters the oceans as runoff (**Fig. 9.14**). Sulfur also enters the ocean from underwater geothermal vents. Hydrothermal sulfur vents

Fig. 9.14

At this sulfur vent in Lassen Volcanic National Park in northeastern California yellowish sulfur deposits are visible near the mouth of the vent.

Source: Yellowstone National Park under Public Domain

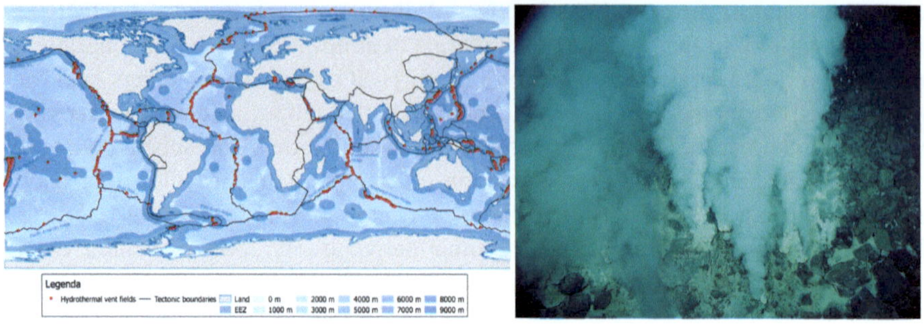

Fig. 9.15

Distribution of the world's hydrothermal vents and white gaseous plumes being released from Champagne Vent in Eifuku, Japan (> 100°C [212°F]).

Source: Modification of DeDuijn under CC BY SA 4.0 and NOOA, Public Domain

in oceans run along Earth's plate boundaries (**Fig. 9.15**). Approximately 250 active hydrothermal vent fields are visible at the seafloor. The dissolved sulfates present in marine waters and sediment represent the largest reservoir of sulfur. Here marine phytoplankton and aquatic plants metabolize sulfur, which is essential to protein synthesis. During anaerobic conditions sulfates are one possible source of metabolic energy for chemoautotrophs.

Human activities have played a major role in altering the balance of the global sulfur cycle. This occurs by the same **anthropogenic** activities that have altered the natural carbon, nitrogen, and phosphorus cycles. Burning of large quantities of fossil fuels, especially from coal, release far more hydrogen sulfide gas into the atmosphere than would occur by natural means. Nitrogen oxides and sulfur dioxides, from the combustion of fossil fuels, react with oxygen and rainfall to form nitric acid and sulfuric acid in a phenomenon known as acid rain. **Acid rain** is corrosive to surfaces and toxic to many species. The acidity causes damage to aquatic ecosystems (particularly shell builder species) but also damages the entire aquatic ecosystem by lowering the pH of lakes below a sustainable range for many of the resident fauna. The low pH causes leaching of aluminum into waterways, itself extremely toxic to aquatic organisms. In terrestrial environments acid ran causes damage to man-made structures, chemically degrading concrete and stone buildings and statutes. For example, many marble monuments in Washington, DC, including the Lincoln Memorial and the statue of George Washington, have incurred significant surface damage due to acid rain precipitation (**Fig. 9.16**). These examples are part of the wide-ranging anthropogenic alterations of our environment and a major challenge to conservation and restoration efforts. The removal of fossil-fuel burning stands as the major resolution to human-caused sources of acid rain.

9.4 Carbon Dating

Atmospheric carbon is in the form of carbon dioxide and methane. Carbon-14 (^{14}C) is a naturally occurring unstable radioisotope that is created in the atmosphere from

Fig. 9.16

Effects of acid rain on statues on marble and limestone.

Source: Nino Barbieri under CC BY 2.5

atmospheric nitrogen, ^{14}N (by the addition of a neutron and the loss of a proton) because of cosmic rays. This is a normal, continuous reaction, and the isotope ^{14}C is always being created. But it is in a very small proportion compared to stable carbon (^{12}C). There are approximately one ^{14}C to one trillion ^{12}C atoms in the atmosphere. As living organisms incorporate ^{14}C daily, by carbon fixation from photosynthesis into the food web, the relative amount of ^{14}C in a living organism is approximately equal to the concentration of ^{14}C in the atmosphere. When an organism dies it stops ingesting carbon. Since the ^{12}C atoms in the body tissue are stable it will not degrade. But the radioisotope ^{14}C begins to decay in a defined fashion (according to isotope properties, half-life decay rates). Now the ratio changes between ^{14}C and ^{12}C and will continue to do so over long periods of time as less and less ^{14}C exists. For the radioisotope ^{14}C its half-life is 5730 years, which means half of the incorporated concentration of ^{14}C will have been converted back to ^{14}N by year 5730, whereas none of the ^{12}C will have decayed in that same time period. Because the half-life of the radioisotope ^{14}C is long it can be used to date bones or wood from formerly living objects.

By comparing the ratio of the ^{14}C remaining in an object to the amount of ^{14}C detected in the atmosphere it is possible to give a reliable age since an organism stopped living (**Fig. 9.17**). Based on carbon dating the age of the pygmy mammoth was calculated with accuracy up to 50,000 years and was corroborated by other archeological and geologic methods. Several elements have isotopes with distinct half-lives. For example, ^{40}K (potassium-40) has a half-life of 1.25 billion years, ^{235}U (Uranium-235) has a half-life of approximately 700 million years, and they are used to determine geologic events. Using radiometric dating scientists can study the age of fossils or other remains of extinct organisms to understand how organisms have evolved from earlier species.

Fig. 9.17

Carbon dating a pygmy mammoth, which is less than approximately 50,000 years old.

Source: Bill Faulkner, OpenText, BC, CA under CC BY 4.0

Glossary

Matter
Elements
Closed system
Catabolism
Watershed
Sessile
Anthropogenic

Citations

1. R. Jacoby, The role of soil microorganisms in plant mineral nutrition—current knowledge and future directions, Front Plant Sci 19 (8) (2017) 1617. https://doi.org/10.3389/fpls.2017.01617
2. H.M. Griffiths, The impact of invertebrate decomposers on plants and soil, New Phytologist 231 (6) (2021) 2142–2149.
3. D. Batzer, D. Boix, An introduction to freshwater wetlands and their invertebrates, in: D. Batzer, D. Boix, (Eds.), Invertebrates in Freshwater Wetlands, Springer, Cham, 2016. https://doi.org/10.1007/978-3-319-24978-0_1
4. M. Hayer, et al., Microbes on decomposing litter in streams: entering on the leaf or colonizing in the water, ISME J 16 (2022) 717–725. https://doi.org/10.1038/s41396-021-01114-6
5. Y. Fukasawa, K. Matsukura, Decay stages of wood and associated fungal communities characterise diversity–decomposition relationships, Sci Rep 11 (2021) 8972. https://doi.org/10.1038/s41598-021-88580-2

6. Shizing Zhou, Decomposition of leaf litter mixtures across biomes: the role of litter identity, diversity and soil fauna, J Ecol. Special Feature: Citizen Science 108 (6) (2020) 2283–2297. https://doi.org/10.1111/1365-2745.13452
7. C.T. Kraamwinkel, et al., Planetary limits to soil degradation, Commun Earth Environ 2 (2021) 249. https://doi.org/10.1038/s43247-021-00323-3
8. http://paleoportal.org/index.php?globalnav=time_space&sectionnav=period&period_id=12 Accessed 3.2.24
9. https://ralucapalade.wordpress.com/the-carboniferous-period-2/ Accessed 3.2.24 https://www.nationalgeographic.com/science/article/carboniferous
10. D. Bačėninaitė, K. Džermeikaitė, R. Antanaitis, Global warming and dairy cattle: how to control and reduce methane emission, Animals (Basel) 12 (19) (2022) 2687. https://doi.org/10.3390/ani12192687.
11. A. Bernhard, The nitrogen cycle: processes, players, and human impact, Nature Education Knowledge 3 (10) (2010) 25.
12. https://www.unep.org/facts-about-nitrogen-pollution Accessed 2.25.24.
13. L.Q. Shen, et al., Estimating nitrogen and phosphorus concentrations in streams and rivers, within a machine learning framework, Sci Data 7 (2020) 161. https://doi.org/10.1038/s41597-020-0478-7.
14. Xinning Zhang, Global nitrogen cycle: critical enzymes, organisms, and processes for nitrogen budgets and dynamics, Chem Rev 120 (12) (2020) 5308–5351. https://doi.org/10.1021/acs.chemrev.9b00613.
15. Natalie Mahowald, Global distribution of atmospheric phosphorus sources, concentrations and deposition rates, and anthropogenic impacts, Global Biogeochemical Cycles 22 (4) (2008). https://doi.org/10.1029/2008GB003240.
16. M.F. Chislock, E. Doster, R.A. Zitomer, A.E. Wilson, Eutrophication: causes, consequences, and controls in aquatic ecosystems, Nature Education Knowledge 4 (4) (2013) 10.
17. J. Carey, News feature: the complex case of Chesapeake Bay restoration, Proc Natl Acad Sci 118 (25) (2021). https://doi.org/10.1073/pnas.2108734118 e2108734118.
18. University of Maryland Center for Environmental Science, Chesapeake Bay Watershed report card. https://ecoreportcard.org/report-cards/chesapeake-bay/. Accessed 2 March 2024.
19. R.L. Mann, E.N. Powell, Why oyster restoration goals in the Chesapeake Bay are not and probably cannot be achieved, J Shellfish Res 26 (2007) 905–917.
20. Katz, Harvey M. (2009). "Chesapeake Bay Executive Order | Targeting Resources to Better Protect the Chesapeake Bay is Aim of U.S. Department of Agriculture's 202(b) Report." https://web.archive.org/web/20140305050928/http:/executiveorder.chesapeakebay.net/post/Targeting-Resources-to-Better-Protect-the-Chesapeake-Bay-is-Aim-of-US-Department-of-Agriculture.aspx. Accessed 2.2.2024.
21. Sarah Gibbens, Massive 8,000-mile "dead zone" could be one of the gulf's largest, Nat Geo Environ (2019). https://www.nationalgeographic.com/environment/article/massive-dead-zone-predicted-to-be-one-of-largest-gulf-of-mexico.
22. H. Bao, S. Yu, D. Tong, Massive volcanic SO2 oxidation and sulphate aerosol deposition in Cenozoic North America, Nature 465 (2010) 909–912. https://doi.org/10.1038/nature09100.
23. A. Jasińska, et al., Oceanological and hydrobiological studies. Intl J Oceanography. 41 (2) (2012). https://doi.org/10.2478/s13545-012-0019-x.
24. S. Beaulieu, An authoritative global database for active submarine hydrothermal vent fields, Geochem Geophys Geosyst 14 (11) (2013) 4892–4905. https://doi.org/10.1002/2013GC004998.

CHAPTER

Population Dynamics 10

Chapter Outline

10.1 Properties of Populations

By the end of this section students should be able to:

- Classify organisms as unitary or modular.
- Describe how distribution and density of populations are measured using sampling methods.
- Examine age pyramids to assess the health and future growth of a population.
- Explain how population structure changes temporally and spatially.

Species Abundance and Distribution

A **population** is a group of individual members of the same species that are concurrently inhabiting a given area. A population can be described by its overall structure, whether densely populated or dispersed, and the age distribution of its members. Each of these parameters is dynamic, meaning they can be expected to change with time. Sometimes the size of a population is dispersed due to natural disasters (**Fig. 10.1**), Before counting population size it is necessary to define what constitutes an individual within that population. Individual organisms may be unitary or modular

Ecology for the 21st Century. DOI: https://doi.org/10.1016/B978-0-443-24031-7.00010-6

and sometimes that is not a straightforward answer. **Unitary** organisms exist as discrete individuals. Most animals are categorized as unitary, which is the easiest way to identify and count all the individuals within a population. The origin of each new individual occurs after fertilization events and completion of a series of predictable embryonic stages. Once the zygote has grown into a genetically unique, separate organism it will be counted as an individual.

Fig. 10.1

Mount Etna, Sicily eruption plume on Christmas Eve, 2018, spewing ash across the island impacted all living species in its path. Earth views were taken by the Expedition Five Crew aboard the International Space Station ISS. NASA Identifier: iss005e19024 Public Domain.

Source: BenAveling under CC BY-SA 4.0

Modular organisms are found in each of the kingdoms (plant, fungi, bacteria, etc.) growing in the same location after either asexual or vegetative reproduction from a single ancestor. In the modular organism an individual zygote develops into module units of construction that increase in size by producing *multiple* modular units. Most plants are modular. The plant might have suckers, or new stems, that sprout from surface roots to form new individuals, yet the whole plant is considered one *genetic* individual.

New functional modules are known as **ramets** whereas the parental is known as the **genet**. The genet was produced from a zygote, the ramet was produced by asexual reproduction (**Fig. 10.3**). By producing ramets the genet can cover a large spatial area to extend its life. A **clonal colony** (preferred name for genet in animal or microbial species) either remains physically connected or separates from its parental body. An interesting paradox is that modular organisms can be uniquely different in form, whereas a unitary organism follows a determinant, uniform plan. That is, humans will always have two legs, cats will have four legs and whiskers, etc. Modular organisms are less predictable, with plasticity in growth habits and timing of reproductive maturity based on their interactions and environmental conditions.

Plants are the easiest to recognize as modular, originating through branching, repeating structural units that remain attached to the body. The above ground vegetative leaves, with their axillary bud and **internodal** stem, produce more leaves, more axillary buds, etc. Roots also reflect modular growth characteristics. Modules may spread vertically as stems or laterally as a stolon, which is a specialized stem that lies just above the ground surface. Rhizomes are specialized stems that grow and reside just below the ground surface. Lateral rhizome structures are versatile and capable of spreading to establish a new root system. Other examples of modular organisms include corals, sponges, bryozoans, many protists, and fungi.

Fungi are an example of a modular system through lateral underground growth, though they appear to be unitary. This is because the distinct mushrooms growing aboveground appear to be individual unitary organisms. However, underground there can be miles of lateral growth, connected as one modular fungal system, with scattered visible mushroom caps aboveground. In fact the largest known organism on Earth is not the whale as one might first guess but an extensive fungus (*Armillaria ostoyae*) located in Oregon's Blue Mountains (**Fig. 10.2**).

The Oregonian humongous funguses of the genus *Armillaria* encompass nearly 4 mi^2 (1665 football fields)! This massive structure is a *single specimen*. It is believed to be 2400 years old, based on DNA testing and by calculation of its annual spread. *A. ostoyae* lifestyle is parasitic growth classified as pathogenic. Conifer trees invaded by these fungi can develop a disease called **Armillaria** (white rot root disease). However, most species of fungi are neither massive in size nor detrimental to tree populations. There are approximately 50,000 mycorrhizal fungal families that grow through spreading hyphal threads underground, with occasional sprouting mushroom revealing their presence. These underground ecosystems are essential for soil fertility and are another great source of carbon storage.

The *Armillaria* species has a unique adaptation to extend its geographic range to an even wider range than most fungi: thick cords of hyphal threads called

Fig. 10.2

The honey mushroom of the "humongous fungus," *Armillaria ostoyae*.
Source: USDA Forest Service, Pacific NW Research Station, Public Domain

rhizomorphs. *Armillaria* is classified as a **facultative saprophyte**, meaning it acquires nutrients by absorbing dead plant material. For their own survival parasites often moderate resource extraction to keep their host, or feeding supply, alive. But the parasitic *Armillaria* is not adversely affected by its host's death. After feeding on the tree root's dead mass *Armillaria* is capable of using its extensive rhizomorph structure to reach a new host. Aboveground the *Armillaria* sprouts the honey mushroom. Honey mushrooms are considered a delicacy, an acquired taste in their native habitat situated in the cool northern hemisphere of British Colombia, Russia, Poland, and Ukraine. As a wild mushroom more prized than chanterelles and morels, the honey mushroom is mildly poisonous if not properly cooked.

Fungi are modular, with a highly branching, varied form influenced heavily by their environment. Each mushroom is a fundamental unit, or ramet.

So the question arises, should ramets be counted as separate individuals in a population study? The ecologist will need to carefully consider both the clonal colony (genet) and its modular units (ramet) according to the scope and purpose of the population study. For practical purposes ramets are mostly counted as individual members because in the natural world they function physiologically independently, even though they are genetically identical. But in some cases it would be impossible

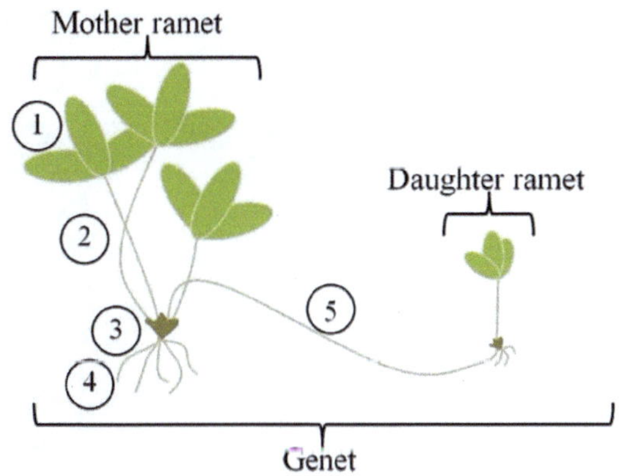

Fig. 10.3

Genet vs Ramet.

Source: Waters, EM and Watson, MA under CC BY SA 4.0

to separate ramets for counting purposes. Therefore, for some modular organisms, the only way to accurately assess population size would be to measure the species' total biomass.

Geographic Range

The distribution of a population refers to its spatial location, or area that it occupies. To determine the area where all individuals are present, or absent, in a region under study a map is constructed covering the site of each with a line drawn around all positions where individuals were found. The map is sometimes divided by grid lines (common in human population densities) to spatially organize the data. The **geographic range** of a species considers all its individual members, including those members that migrate or hibernate. The primary factor that influences the distribution of a population is **habitat suitability**. Geographic range therefore includes individuals found in high concentrations in more suitable habitats and regions that have fewer members present. Low density reflects **species tolerance** levels under varying conditions within the overall range (temperature, soil moisture, pH, elevation, etc.). Some species, for example, the plant genus *Aloe*, have widespread geographic ranges and can be found in every country. The leopard (*Panthera pardus*) also covers a broad range, in excess of 20 mil square kilometers (7.7 mil square miles) throughout much of Africa, the Middle East, and Asia. Other species have very narrow ranges, for example, the rare wild cat, Iriomote (*Prionailurus iriomotensis*), which resides only on Japan's Iriomote Island, with a range of approximately 292 square kilometers (113 square miles) (**Fig. 10.4**).

A good example of a restricted distribution pattern is found in the Allegheny Mountains, where the shale-barren evening primrose (*Oenothera argillicola*) only

Fig. 10.4

Iriomote cat (*Prionailurus iriomotensis*).

Modification of Cougar Hill Nat'l Wildlife Humane Society (NWHS)

grows in open woodlands of the south- or southwest-facing slopes of the hot, shale-barren Allegheny terrain. The direct sunlight's surface temperatures are lethal to many plant seedlings in the forest, which minimizes competition for the shale-baren primrose in the limited soils on the shale surface.

Species with broad ranges covering most of Earth are considered to have a **cosmopolitan** or **ubiquitous distribution**. For example, the blue whale (*Balaenoptera musculus*) inhabits every ocean on the planet. The aloe plant and the leopard are cosmopolitan species. Endemic species have a geographically restricted distribution in a localized habitat and may have acquired specialized habitat requirements. The shale-barren evening primrose is an example of an endemic distribution pattern. Quite frequently species listed as endangered or threatened have endemic distribution patterns and specialized habitat restrictions. This includes the northern spotted owl (*Strix occidentalis caurina*) of the old growth forest of the American northwest and the panda bear (*Ailuropoda melanoleuca*) of the temperate high forests of southwest China.

The distribution pattern of species is determined by ecological and evolutionary processes. The inability of species to expand their geographic range is due to constraints on the species that prevent them from overcoming limits on their distribution. In some cases the factors are simply physical barriers such as a large body of water is inhibitive for most terrestrial species. The common geographic barriers that are likely to reduce or prevent individuals from moving and colonizing new areas are:

- Bodies of water, including rivers
- Mountains
- Large areas of unsuitable habitat such as deserts
- Interactions with other species, including competition and predation

Individuals are generally not evenly distributed across their geographic range. They will be found clustered in the most suitable habitats within their geographic range with the highest species tolerance. An optimal habitat is dependent upon having all environmental factors conform to the range of tolerance of the species. In actuality the environment will have heterogeneous populations divided into subpopulations appearing in the suitable habitat patches surrounded by unsuitable habitats with fewer or no members present. The **metapopulation** consists of a collection of the local populations connected to one another by the movement of individuals among them (**Fig. 10.5**). Movement between local populations also retains the gene flow. It is often easier and more economically feasible for ecologists to study

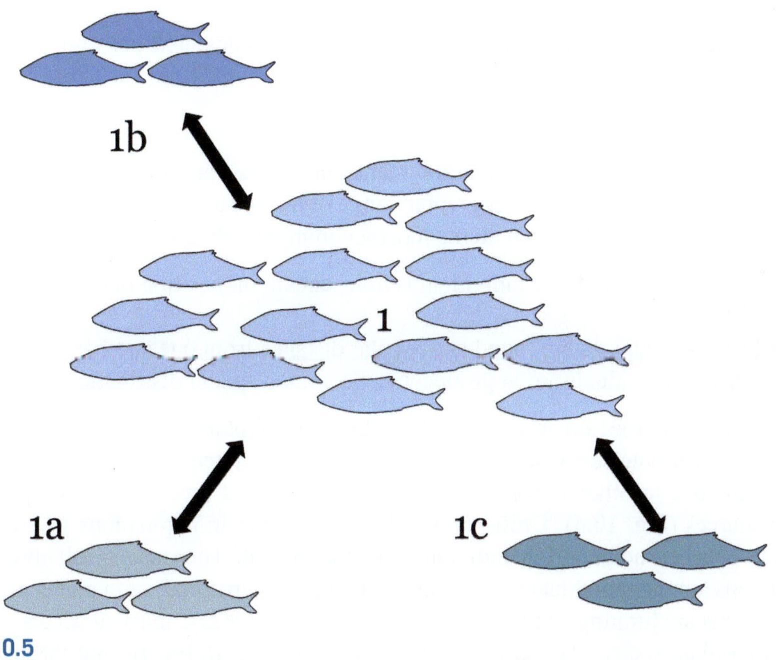

Fig. 10.5

Metapopulation, a "population of populations," is useful when studying fisheries. For example, local population (1) is the source for interbreeding with surrounding subspecies populations (1a, 1b, and 1c). Independent populations overlap only during breeding times.
Source: Image modification of Harrison and Taylor (1997) and Stith et al. (1996)

subpopulations rather than collect data across the entire geographic range and the metapopulation.

Ecotypes of a species are distinct varieties that exhibit behavioral, structural, or physiological differences from other members within the same species. Ecotypes result from modes of natural selection, as distribution of a trait in a population may change during the adaptation process. Wherever populations are explored the inclusion of existing ecotypes (which ones, separate counts for each ecotype, etc.) would need to be resolved and quantified.

The commonly acting modes of selection on ecotype are:

- Stabilizing—selected for optimized trait
- Directiona—shifts to right or left of original trait range
- Disruptive—selected away from original optimum toward extremes (bimodal evolution)

Abundance defines the size of a population, or number of its individuals. Abundance is a function of population density and the area over which the population is distributed (density x area). The **crude density** value is the number of individuals per unit area, using the formula:

- # individuals/m^2 or km^2

Crude density of aquatic individuals is sometimes calculated per unit volume, using the formula:

- # individuals/m^3 or liter (0.001 m^3)

Given the variability in habitat requirements, individuals are often not evenly distributed across the grids in the geographic range of the population. Density can vary widely, with three distribution patterns mostly commonly observed in organisms.

- Random—position of one individual is independent on position of other individuals
- Uniform—organisms are found at a regular distance from one another
- Clumped—individuals are in groups, most common spatial distribution

Random distribution is seen in the scattering of plant seeds through wind. Pollinators distribute seeds in patterns that are not considered random but randomness results due to other factors, for example, presence of predators or competing pollen sources (**Fig. 10.6**). **Uniform distribution** is seen in populations where the distance between neighboring individuals is maximized. This may result through interactions among individual species members that are territorial, or in plants, when soil nutrients are limiting so that spreading out allows favorable nutrient access. The California plant species *Salvia leucophylla* grows with uniform spacing through a strategy whereby its flower releases chemicals that inhibit the growth of any other plants nearby (1,8-cineole and camphor oils). These monoterpene compounds inhibit seed germination in proximity to the salvia plant and are an example of a type of adaptation called **allelopathy**. **Clumped distribution** occurs where suitable habitat

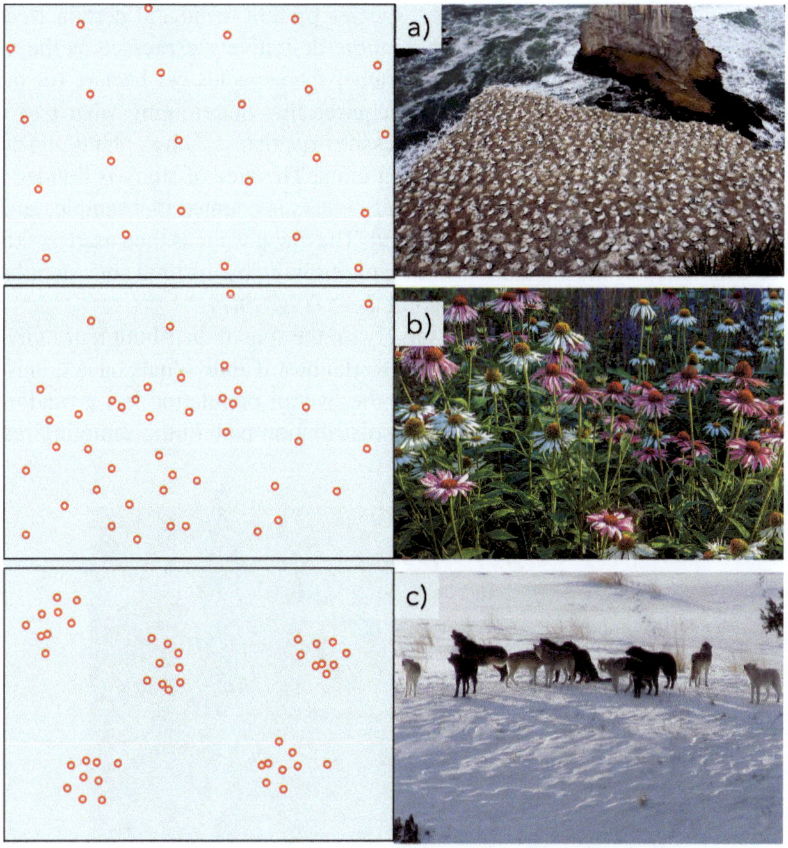

Fig. 10.6

Sampling methods used to determine population density are dependent upon distribution patterns. (a) Uniform distribution in a population of gannets on a rocky shoreline. (b) Random distribution in wildflowers. (c) Wolves cover large ranges in packs.

Source: Modification of Yerpo under CC BY SA 4.0

or resources limit species' presence into patches, or when species form social groups (flocks or birds, plant ramets, or the wolf pack).

The specific distribution pattern—random, clumped, or uniform—represented by the population of interest would need to be considered before determining the most accurate sampling method.

Sampling Methods

The abundance of a species reflects both population density and distribution. To account for the patchy distribution of some species ecologists sometimes use **ecological density**, a measure of the number of individuals per unit of *available* living space rather than the total range. If the species prefers to inhabit certain trees or shrubs in its geographic region, density is more definitively expressed as the number of species per mile of tree (or shrub) rather than species per hectare (or mile). Measuring ecological density accurately requires first determining what part of a habitat is suitable for a particular species. Sessile organisms, such as plants and coral, can be sampled using quadrats, or sampling units. The area of study is divided into subunits of a defined size, the number of individuals is counted in a sample, and the mean density of all the samples is determined. The mean value is then used to extrapolate an estimation of the population size by multiplying by the total area: population mean = (quadrat mean X total area/quadrat area) (**Fig. 10.7**).

The accuracy of sampling depends largely on the spatial distribution of individuals in the population. A sampling method works best if individuals have a uniform distribution and would be less accurate if the overall population has a random or clumped distribution. Even with a uniform distribution pattern the sampling results

Fig. 10.7

Quadrat sampling method to estimate population size of sessile species. Quadrats can be placed randomly or systematically (shown above) on the area to be sampled.

Source: Quadrant Sampling under CC 0

can be influenced or biased for a few reasons, such as choice of boundaries or favoring the collection samples only with easy access, inducing a significant environmental bias.

A commonly used technique to measure mobile populations is the **mark-recapture** method. Essentially this method is based on capturing several individuals in a population, tagging them with a mark that will not be lost during the course of the study, and releasing a known number of marked individuals (M) back into the population, N *(unknown)*. After an appropriate period of time for M to reintegrate back into the population, a second capture is collected from the population (n). Those that were marked in the first sampling are designated now as recaptured (R). Assuming the ratio of marked to total sampled individuals in the second sampling (n/R) represents the entire population (N/M) it is possible to calculate an estimate of the population, as follows:

$$\frac{N}{M} = \frac{n}{R} \text{ To solve for N(population), rearrange the equation: } N = \frac{nM}{R}$$

The assumptions applied to this index are that: (1) the samplings are random, with all individuals having an equal probability of being captured and (2) marked individuals will distribute themselves randomly back into the population. Another consideration is the impact of markings on future survival of an individual released back to the population. Visible marks may lead predators to an individual, or the capture process may otherwise alter its fitness. The mark-recapture method assumes the ratio of marked to unmarked individuals remains the same across sampling periods. If the recapture is done without bias based on redistribution or fitness wildlife biologists favor this method for small mammals, animals with shells (snails, limpets), or invertebrates with exoskeletons.

Population numbers are a vital measure for the success or failure of conservation efforts. Methods used to estimate species' abundance invariably include indirect observations that indicate the presence of organisms rather than actual direct counts.

Types of observations and signs used to determine presence of target species include:

- Counts of vocalization (i.e., bird song)
- Counts of animal scat (feces) observed
- Counts of animal tracks

If there is a relatively constant relationship between observations and total population size an index of abundance can be compiled. For example, the observation data can be converted into number of individuals heard per hour (or seen per kilometer), which represents indices of abundance for the target species. While abundance estimates alone do not function as actual density, if observation periodicity is compiled over several years it will provide useful trends on target species abundance.

This approach is used to estimate lion density in sub-Sahara Africa (**Fig. 10.8**). Historically lions had a geographic range throughout Africa and Southeast Asia to India. Today lions exist in much fewer numbers and are mostly

Fig. 10.8

Six-year-old male lion in Phinda Private Game Reserve, South Africa.

Source: Charles J. Sharp under CC BY SA 4.0

restricted to sub-Saharan Africa or dedicated conservation areas in India. Lion population numbers are difficult to estimate, leading to debate over their true population status throughout Africa. Surveying their habitats and population sizes is difficult to assess because they are mobile over long ranges.

The assertive male lion will claim huge territories from 30 km^2 to 400 km^2 shared with several females. Conservation biologists have traditionally used track surveys to count lion populations by recording their footprints and using call-up surveys broadcasting sounds to attract lions for counting. Some conservation experts argue there are 20,000 lions left on the African continent while others claim the estimate is more likely 30,000. In 2016 researchers at Oxford University adapted a method to document tigers and cheetahs to obtain an accurate lion count in Africa. The goal was to quantify lions in the Kenyan conservancies surrounding the Maasai Mara National Reserve. The case study was performed at Oxford's Wildlife Conservation Research Unit (WildCRU), which estimated 420 lions existed in total in the Maasai Mara. This protected region represents one of the highest remaining lion densities recorded anywhere on the African continent. By using a "spatially explicit" approach researchers carefully surveyed densities over 90-day periods by high-resolution image captures of individual lions. By capturing high-tech pixel-specific lion densities researchers were able to log individual lion's unique whisker spots. A total of 203 lions were identified within a 2400-square-kilometer range (**Fig. 10.9**). Extrapolating their data across the entire geographic range, lion concentration in the Maasai Mara was determined to be approximately 16.85 lions (over the age of 1 year) per 100 km^2. This new data reflected a surprisingly productive ecosystem for the lion. Today the total population size of the "king of the jungle" in Maasai Mara National Park and surrounding conservatories is estimated to be 850 to 900 lions.

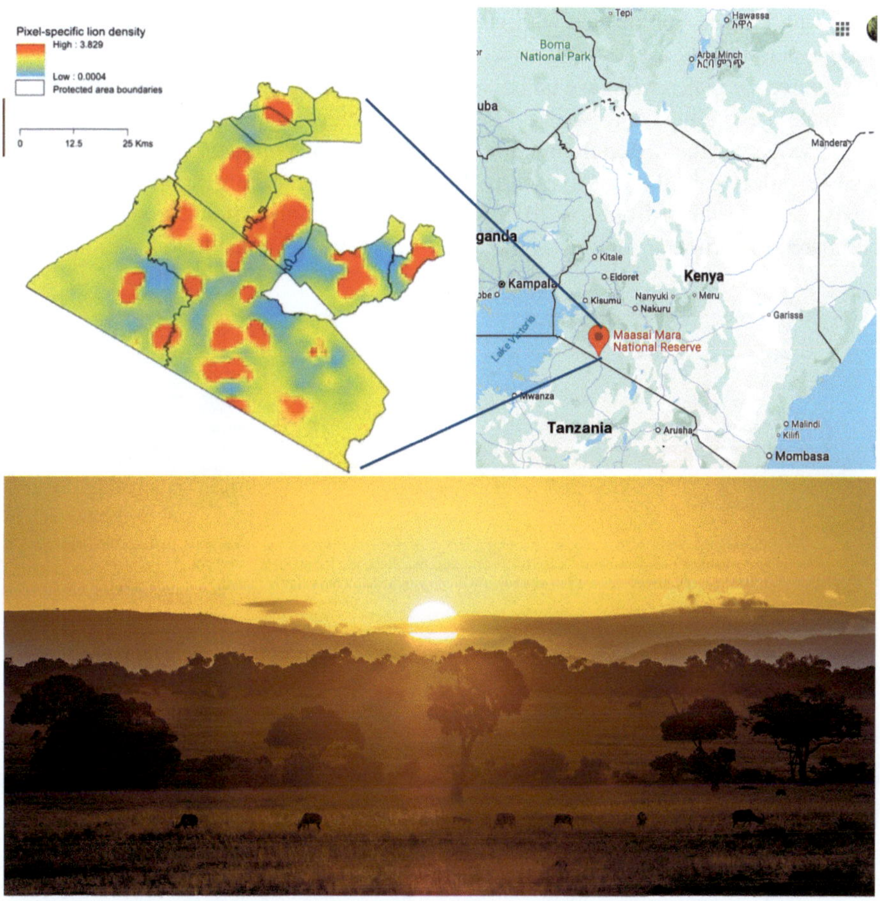

Fig. 10.9

Maasai Mara National Reserve in southwest Kenya (red dot).
Source: Modification of Elliot and Gopalaswarmy (2016)

Structural Characteristics of Populations

All species populations have an **age structure**, with a proportion of individuals in several different age classes. Ecologically age structure is classified into three groups: prereproductive, reproductive, and postreproductive. Several techniques are used by ecologists to determine age, including the wear and tear of teeth or examining carcasses to determine age at death, growth rings in horns, scales, or plumage changes in birds. In trees age will be determined by counting annual growth rings or trunk diameter.

Age structures are often depicted in **age pyramids**, which provide a graphic illustration of the different age groups within the population (**Fig. 10.10**). This information is useful to predict the health of a population and its potential for growth, particularly if the population's age of first reproduction and its reproductive strategy is already known. For species with variable generations the length of the prereproductive period will have a strong impact on the population's growth rate. And the age and number of males will have an impact on the breeding dynamics (more breeding-age males generally lead to earlier, less-disruptive breeding periods). The movement of individuals within a population directly influences its population density and will also be reflected in the age pyramids. **Fig. 10.10** depicts the general age pyramid types and data taken from the 2015 US Census to graph national population size compared to particular regions of the country. Would you consider the national age

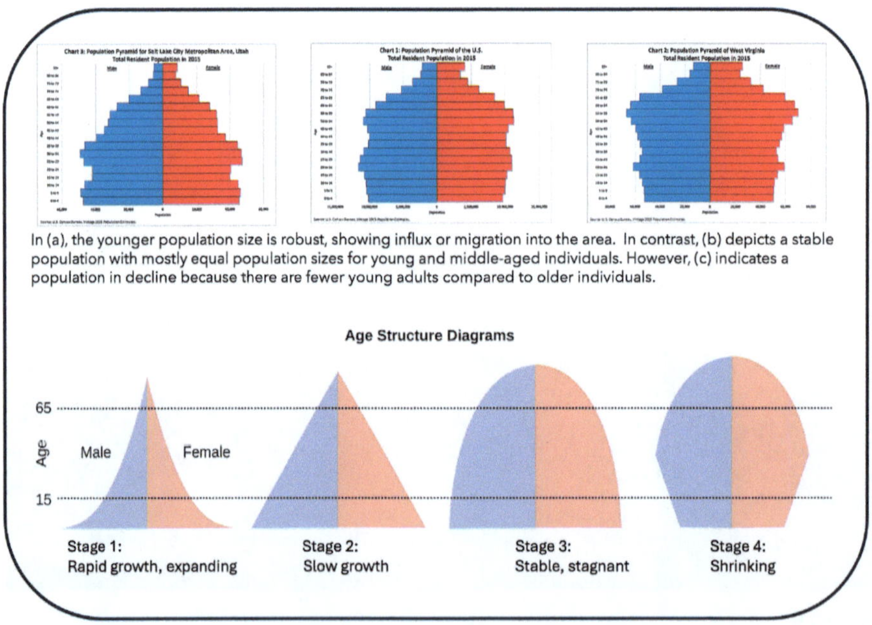

Fig. 10.10

Age structure diagrams in humans.

Source: Modification of Census.gov, Public Domain

pyramid indicative of rapid or slow growth? Is it possible to project whether there will be significantly more females or males from any of the diagrams?

Age pyramids show population as a bar chart for each age group. Males and females are usually shown separately on either side of the meridian. In 1958 an early pioneer in the field of ecology, Dr. F.S. Bodemheir, suggested the use of three major age groups—prereproductive, reproductive, and postreproductive—to represent "young," "mature," and "old" ages, respectively. The length of time in each category varies depending upon the species, for example, some species spend more years in prereproductive years. The shape of the pyramid is informative about the population and life histories of the species. **Life histories** are unique to each species, including its pattern of survival and reproductive strategies. We will go into more depth on life histories in the next chapter.

A pyramid with a wide base (young population) indicates the population is growing fast. If the pyramid base is narrow, resembling an urn or diamond shape, the population size is in decline (declining birth rates). If a wide base pyramid declines rapidly to a narrow tip the population has experienced high infant mortality and short life expectancies (few survivors in older ages). A rectangular shape with broader tip indicates a stable population with even distribution and longer life histories (has more survivors in older ages). By inspecting specific age bars detailed information can be ascertained to explain changes occurring within a specific population.

For example, sex ratios in a population may shift with age. Populations of sexually reproducing organisms usually have a sex ratio approximating 1:1 (M:F). Since the male parent has a 50:50 chance of contributing an "X" or "Y" chromosome at conception the sex of the child will have a probability of 1:1 (50% each sex in total population). This is known as the **primary sex ratio**. In most mammalian populations the sex ratio *at birth*, or **secondary sex ratio**, will be slightly weighted toward more males. The secondary sex ratio in humans is approximately 1.07:1 (M:F). The sex ratio in older age classes will again change, normally shifting toward females since males in many species experience a shorter life span than females. This is a predictable outcome given that many species are characterized by rivalries among males for resources or mates. An exception is found in bird populations, where the number of older age males tends to be higher than females. This contradiction is explained by the fact that in many bird species females are susceptible to attack and predation during nesting periods.

Population Migration Patterns

Dispersal is the movement of individuals in space (location) and generally implies the movement of individuals away from one another.

- **Emigration**—when individuals leave a subpopulation, leading to a decline in population size
- **Immigration**—when individuals enter a subpopulation, increasing the population size

Movement of individuals between subpopulations is an important factor to maintain gene flow. The means of dispersal can be wind, water, gravity, or hitching a ride on an animal. Seeds generally fall nearby the parent plant but lighter seeds can be carried longer distances by wind. Animals are important in dispersal of plant seeds whenever they eat fruits or berries and indirectly the seeds that they contain. Seeds pass through the gut and are deposited by defecation back into the soil where they have a chance to germinate. Some seeds have adaptations for spines or hooks that attach to animal fur or bird feathers and are carried off passively until rubbed or otherwise break free to new soils. Individual animals may themselves disperse passively, for example, wind carrying spider juveniles or water current carrying the larvae of invertebrates downstream.

Active dispersal in mobile animal populations occurs according to the animal's age, class, or sex. For example, bird dispersal occurs in the younger age groups. There are many factors that may affect dispersal patterns and intensity, including crowding, food availability/quality, and temperature change.

Population distribution and density will experience changes both temporally and spatially. Species geographic range can be highly plastic, contracting or expanding to corresponding shifts of their ecological niches.

Migration is the round-trip movement of organisms, occurring daily or seasonally, and even in age-specific patterns. The desert locust travels up to thousands of miles annually. For example, in 1988 a swarm of desert locust traveled from West Africa to the Caribbean (3100 miles) over a 10-day period. Bats are the only mammal capable of sustained flight. They depart from their hibernaculum each dusk in search of feeding areas and return by dawn. Earthworms move deep into the soil during the winter to avoid freezing and travel back to the surface in the spring. Gray whales feed in the Arctic during the summer and winter off the Californian coast, where they birth their calves (NOAA Fisheries). Young Pacific salmon, called smolts, depart from the freshwater rivers of their birth when just a few inches long and begin their travel to the open seas of the Pacific Northwest to complete their young adult life (**Fig. 10.11**). Salmon are dependent on their biological clocks to return to their rivers of origin to reproduce. As juvenile Pacific salmon pass from Puget Sound through the man-made locks of the Lake Washington Watershed they are confronted with a mix of salinity, temperature, and changes in dissolved oxygen concentrations. Salmon species represent an example of a **single life-span migration**. This is because the adult salmon only returns to its origin location to spawn (reproduce) at the end of their life. For indigenous people throughout the Pacific Rim where salmon is their primary protein source the salmon's life cycle is an important part of their culture. Native groups perform rituals to celebrate the first return of salmon, which are harvested and preserved to provide their communities with a storage food for the long winter. This is an example of sustainable fishing, as the salmon's reproductive stage has been successfully completed before harvesting the adults.

Dispersal patterns can alter spatial distribution of individuals within a population. Emigration leads to a decline in density. Immigration leads to an increase in density or may establish a new subpopulation within a previously unoccupied habitat.

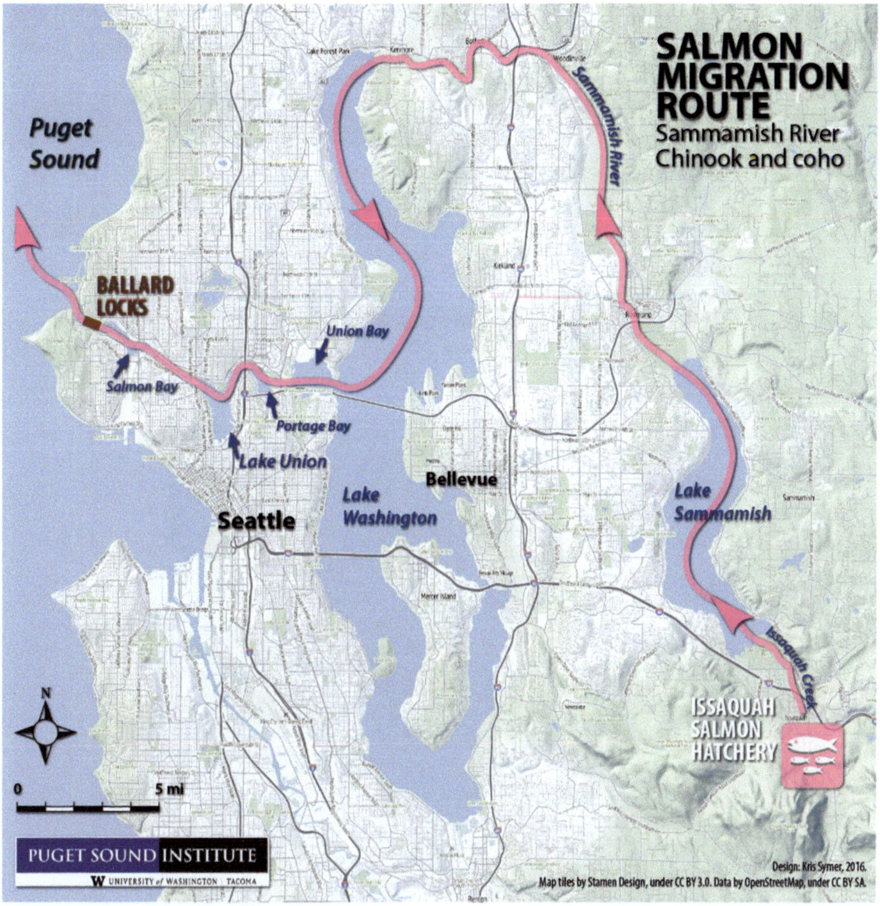

Fig. 10.11

Salmon migration route of Pacific Chinook, Coho, and Sockeye salmon between Lake Washington watershed and open waters of Puget Sound.

Source: Chris Symer/PSI Stamen Design, under CC by SA

Species introduced into an area where they did not previously live have the effect of expanding their geographical range. These introductions may be intentional or random and unintentional (**Figs. 10.12a, b**).

Shifts in climate or local environmental conditions may lead to an increase in species' dispersal. The tiny 4.5-inch bog turtle is endemic to the eastern United States. As a semiaquatic species the bog turtle only thrives in nature in temperate habitats where it has access to cooler, shallow, and slow-moving waters, or boggy meadows. The historic bog turtle sites have experienced habitat degradation or loss, with only 25% turtle habitat sites in New York still extant. Loss of habitat has forced the

Fig. 10.12a

Bog turtle natural habitat in wet meadows and slow streams.

Source: NYS Dept. Environmental Conservation, Public Domain

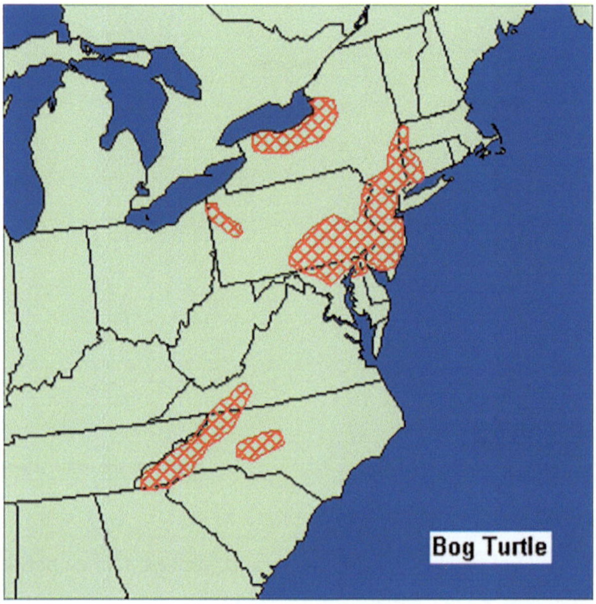

Fig. 10.12b

Bog turtle distribution in the United States.

Source: NYS Dept. Environmental Conservation, Public Domain

bog turtle to migrate to find suitable areas where a new niche may have developed. Successful locations for migration of bog turtles sometimes occur as the result of fire, beavers, agriculture, or other disturbances that tend to create open, wet meadow-type habitats that are suitable for the bog turtle.

For the most part sprawling urban development inhibits species' successful immigration. When climate change forces a species to migrate just to remain in their environmental niche or make adaptations to avoid extinction, modified landscapes add an additional layer of complexity. Anthropogenic-related habitat loss and fragmentation reduce successful dispersal. For the bog turtle populations become more isolated in the search of new suitable habitats.

Temporal environmental conditions affect flora and can shift the spatial distribution of a population. For example, the climate in North America has changed significantly since the last glacial maximum (LGM, approximately 20,000 years ago). The populations of many species of trees have shifted further north with the glacier retreat as the North American continent warms. Plant distributions have been mapped and quantified to access the effects of climate and establish a baseline to accurately predict future flora dispersal. A predictive model, based on distribution of species and its environmental requirements, is the Climate Envelope Models (CEM). This approach was applied to tree species in the North American biogeographic range to predict tree species future distribution along climatic gradients. Studies that assess altered temperature and precipitation include data collected over a long period of time (**Fig. 10.13**), for example:

- Annual mean temperature
- Minimum temperature of the coldest month
- Maximum temperature of the warmest month
- Annual precipitation
- Precipitation in the warmest quarter
- Precipitation in the coldest quarter

Fig. 10.13

Birnam Wood comes to Dunsinane in the form of branches carried by the soldiers closing in on Macbeth. In North America many tree species are "moving" northward toward the tundra, analogous to Birnam Wood.

Source: After McKenney et al 2011, Natural Resources, Canada

Ongoing field studies corroborate the predictive models. Researchers in Alaska have tracked the boreal forest tree line moving closer to the artic tundra. Trees relocating into higher latitudes and altitudes have been compared to the Shakespearean play, "Macbeth." In the Birnam Woods Scottish soldiers hid behind tree boughs as they slowly advanced upon Dunsinane. The effect of their subterfuge was an image of trees literally moving and closing in on Macbeth. With today's climate change threat, North American pine and deciduous trees are experiencing successful seed germination toward more favorable temperatures and precipitation (northerly and westerly), while the now warmer, drier climates they once densely populated have become less favorable to their seed dispersals.

10.2 Population Growth

By the end of this section students should be able to:

- Apply birth and death rates to calculate a population's growth.
- Explain how carrying capacity impacts population growth.
- Use Life Tables to determine age-specific mortality and survival.
- Discuss relationships between net reproductive rate, birth rates, and survivorship.
- Outline one or more factors contributing to population extinction.

Populations grow through births and immigration. Populations decline through deaths and emigration. If the population experiences immigration and emigration it is considered an **open population**. In a **closed population** the levels of immigration and emigration are nonexistent or occur at very low levels and do not influence overall population growth. In the study of population dynamics population growth is explored through demographic processes related to birth and death, using geometric and logistic population growth formulas.

Geometric Population Modeling in Closed Populations

Population dynamics are simplest in a closed population, or one in which the rates of immigration and emigration are equal. The difference between rates of birth and death reflects the population growth. Using an example of a fish species within a tank we define the closed population as:

- N = # individuals in the population
- t = time
- N_t = # individuals in the population at a given time
- N_0 = the initial population size at time zero

In this example the assumption is that the population size is lower than available food supply. In other words food resources are not a constraint. All births will be added and all deaths will be subtracted. Using a population size of $N_0 = 100$, with 50 new births and 8 deaths, calculate the size the population at the end of day one (N_1):

- $N_0 + B - D = N_1$ or $100 + 50 - 8 = \mathbf{142}$

What is the population size at the end of day two, where there were 40 births and 8 deaths?

- $N_1 + B - D = N_2$ or $142 + 40 - 8 = \mathbf{174}$

An absolute value, based on number of births and deaths, is dependent on population size. Often geometric population models are derived from an estimate that is independent of population size and can be expressed as a rate. This is because growth rates are more useful to compare different sets of data, such as comparing the accidental death rate of humans in one county compared to accidental death rates across the entire state. To obtain a population growth rate, known as the per capita birthrate, ***b***, using the day one data above, apply the following formula:

- $b = B/N_0 = 50/100 = 0.5$
- The per capita birth rate is **0.5**.

It is also possible to calculate the per capita death rate, ***d***. Determine the per capita death rate for the data from day one above, using the following formula:

- $d = D/N_0 = 8/100 = 0.08$
- The per capita death rate is **0.08**.

By assuming ***b*** and ***d*** are constant rates it is possible to predict the growth of a population over time, regardless of population size. Determine the population growth after one day on a beginning population size of 100 and the per capita birth and death rates:

- $N_{t+1} = N_t + bN_t - dN_t$
- $N_1 = 100 + 0.5(100) - 0.08\,(100) = \mathbf{142}$

Closed populations reflect a **geometric growth model** and have a finite rate of increase. The data predicts changes in population size by discrete time intervals (i.e., one day). Try to work out the above three problems yourself before viewing the given solutions.

Solutions.

1. In the first problem set you simply take the total population and divide by the total area (300 bog turtles/20 hectares = 15 bog turtles/hectare).
2. In the second problem set you are finding the total amount of salmon in the area. First, find the total square miles (20 × 200 = 4000 square miles). Since the distribution was given as uniform multiply the amount found in each square mile by the total square miles (4000 × 50 = 200,000 salmon in the river tributary). Do you think it is likely that salmon are uniformly distributed across the tributary?

3. In the third problem set find the per capita growth rate, $\boldsymbol{r} = \boldsymbol{b} - \boldsymbol{d}$. For the Iriomote cat it is $\boldsymbol{r} = 0.6 - 0.12$, or: –0.6, which indicates a declining growth rate. Next find how many are dying during the period being measured. To accomplish this multiply the total population by the death rate (600 × 0.12 = 72 cats). Finally, determine the births during the same period by multiplying the total population by the birth rate (600 × 0.06 = 36 cats).
4. The only correct answer would be the Iriomote cats since they are island-bound and cannot immigrate.

In most cases it is necessary to consider the growth rates of mobile populations without barrier to egress or entry of time, which is defined as an open population.

Exponential Population Modeling in Open Populations

In open populations the growth rate is determined by considering not only birth and death rates but also immigration and emigration rates. In this model $\boldsymbol{I}$ = immigration, $\boldsymbol{E}$ = emigration, and the change in population (ΔN) is determined by factoring in both birth/death and immigration/emigration values, which can then be expressed by the intrinsic growth rate, $\boldsymbol{r}$.

- $\Delta \mathrm{N} = [b - d] + [\mathrm{I} - \mathrm{E}]$
- $\Delta \mathrm{N}/\mathrm{N}_0$ = intrinsic growth rate ($\boldsymbol{r}$)

A second formula, the **exponential growth model** ($N_t = N_0e^{rt}$), is used to predict changes over time, using the growth rate $\boldsymbol{r}$ and the year since N_0 as $\boldsymbol{t}$ and the mathematical exponential constant, 2.71, as $\boldsymbol{e}$. Exponential growth is characteristic of newly colonizing populations with favorable environments at low population densities so that populations are density independent. Exponential growth accounts for individuals reproducing *continuously*, and in this model generations can overlap. Population growth charts of exponential growth will display a rapid growing J-shape. Note that the exponential population model is continuous and more realistic in nature but with the caveat of no limiting factors.

Some of the best examples of exponential growth are seen in prokaryotes (bacteria) that reproduce rapidly, sometimes within an hour. For example, if 1000 bacteria are placed in a flask supplemented with an abundant supply of nutrients, bacteria will have doubled from 1000 to 2000 in 1 hour. By the end of the second hour each of the 2000 bacteria will have reproduced so there are now 4000 bacteria. By the end of one day, or 24 reproductive cycles, the bacteria population will have increased from its initial population size of 1000 to more than 16 billion. In **Fig. 10.14** when a population, N, is plotted over time it reflects the characteristic exponential, J-shaped growth curve.

- $N_t = N_0e^{rt}$ where,
- N_t = population size at time t
- N_0 = original population size. $e = 2.71$ (constant)
- r = intrinsic rate of increase
- t = time

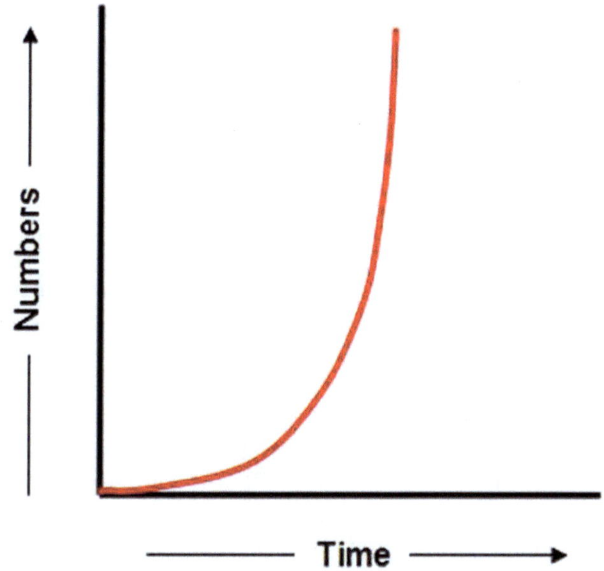

Fig. 10.14

Exponential growth chart with no limiting factors.

Source: Images by author

Various factors alter the predicted exponential growth rate. Ultimately a growing population reaches **carrying capacity**, which is the maximum density a region can support indefinitely, given adequate resources.

Darwin recognized the natural limitations placed on population growth when he wrote of the "struggle for existence." That is, individuals compete among their own species or with other species for use of limited resources within their environmental niche. Limiting resources include inadequate food resources or a shortage of space for the species' population. Both factors may lead to decreased birth rates, increased death rates, and mass emigration to new locations. Whenever limiting factors impose a greater constraint as population density increases the population follows **density-dependent growth**. Typical density-dependent factors include disease, competition, and predation. When limiting factors are more accidental, not related specifically to population's density (i.e., flooding or fires), the population is following **density-independent growth**. Some other examples of density-independent factors are weather, natural disasters, and pollutants in the environment. A species' carrying capacity is the maximum population size that a particular environment can sustain under its combination of various density-dependent and density-independent limiting factors. Carrying capacity values are unique to each species present and are dynamic, necessarily changing as the conditions change.

So far we have reviewed closed populations under the geometric model and open populations under the exponential model. But since resources are not unlimited exponential growth is not reasonable and cannot be sustained.

Carrying Capacity and the Logistic Population Model

In the real world exponential growth occurs as long as there are few individuals and unlimited resources. As soon as the population grows to the point its resources are depleted the growth rate will slow down and eventually plateau or level off, reaching carrying capacity. The final population growth model we will discuss takes into consideration a potential shortage of resources. Open population models that factor in carrying capacity are called logistic.

Carrying capacity, $\boldsymbol{K}$, represents the maximum population size that can be sustained in a particular environment. As environmental conditions fluctuate $\boldsymbol{K}$ will fluctuate. **Logistic growth** is the continuous population growth rate experienced in an environment when resources are limited. Logistic growth falls under density-dependent growth, characterized by a sigmoidal, or S-shaped, growth curve (**Fig. 10.15**). The population is expected to increase rapidly and then stabilize. The growth rate begins to flatten as the population nears carrying capacity and resources begin to run low. At carrying capacity the growth rate will reach zero with no further change in population size. Note that the first half of the logistic growth curve looks like the J-shaped exponential growth curve, and that it is carrying capacity that alters the rate of growth.

Recall that the change in population abundance over time is a function of birth and death rates, represented by the per capita growth rate $\boldsymbol{r}$. For the human population the current growth rate is estimated as 1.18%, so $\boldsymbol{r} = 0.0117$. If the year 2010 is time $t = 0$ and $N_0 = 6.8$ billion, population size at year one is $N_1 = 6.8 \times e^{0.0117(1)}$. And,

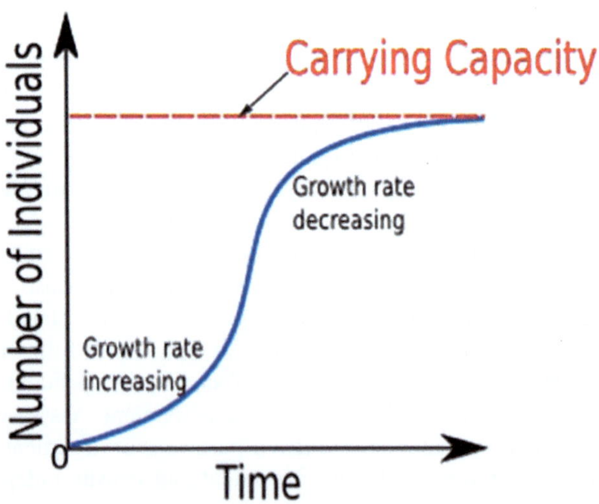

Fig. 10.15

Logistic growth chart.

Source: Images by author

if ***r*** remained constant the human population would reach more than 80 billion in 215 years! But many events, including warfare and famine, will drastically reduce *r* before the population ever reaches 80 billion. Those factors are directly or indirectly related to the human carrying capacity.

Species historically fall into two growth patterns: ***K*-selected species** increase in population size until they reach carrying capacity. These species are known to expend a great deal of their energy for parental care, have fewer offspring, and their total population size fluctuates around carrying capacity. The ***r*-selected species** are short lived and experience maximum growth rates with corresponding large swings up and down. Species with *r*-selected life strategies are governed by their biotic potential. These species expend little energy for parental care, instead producing large offspring litters with higher risks of infant mortality. This is a successful strategy whenever strong predator presence threatens juveniles. The population will have episodic crashes. And if the population under study is the primary food source of a predator they will be linked in such a way that their predator–prey relationship correlates with the rising and falling population growth rates in both species.

Which growth pattern, *K*-selected or *r*-selected, would you say describes the human species? *K*-selected growth describes humans and other mammals. What is an example of a species with a *r*-selected growth pattern? The invasive Asian carp fits the *r*-selected growth pattern. In their normal habitat the many species of Asian carp face heavy predation. To improve their chances of survival they produce huge numbers of offspring. Therefore *r*-selected growth describes the Asian carp and most aquatic species.

Density-dependent factors include food resources, space, migration, disease, and predation. Each has a role in regulating growth of the population according to its density. As population density increases resources are consumed at a faster pace, followed by birth rates decreasing while death rates increase. Species dispersal (emigration) increases, as species seek habitats with more plentiful resources. The overall result is decreasing population size. What are density-dependent factors in this case?

- Competition for resources between and among species
- Possibly disease and waste accumulation.

Density-independent factors include effects on birth and death rates independent of the actual number of individuals in the population. For example, temperature, precipitation, or catastrophes such as floods or hurricanes which impact both large and small population sizes equally.

Species will exhibit fast reproductive capacity if environmental conditions permit. If a species enters a previously uninhabited region its population will go through a typical pattern of growth represented by the population growth curve (**Fig. 10.16**). A population growth curve has four distinct phases.

1. lag
2. exponential growth
3. deceleration
4. stable equilibrium

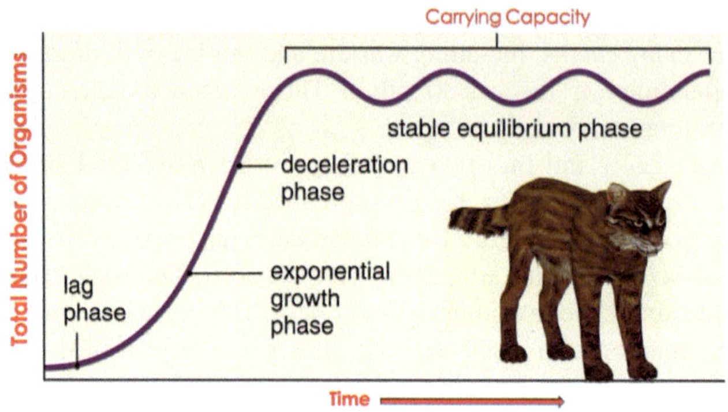

Fig. 10.16

Four phases of population growth with carrying capacity.

Source: Images by author

In the **lag growth phase** immediately following establishment of a population the population is small and relatively constant with low birth and death rates while reproduction follows a mating period. Depending on the plant or animal lag growth phase may occur over many years. The **exponential growth phase** is the time period when the population is exponentially growing (increases by the same percent each year). Eventually the rate of growth will level off to the **deceleration phase**, in which the rate of population growth will begin to slow down. This is because the death rate has begun to approach the birth rate. Once birth rates have stabilized and are equal to death rates, and individuals begin to exit by migration, the population is said to have reached its **stable equilibrium phase**. While the population size is stable individual members are still fluctuating. A new mixture of individuals occurs because the rates of birth, death, and migration are ongoing. Most populations fluctuate at carrying capacity in an undulating fashion.

Life Tables

Ecologists and conservationists often need to investigate factors that limit or promote population growth. Gathering information to anticipate birth and death rates of populations is essential to predicting their future population size. **Life tables**, first developed by life insurance companies to predict human life expectancies, have been adopted by ecologist as a tabulated expression of species life expectancies. The data compiled in a life table can be used to predict age-specific survivorship and basic reproductive rates or to make comparisons between different groups. To assess the plight of an endangered species in the wild ecologists might use a Life Table to predict the growth and decline of an existing population. Three important parameters in the Life Table are:

- S_x = **survival rate**: chance an individual of age x survives to age $x + 1$
- l_x = **survivorship**: proportion of surviving individuals from birth to age x
- F_x = **fecundity**: average number of offspring a female will produce at age x

Populations are dynamic, increasing or decreasing depending on factors like reproductive rates and longevity, particularly how long the females can produce offspring. The life table tabulates an age-specific account of mortality, summarizing birth and death rates for individuals at the different stages of their lives (egg, hatchling, juvenile, and adult stages). As birth and death rates vary with age the life table will often be set up to follow the life history of a cohort, or group of individuals from the population, born in the same time period. The cohort provides researchers a systematic picture of the population's mortality and survival. Life expectancy, $\boldsymbol{e}$, is the average number of years an individual of a species lives. The age-specific life expectancies, $\boldsymbol{e_x}$, is the average number of years an individual of a certain age is expected to live into the future.

There are two basic types of Life Tables:

- dynamic
- time-specific

From a population sample, a **time-specific life table** is constructed to obtain the distribution of age classes during a single period, especially useful for short-lived species, like insects, with one breeding season. The static time-specific life table is, of course, the easiest to construct. The time-specific life table assumptions are:

- Sampling from age classes will be proportional to their numbers in population as a whole.
- Age-specific birthrates are constant over time.
- Age-specific mortality rates are constant over time.

The cohort, or **dynamic life table**, shows the fate of a single group of individuals born at a given time (the cohort) from birth to death. Dynamic life tables are useful for long-lived vertebrates with overlapping generations.

The first row in the dynamic life table represents the birth year of the cohort and each subsequent row typically represents data compiled in the following year. Assuming a unit of age $\boldsymbol{x}$ in years the column $\boldsymbol{n_x}$ is the number of individuals still alive in following years. **Survivorship** $\boldsymbol{l_x}$ is the proportion of individuals still living from the original cohort ($l_x = n_x / n_0$) in each subsequent year. **Age-specific fecundity** $\boldsymbol{m_x}$ is the average number of offspring born to individuals in each age. Some life tables use the notation $\boldsymbol{b_x}$.

Life tables provide useful data to answer questions on whether a population is shrinking or expanding. A population growth parameter, the net reproductive rate $\boldsymbol{R_0}$ (pronounced R-naught), is the product of survivorship and fecundity ($\boldsymbol{l_x}$, $\boldsymbol{m_x}$) calculated at each age. When the $\boldsymbol{R_0}$ value is greater than 1 the population is growing, producing more offspring than its death rate. In **Table 10.1** the sample life table depicts a population in decline (last column). While a life table is a good indicator of survivorship, assessing and interpreting its patterns by reading columns of data can be less intuitive and cumbersome. A graphic chart constructed from the data is useful to describe the dynamics.

Table 10.1 Dynamic Life Table

Age (X)	Number alive (*nx*)	Survivorship (*lx*)	Age-specific fecundity (*mx*)	Survivorship × fecundity (*lxmx*)
0	100	1	0	0
1	25	0.25	0	0
2	12	0.12	2	0.24
3	9	0.09	5	0.45
4	6	0.06	5	0.3
5	4	0.04	5	0.2
6	3	0.03	2	0.06
7	2	0.02	2	0.04
8	0	0	0	0

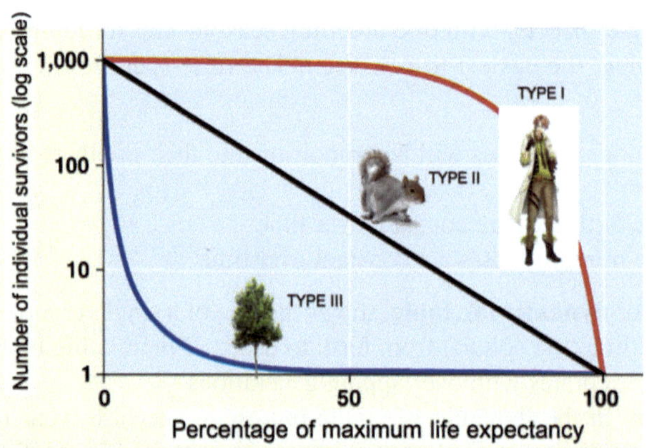

Fig. 10.17

Survivorship curves are idealized models.

Source: After Crouse et al 1987; Ukanda under CC By 2.0

Survivorship Curves

To predict if a population will grow or shrink ecologists often create **survivorship curves**, which are derived by plotting survivorship against time, with l_x on the *y*-axis of the graph in logarithmic scale. This technique provides more intuitive data when birth and death rates at different ages are known along with the current age and sex makeup of the overall population. A survivorship curve plot predicts the number of individuals from a hypothetical cohort to survive to reach different ages (**Fig. 10.17**).

Survivorship curves are classified into three general types:

- **Type I**: most individuals survive to old age, *big mammals (i.e., humans)*
- **Type II**: survival rates constant throughout the individual's lifetime, *birds, mice*
- **Type III**: high death rates in youth, individuals reaching adulthood survive well, *small mammals, fish, invertebrates (i.e., loggerhead turtle)*

Birth rates are often expressed as the number of births per 1000 individuals in a population/unit of time. The **crude birthrate** is the number of births during a time period divided by the starting population size multiplied by 1000. For example, if 5663 births occurred in 1 year within a total initial population size of 149,442, the crude birthrate is 37.9 (5663/149,442 × 1000 = 37.9) Crude birthrates won't account for substantial variations, such as sex ratios and age distributions. To increase the accuracy of birthrate estimates the male and female populations can be separated. This is only possible in cases where the population is sexually dimorphic (only females give birth). Ecologists sometimes use an **age-specific birthrate** to determine the birthrate for females within a specific age class. Note that while fertility infers an *ability* to reproduce **fecundity** is related to the reproductive output. Fecundity is the preferred measure for population studies. Reproductive rates can be measured as the number of eggs, seeds, or asexual propagules, etc.

Age-specific mortality rates and age-specific birthrates can be combined to project future changes within a population. The survivorship and fecundity rates would be determined in the same manner for each successive year (i.e., *Assumption #2:* age-specific mortality rates remain constant). In some populations survivorship and fecundity will be described by size, for example, in fish or perennial plants (**Fig. 10.18**).

Up to this point population growth has been viewed as a deterministic process. In the exponential growth model populations were described as increasing in size without restriction or limits to growth. In the logistic growth model population studies accounted for limits to reproductive growth after reaching carrying capacity. But neither model fully describes natural populations. The logistic model is generally

Life Table and Fecundity in the Loggerhead Sea Turtle

Stage #	Class	Size	Approx. Age	Annual survivorship	Fecundity (eggs/yr)
1	Eggs, hatchlings	<10	<1	0.6747	0
2	Small Juv.	10.1 – 58.0	1-7	0.7857	0
3	Large Juv.	58.1 – 80.0	8-15	0.6758	0
4	Subadults	80.1 – 87.0	16-21	0.7425	0
5	Novice Breeders	>87.0	22	0.8091	127
6	1st year remigrants	>87.0	23	0.8091	4
7	Mature breeder	>87.0	24-54	0.8091	80

Fig. 10.18

Fecundity table showing fertility in the novice breeders of the endangered loggerhead sea turtle in age-specific classes.

considered to be valid and applicable to many population studies; it is also an oversimplification of real-world population dynamics. This is because the logistic model implicitly functions under a carrying capacity that does not change in the given environment. Yet, in nature, the unpredictable has a way of altering the status quo, including the circumstances under which the carrying capacity was established.

But the carrying capacity does vary annually as Darwin's struggle for existence implies, for example, hotter- and drier-than-average summers one year followed by colder- and wetter-than-average summers the next. And the carrying capacity in some areas will be lower during their winter months than during their summer months. Population ecologists suggest that characteristics evolve within a species according to its experiences in their environment and subsequent adaptations, which in turn impacts population size. **Population dynamics** model population growth to predict one exact outcome under investigation and account for annual variations in climate, temperature, precipitation, natural disasters, fire, flood, and drought. Unpredictable fluctuations in population modeling are called **demographic stochasticity** and **environmental stochasticity**.

A population is declining when deaths exceed births. Populations that are declining in size can eventually become extinct. Many situations could make a population more likely to become extinct, for example:

- Small populations are more vulnerable to extinction
- Loss of genetic variability
- Increased vulnerability to demographic and environmental stochasticity
- Extreme environmental events can increase mortality rates and reduce population size (drought, flood, heat waves, extreme cold snaps)

If environmental conditions exceed the bounds of tolerance for the species an extreme event could lead to extinction, for example, ongoing extinctions are occurring today due to changes in regional and global climate. According to the latest data on Earth.org North America is experiencing the greatest number of species extinction, with 33% biodiversity loss. Severe resource shortages, through extreme environmental conditions or overpopulation, will result in population decline. Extinction is possible whenever the resource base does not recover in time for survivors to reproduce.

Gains and Losses in Species Populations

The diminishing bumblebee. In the North American continent more than 700 bee species are in decline, according to the Center for Biological Diversity. More than half of those bumblebee species (*Bombus sp.*) are at risk of extinction. In 2015 a UN group reported bee populations were declining worldwide by 37%, with 9% of all butterfly and bee populations facing extinction. By January 2018 the bumblebee was officially added to the ever-growing list of endangered species. The endangered list includes many popular species whose plight and imminent decline have proven newsworthy: the grizzly bear, the northern spotted owl, and the gray wolf. But there are many at-risk species that never make the news cycles. According to the 2020 WWF Living Planet Report nearly 3 billion birds have been lost in North American in the past 50 years. And the endemic bumblebee was abundant in grasslands and prairies across the US's eastern and midwestern territories. Today the bumblebee is being restricted to conservation protection sites due to the alarming rate of population decline. As the nation's primary agricultural pollinators these insects play an important economic role to sustain crop propagation, which is valued at billions of US dollars annually. Because the decline of the bumblebee represents a global threat to agricultural its plight has now captured international attention. The causes of population decline in the bumblebee are complex. In *Science* magazine Goulson et al. pointed to several interacting factors that were indicated in most of bee research: availability of food and habitat loss (nest sources), exposure to agrochemicals, and climate change. This research suggested that growing susceptibility to parasites and disease in bee populations (a direct cause of death) may be the indirect result of stressors leading to weakened immune systems. The weakened physiological response was attributed to the energetic costs of chronic coping with pesticide exposure, climate change, and other threats like habitat loss (**Fig. 10.19**).

In the early spring solo queen bees that survived parasites, predators, and the winter (some species have an antifreeze-like protein in the blood) will emerge from underground, gather food, and start to lay the first eggs from mating that occurred prior to her winter hibernation. Once the first female worker bees are hatched the queen rests and finishes laying eggs to sustain the colony. Forager and male bees

Fig. 10.19

In July 2013 researchers from University of Maryland and the USDA reported that pesticides had contaminated the pollen that bees harvest during pollination and bring back to their hives.

Source: USDA, Public Domain

have the shortest lifespans as they are susceptible to conditions outside the hive, which can be deadly.

An ecologist studying bee populations might prepare age structures and life tables to answer questions such as which factors are most vital to survivorship, which bee worker class experienced shortened life expectancies, and which bee species are under threat for extinction. Life tables and survivorship curves constructed for the bumblebee provide insights into how to protect their diminished populations and select the best species for conservation efforts.

The Primary Factors That Threaten the Bumblebee

Habitat loss. Land conversion for human needs results in restricted and isolated patches of flowering plants, while intensive farming of mass-flowering crops (monocrops) fails to provide sufficient quality nectar and pollen resources (high-energy food) for bees. The primary food source for bees is pollen, high in nutrients (protein, carbohydrates, vitamins, and lipids) but the ratio of each constituent in pollen is plant-species specific. The crude protein content ranges widely, from 2.5% to 61%.

Climate change. Even undisturbed wild habitats are showing deficits in flowering due to climate change. Bees are adapting to a wider range of plant variety for pollen, through a process called **behavioral plasticity**. In turn bees are increasing their exposure to parasite transfer from contact with new plants. And they need to sustain daily activities with less energy-efficient food sources. Climate change appears to be constricting the bumblebee population's geographic range. This is partly explained by the bumblebee having a lower heat tolerance, although it is unclear which factors are preventing them from migrating out of their decreased geographic range.

Agrochemicals. Neonicotinoids are agricultural pesticides that attack the central nervous systems of insects, including bees. The honeybee, one of the most prominent agricultural pollinators, has exhibited neonicotinoid-induced declines in both their foraging success and navigation (used to find pollen sources and to navigate home). Honeybees apparently prefer neonicotinoid laced nectar and become inebriated! While not all bees are equally susceptible to neonicotinoids, another study compared wild bee to honeybee performance in paired fields of land-use history and applications of the agrochemical. The wild bee population densities and colony success showed declines when localized to neonicotinoid-treated oil seed rape fields.

In the fall of 2021 the US Fish and Wildlife Service announced that the American bumblebee's population had plummeted by nearly 90%. The bumblebee joins the grizzly bear, the northern spotted owl, and the gray wolf on the growing list of endangered species, along with another 700+ animal species already listed as extinct. Once abundant in the East and Midwest, the rusty-patched bumblebee (*Bombus affinis*) is now restricted to protected regions in the continental United States as its population size dwindles at an alarming rate.

The rise of invasive exotic species. Invasive nonnative species are particularly problematic as they are known to take over a geographic region from native species and disturb their local ecology. Invasive or exotic species are nonnatives introduced to a new region where they don't have known predators to keep their population growth in check, leading to unintentional and sometimes devastating consequences. For the bumblebee invasive species can take over native flora and alter the quality of available pollen sources. Chemical analysis studies have shown some invasive plant species led to reduced nutritive intake in the bumblebee. The *Buddleja davidii*, a wild butterfly bush that forms shrubby thickets, is an invasive species in the Pacific northwest. Its pollen was analyzed and shown to have lowered levels of the amino acid proline, which is required for bee flight.

An example of intentional dispersal, followed by an *unintentional* dispersal of an exotic species, is the 1970s introduction of the Asian carp to North America. Southern state fish farmers had initially imported Asian carp from China to help clean their commercial fishponds. This decision was based on the Asian carp's known superiority consuming algae and vegetation overgrowth. Soon the successful establishment of the Asian carp in fishponds was unintentionally dispersed into nearby lakes and rivers (Mississippi and Illinois Rivers) through flooding events. This led to serious decline in endemic fish populations in any lakes and rivers where the more aggressive Asian carp out-competed endemic species for food and space. Once overpopulation of the Asian carp has taken hold of an aquatic system the water quality soon declines because it consumes much of the native aquatic filter organisms, like the freshwater mussels. Currently species like the silver Asian carp are known to be highly detrimental to the environment in many regions of the United States. In 2007 the US Fish and Wildlife Service established a national plan for the management and population control of invasive Asian carp (specifically the bighead, silver, black, and grass carp). The ecological ramifications of invasive species on community dynamics will be further explored in Chapter 13.

Invasive species, whether flora and fauna, have the potential to negatively affect the delicate balance of an ecosystem. They may impact the ecosystem's health through disturbing local population densities by outcompeting for limited resources, which often results in regional biodiversity losses. In the case of the Asian carp the damages are long ranging, affecting the food supply of entire lake ecosystems.

Glossary

migration
allelopathy
habitat suitability
closed population
open population
demographic stochasticity
environmental stochasticity
invasive (exotic) species
time-specific life tables
dynamic life tables
density dependent growth
density independent growth
unitary
modular
genet
ramets
clonal colony
internodal
Armillaria
rhizomorphs
facultative
saprophyte
ubiquitous
abundance
crude density
random v. uniform v. clumped distribution
behavioral plasticity

CHAPTER

Life Histories

11

Chapter Outline

11.1 Life History Continuum

After reading this section students should be able to:

- Describe the life history of an organism through its allocation of resources.
- Understand life's evolution inevitably involves trade-offs.
- Define life history through interactions between genotype and environment.
- Identify characteristics of life history that reflect the influence of natural selection.

Life History

The life history of an individual organism is its lifelong pattern of survival, including growth, development, and reproduction. In contrast to population modeling, the topic is the individual and the focus is traits that affect its survival. Life histories investigate the timing of age-specific patterns and events that make up an organism's life, notably juvenile development, age of sexual maturity, mating patterns, fecundity, and parental investment in offspring. **Life history theory** is at the heart of biodiversity. It seeks to explain the specifics of an organism's anatomy and behavior by an analysis of its life history patterns and how they would have been shaped by natural

Ecology for the 21st Century. DOI: https://doi.org/10.1016/B978-0-443-24031-7.00011-8

selection. Given that individuals have finite quantities of resources and energy, an allocation of time or energy to one aspect invariably reduces those resources available for other aspects. For example, allocating resources to reproduction, in turn, reduces the resources available for food foraging and growth. The strategy individual species adopt will depend on the constraints imposed on their life history. An accumulation of trade-offs shapes the life history of an organism. For species where maturity is delayed until a greater size is reached the parent makes an energetic investment in the offspring. But this trade-off means there are fewer years for adult reproductive value. Alternatively, earlier age of maturity with longer adult reproductive years increases the number of reproductive events, known as **parity**. Wherever life histories experience higher mortality (low survivorship), earlier maturation and greater fecundity would be favored to increase chances of reproductive success into the next generation (**Fig. 11.1**). **The Life History Continuum** illustrates a gradient between those species that have lower fecundity but greater survivorship and those that have greater fecundity but lower survivorship (**Fig. 11.2**).

Whereas there is a wide range of reproductive strategies to stabilize a population size, a balance must be met between maturation and adult reproductive value. This simplistic view of life history as a continuum between fecundity vs survival hints at the much more complex life strategies taken to avoid extinction. Ultimately the trade-offs imposed by environmental conditions affect growth, reproduction, and survival by constraining and optimizing across a wide range of possible life strategies occurring over time on the species' evolutionary pathway. Trade-offs are imposed by physiological, energetic, or environmental constraints.

The main trade-offs are:

- Reproductive mode
- Maturation age
- Parental investment
- Fecundity (#/size of eggs or seeds)
- Timing of reproduction

Constraints can be classified as intrinsic or extrinsic (ecological) factors. **Intrinsic factors** impose life history constraints on the organism. These include evolutionary

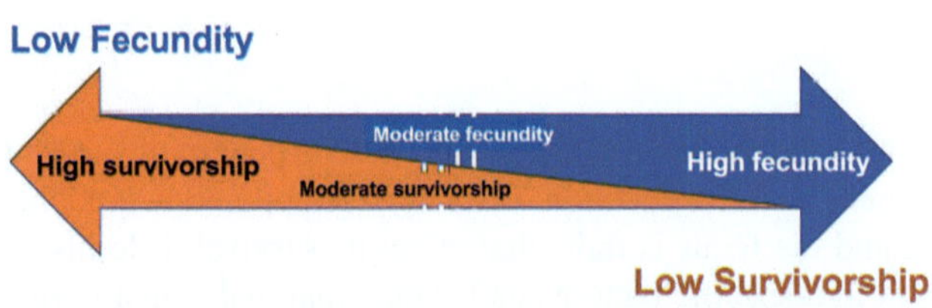

Fig. 11.1

Life History Continuum Model. Life histories reflect a trade-off between reproductivity and adult survivorship.

Source: Depiction of Life History Continuum Model by author

Fig. 11.2

Ocean sunfish produce 600,000,000 eggs vs the shark, which produces on average 1 to 2 eggs per clutch. While there may be several shark fetuses in one egg sack, the strongest eats the others.

Source: Modification of G. Johnson under CC BY-SA 3.0

history, migratory patterns, genetics, physiology, and density dependence. **Extrinsic factors** influence survivorship and mortality of the organism. These include climate, physical environment, and species interactions (predators or competitors).

Life history characteristics and mating systems are products of evolution and reflect adaptations to the environmental conditions. A modification of the Life History Continuum, known as the **Fast-Slow Life History Continuum Model**, emphasizes the *forces* imposed by mortality at different stages during the life of an organism (**Figs. 11.3a**, **b**). As a continuum the model accounts for a range of species including those that experience high adult mortality and those that experience low adult mortality. Note that the Fast-Slow Life History Continuum Model does not contradict the first life history model introduced (**Fig. 11.1**). Rather, it adds more aspects to the individual's life history. The Fast-Slow Life History Continuum covers a gradient from the fast life history of shorter life spans with faster rates of developmental, earlier maturation, and higher fecundity toward the opposite end with aspects of a slow life history (longer life spans, slower developmental growth, late maturation, and lower fecundity). Notably the parental investment is greater in species that take longer to develop and those that will have greater survivorship but fewer offspring.

Environments are sometimes classified by the degree of variation: exhibiting fluctuations and variability or exhibiting stability over time. The species present will need to adapt to whatever fluctuations are present in the environment to maintain a balance between life span and fecundity. Seasonality exerts strong control over biodiversity. Those species better at adapting to variation tend to possess generalist life strategies (also known as cosmopolitans). Endemic species are less adaptable

Fig. 11.3a

Fast-Slow Life History Continuum Model, adapted from Del Giudice and Ellis.

Source: Adapted from Del Giudice & Ellis under CC BY 4.0

Fig. 11.3b

Great Basin Bristlecone pine trees (*Pinus, Balfourianaeare*) live in subalpine extreme conditions. Exact locations of the two longest-living tree individuals found in the White Mountains of California (5068 years and 4850 years) are kept secret. The Bristlecone pine's long life span is offset by low reproductivity.

Source: Desires Photo under CC BY-ND 2.0

to seasons or other types of environmental variation. Areas that house the greatest biodiversity (e.g., tropical rain forests) are predictably more environmentally stable with less seasonal variation.

Genotype and the environment. Natural selection acts upon phenotypic variation among individuals in the population. Phenotypic variation arises from genotypic variation among individuals of a species and their interactions with the environment. An individual's life history is a representation of the interaction between genotype and the environment. Organisms display phenotypic plasticity in life history traits (but not *genotypic* plasticity). For example, fierce predation may lead to camouflage strategies, using color patterns that match the environment to provide some protection for its vulnerable prey. The ability for some species to change color may be a regular physiological event (e.g., during mating) and can occur either slowly or very quickly with alteration of pigment molecules on the organism's surface. Every organism will have its own developmental history, which reflects the environmental constraints it encounters daily. For example, crabs are known to change color diurnally, from darker grays at night to brighter colors that match sun-reflected sandy beaches during the day.

Phenotypic variation is seen in:

- The age at maturity
- Allocations dedicated to reproduction
- Size of offspring produced
- Number of offspring produced
- Level of independence at hatching or birth

A change in temperature during a critical developmental stage may lead to a lifelong alteration, as observed in sex determination in amphibians, a phenomenon known as temperature sex determination (TSD). The gender of developing American alligators (*A. Miuissippiensis, reptile*) varies according to the temperature during the egg incubation period. In cooler weather, below 31.5°C (88.7 °F), females will develop from the egg but if the temperature warms to 32.5°C to 33.5°C (90.5 °F–92.3 °F) males will develop from the same egg. In turtles a similar situation occurs except those cooler temperatures determine male sexual development over females. In 2014 investigative studies on reptilian TSD by Spencer and Janzen showed that one-sex hatchlings had healthier first year survivorships with greater energy efficiencies than mixed-sex hatchlings. This fact provides a rationale for the persistence of TSD in life history strategies. In both alligators and turtles maintaining sex ratios was accomplished by reproductive timing or behavior modification. This is possible by laying eggs in either cool, shaded areas or upon sunny precipices.

Phenotypic Variation Is a Special Form of Acclimation

Acclimation is generally described as a short-term adaptation to a temporary environmental condition. The acclimation may be through a modified physiological or metabolic behavior that is reversible when the altered environmental condition is removed. Phenotypic plasticity describes the *ability* of a genotype to give rise to variable responses to stressful environments. Put together the genotype can generate different phenotypes because of its plasticity in the presence of stressors. The result is enhanced tolerance levels via acclimation. Phenotypic plasticity can be sustained over a longer time frame, spanning generations, even involving changes at the morphological level. It is an important part of life history because those individuals who successfully develop variant features are naturally selected for survival. And those individuals in the population who cannot acclimate will be at a selective disadvantage.

Phenotypic variation within populations is possible through two mechanisms: (1) variation in genotype among individuals and (2) interactions between environment and genotype. In Chapter 10 we discussed how habitat loss compelled bumble bees to feed from a wider range of pollen sources, even if the nutrient content was inferior. The set of phenotypic variations available to species across the *range* of environmental conditions is termed its **norm of reaction**. Norm of reaction is demonstrated in the difference in size at sexual maturity between sexes in a species. For male mating success size is less important than quickly reaching age at maturity. The timeliness of reaching maturity is most important, as it increases the male's probability of reproducing before its end-of-life span. For females size is more important because females must be at their minimum size of reproductive maturity. The differences in the reaction norms reflect the trade-offs for each sex relative to its reproductive success and fitness. Note that the variation in the norm of reaction across the sexes still draws from the same gene pool. The species gene pool is simply reflected differently in the female norm of reaction compared to the male norm of reaction.

A study on *seasonal* phenotypic plasticity in the drosophila fruit fly, *Drosophila suzukii* demonstrated that the flies switch to their winter phenotype (darker pigmentation and longer wing length) once the weather cooled from summer to winter. The winter phenotype is an important acclimation in the fruit fly because it is thought to increase its survival at freezing temperatures (1°C [32°F]) (**Fig. 11.4**).

Another example of phenotypic plasticity in shown in the relationship between sessile stony coral and their algal symbiont microorganisms of the genus *Symbiodinium* and marine bacteria. The coral and microbes together comprise the holobiont (**Fig. 11.5**). The composition of coral holobionts varies across environmental changes, which in turn influences adaptability of the holobiont to environmental stressors. Some marine bacteria release an antibiotic to heal coral infections and establish a network of communications. In one study coral populations in the Red Sea between Africa and Asia were found to be acclimating to warming waters. This was evidenced by the northernmost coral populations' continued growth compared to limited ability to adapt in southernmost coral, already existing at their upper thermal limits. The report suggests warming ocean temperatures in the southern region of the Red Sea were outside the norm of reaction for coral reefs and that in the north phenotypic plasticity experienced by the holobiont occupants was instrumental in adapting to warming climates.

Fig. 11.4

Phenotypic variation in the adult fruit fly, *D. suzukii*, raised in laboratory conditions to emulate seasonal cues for summer vs winter climates. Summer adults (*top*) have very different morphologies than winter adults (*bottom*).

Source: Image courtesy of Joanna Chiu

Fig. 11.5

The coral holobiont. Polyps in high magnification of cold-water coral (*Lophelia pertusa*).

Source: USGS, Public Domain

Plant Life History Characteristics

While some animals are known to live a sedentary lifestyle (barnacles and sponges) most are able to move to new locations when conditions become tenuous or threaten their existence. In contrast plants are obligate sedentary and must deal with any challenges in their immediate environment. For example, if an insect feasts on the leaf of a plant the plant cannot simply run away. Plants are fortunately equipped with an extensive chemical arsenal, anatomical features, and sticky terpene oils to deal with noxious pests.

Many plants developed adaptive strategies to deal with fluctuating shortfalls or excesses in temperature or water (**Fig. 11.6**). Besides producing a thick, protective waxy cuticle layer and reflective hairs designed to reduce overexposure to light and heat plants can mobilize their internal water supplies to help cope with heat imbalances.

Recall that water has the capacity to soak up excess heat (high heat capacity). The heated water can then be released as water vapor (analogous to the sweating mechanism in mammals). Adaptations for water and temperature imbalances are examples of plant trade-offs. Drought-tolerant and desiccation-tolerant plants are acclimated to improve fitness to their environments. Desiccation-tolerant plants can survive dehydration if the water concentration drops to 0.1 g H_2O per 1 g dry biomass (10% of its biomass). Most plants have relative water concentrations closer to 85% biomass under actively growing conditions and cannot survive concentrations below 60% to 30% biomass. This plant acclimation is accomplished partly through altering cellular proteins and sugars to increase osmotic pressure and retention of cellular water.

Fig. 11.6

Northern plains, US corn fields in summer 2017 experienced reduced precipitation and drought conditions.

Source: South Dakota State Extension Service, US Drought, Public Domain

The drought-tolerant plants can physiologically acclimate to drought conditions by developing longer root structures, denser stomatal openings, and slower rates of transpiration.

Ecologists often categorize plant life histories according to the environment in which they thrive. Plant ecologist and Professor Emeritus J. Philip Grime (1935–2021) developed a framework for classifying plants. His model was based on the assumption that any plant habitat can be classified as under varying degrees of stress or disturbance. **Stress habitats** will restrict plant growth and productivity due to resource shortfall (light, moisture, minerals, suboptimal temperatures). **Disturbance habitats** refer to the partial or complete destruction of plant biomass due to predation or natural disasters such as flooding. In a framework of high- vs low-*stress* environments, coupled with high vs low *disturbance* environments, only three permutations are capable of creating a suitable habitat for plant life to exist. This is because a high-stress–high-disturbance combination does not provide sufficient conditions for even the hardiest of plant species. The three possible permutations of stress/disturbance habitats for plant life strategies are classified as "C" for low stress–low disturbance, "S" for high stress–low disturbance, and "R" for low stress–high disturbance.

- **"C" (competitive)**—Plants compete effectively in predictable habitats high in resource allocation. These plants are found in highly productive and undisturbed habitats.
- **"S" (stress)**—Plants with acquired adaptations are able to survive in stress environments. These plants are found in stressful habitats like limiting resources or extreme conditions.
- **"R" (ruderal)**—Plants include those with rapid rates of reproduction, short life spans, and commonly short height and wider seed dispersal patterns. These plants are found in productive but disturbed habitats.

In the **Grime Model** three points of triangulation are used in the classification system and show varying intermediate strategies, depending on changing resource availability and habitat (**Fig. 11.7**). The Grime Model also accounts for the phenotypic plasticity expressed in plants under varying habitats.

In summary, patterns of life history characteristics across organisms reflect external selective forces and inherent plasticity in survival strategies. Some species have acquired characteristics associated with their evolving life history with respect to size, age at maturity, number of offspring, number of reproductive events, and life span.

11.2 Reproductive Modes and Sexual Selection

After reading this section students should be able to:

- Identify a reproduction mode as asexual or sexual.
- Relate reproductive success to trade-offs between fecundity and survival.
- Provide examples of how mating systems can differ in pairing selection.

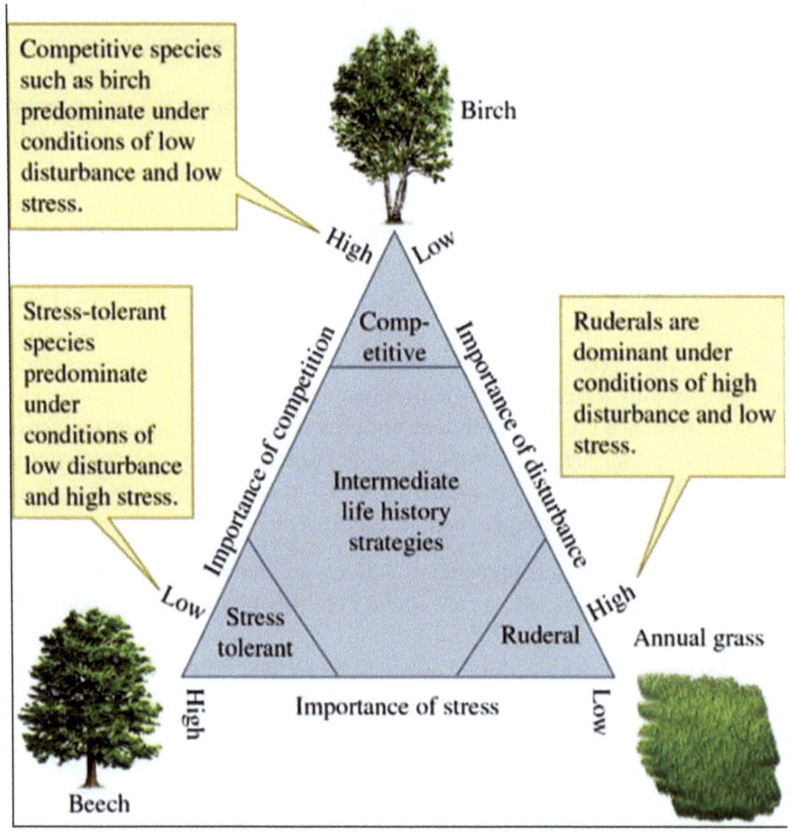

Fig. 11.7

Grime's classification of Life History Strategy in Plants.

Source: After Grime, 1979

Asexual Reproduction

In asexual reproduction offspring are genetically identical to the parent. The process does not involve fusion of distinct sex cells (gametes) and there is no change in the number of chromosomes. Across the kingdoms there are numerous examples of unicellular and multicellular organisms whose mode of reproduction is asexual, or facultatively asexual. Bacteria typically reproduce by binary fission. Many plants reproduce asexually by a stolon, horizontal stems that may root into new plants. Some protists reproduce by simply dividing in half, and the hydra, a type of animal, reproduces by budding.

Forms of asexual reproduction include any of the following:

- **Budding**—Offspring are small buds, or outgrowths, that break away from the parent.

- **Fragmentation**—Organism will fragment, with each piece growing into a new individual.
- **Fission**—The organism duplicates its genetic material and divides into two or more daughter cells.
- **Vegetative propagation**—Tubers or bulbs from an adult will grow into its offspring.
- **Spores**—Offspring arise from reproductive cells to develop a new life phase of the same organism.

The benefits and costs of asexual reproduction include the following:
Benefits of asexual reproduction:

- Genetically identical offspring are already adapted to the environment of their parents.
- They quickly reach high population growth as no mating steps are necessary to reproduce.
- They require few physiological processes to reproduce, therefore are energetically favorable.

Costs of asexual reproduction:

- There is loss of genetic diversity without genetic recombination occurring from two parents.
- Species has small genetic repertoire to respond to changes in environmental conditions.

Parthenogenesis

Female aphids give birth to all-female offspring without fertilization, in a process known as parthenogenesis. **Parthenogenesis** is a derivative form of sexual reproduction that arose during evolution among **dioecious** organisms (separate male and female individuals). Parthenogenesis provides a low-energy investment alternative to the more common high-investment sexual reproduction mode. Some species are entirely parthenogenesis without male involvement in the reproductive process but most are facultative parthenogenesis. **Facultative** means organisms retain the ability to turn on parthenogenesis during unfavorable environmental conditions. It allows them to produce offspring through unfertilized "virgin" eggs (e.g., when there are no viable males). But these organisms will revert back to the normal sexual reproductive modality when conditions are more favorable. The ability to switch back and forth, dependent upon existing conditions, provides considerable advantages.

One of the most important survival values of facultative parthenogenesis is the accelerated reproductive rate that is possible because all new individuals will be capable of having offspring. Note that parthenogenesis is not as low energy as asexual reproduction (budding or fission) because the egg still needs to be produced. The downside of producing offspring from unfertilized eggs is diminished genetic diversity, especially important during times of environmental stress. A few animals and many plant lineages have members known to utilize either obligate or facultative parthenogenesis. But this mode of reproduction is not found in mammals.

Note that some sources list parthenogenesis as a mode of asexual rather than sexual reproduction. In this textbook we categorize the process as an abnormal mode of sexual reproduction because the mechanism of egg production is much more involved than derivative asexual budding or fission.

Seasonal parthenogenesis occurs in the tiny aphid insect (*Aphididae*), which switches between low-energy parthenogenesis during spring and summer months, a time with sufficient resources to support rapid population growth, and sexual reproduction during the fall and winter months when resources are scarce. The greater genetic repertoire (DNA mixed from two parents) aids its chances of survival during the harsher winter months (**Fig. 11.8**).

The filter-feeding, near-microscopic aquatic organisms called rotifers are facultative. They reproduce by parthenogenesis in the spring and summer, producing 100% females. In the fall they produce an incomplete male offspring, which lacks a digestive tract but can contribute sperm for sexual reproduction. The deficient male rotifers exist only for the purpose of supplying the genetic material to fertilize eggs, which will hatch a new generation of females in the spring.

Another species that uses facultative parthenogenesis is the queen honeybee, capable of storing and releasing sperm to control the fertilization of her own eggs. If she releases the stored sperm gametes and its contribution of DNA then fertilized eggs will develop into worker female bees or other queens. If eggs are developed without fertilization from sperm (i.e., parthenogenesis) they will develop into male drones. The needs of the bee colony will dictate the queen bee's reproductive selection of worker females or drones.

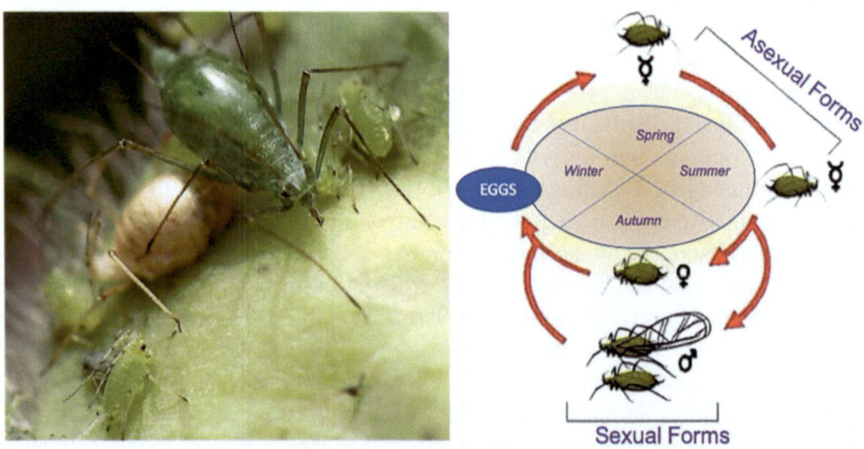

Fig. 11.8

Seasonal change sexual modes is common in insects.

Source: Modification of peccoud, CC BY 3.0

Fig. 11.9

Komodo dragon (*Varanus komodoensis*) at the Cincinnati Zoo.

Source: Mark Dumont under CC 2.0

In an unorthodox form of parthenogenesis reproduction is completed in strictly bisexual species. This unusual mode of reproduction was recently shown to exist in the wild in the North American pit viper snake species, copperhead, and cottonmouth. Several other species have been shown in captivity to utilize parthenogenesis, including sharks and the endangered Komodo dragon (**Fig. 11.9**). It remains unclear if the instances were developed naturally by meiotic errors in the egg or as an outcome of the abnormal living conditions of captivity. These examples highlight the expansive capacity for species to respond to environmental stressors.

Sexual Reproduction

The most familiar form of sexual reproduction is dimorphic, involving separate male and female individuals with distinct sexual organs. It is characterized by the process of meiosis in which gametes receive half of their genetic information from each parent cell and then fuse into a fertilized egg (zygote) that develops into the new individual.

Benefits of sexual reproduction:

- Genetic recombination rearranges the DNA, so each individual is genetically unique, which increases the gene pool of the population.
- There is an increased ability to respond to environmental challenges.

Costs of sexual reproduction:

- Only 50% of the individuals produce offspring.
- There is an investment in developing specialized reproductive organs.

- There is the energy cost of producing sex cells and mating behaviors.
- An investment in producing offspring is not equally shared between the sexes.

Sexual reproduction has developed into several different forms in plants. This makes sense given that plants are stationary and successful reproduction must rely on other actors, for example, abiotic means in their environment (i.e., wind, fire). The floral parts are the reproductive organs of seed plants. They have variation between flowers containing both sex organs or separate flowers for the male and female plant parts. Dioecious plants have separate male and female individual flowers (**Fig. 11.10**). When plants contain both the male (stamens) and female (ovaries) reproductive organs on the *same* flower they are **hermaphrodites**. When plants develop separate male and female flowers on the same plant they are **monoecious**. And finally there is variation on whether dioecious and monoecious plants are capable of self-fertilization.

Hermaphrodites are a unique form of sexual reproduction in which individuals express both male and female sexual organs. A commonly studied model organism is the transparent, microscopic (1 mm) hermaphrodite nematode, *Caenorhabditis elegans* (**Fig. 11.11**). Animals that are hermaphrodites have both male (testes) and female (ovaries) reproductive organs, enabling another form of sexual reproduction. The common sex mode in plants is hermaphrodites with both male (stamens) and female (ovum) reproductive organs.

On rare occasions in sexual reproduction it is the male of the species that gives birth to offspring. The more than 40 marine species known as the seahorse,

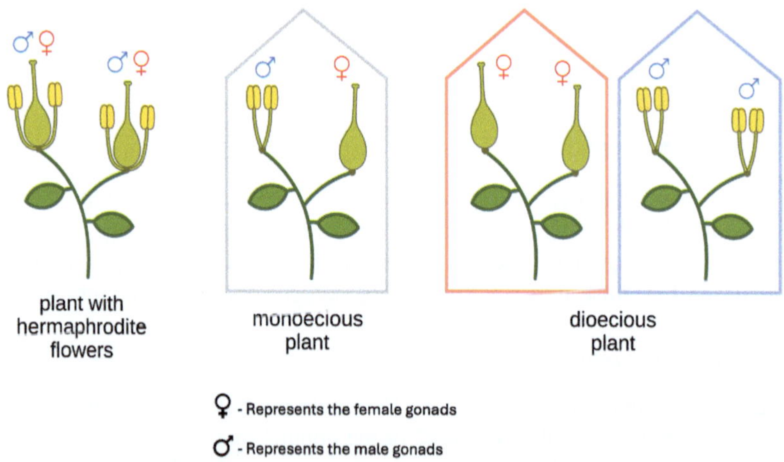

Fig. 11.10

Parts of a dioecious flower. The stamen contains the male gonads and the pistil contains the female gonads.

Source: Holger Casselmann under CC BY SA 4.0 and Ram-Man under CC BY SA 3.0

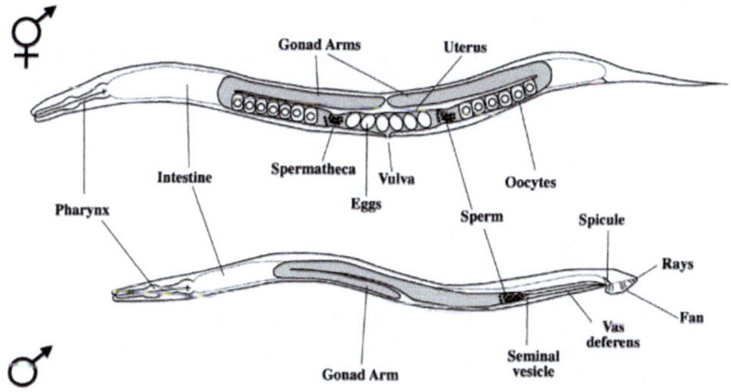

Fig. 11.11

Adult anatomy of *Caenorhabditis elegans* depicted as a hermaphrodite (*top*) and as a male (*bottom*). The gonads (sexual organs) are portrayed in gray shading.

Source: ProFlowers.com under CC BY 2.0

of genus *Hippocampus*, are an example of this variant mode of sexual reproduction. In addition to their energetically costly role in reproduction male seahorses are known as aggressive and will resort to fighting for female attention. The female contributes to the reproductive process by producing the egg. This is also energetically costly as the egg comprises one-third of her total body weight. Researchers measured the energy burden (by respiratory oxygen consumption) and concluded the female's energy expenditure in the reproductive process was still greater than the male's role in child birthing. Even with this unusual life history adaptation rates of survival for newborn seahorses are extremely low, requiring a long gestation period for offspring survival. It is hypothesized that male seahorses have taken on the role of impregnation because of the lengthy period required for females to produce their eggs. By splitting the workload the sea horse is able to increase parity by shortening the birthing intervals. Bond pairing throughout the breeding season, sometimes even in community social groups, has been demonstrated to further increase probability of survival of this unusual genus.

Natural selection favors individuals whose age at first reproduction results in the greatest number of offspring over a lifetime (refer to Life History Continuum). The objective of reproduction with respect to the relative fitness of an individual is not always obvious. There are many activities involved in reproduction that actually increase an individual's probability of dying. The seahorse appears to have resolved some of those risks by sharing the burden of reproduction more equitably.

Reproductive Fitness

Reproductive fitness is a product of fecundity and survival. It underlies the trade-off between the timing and frequency of reproductive effort and the future survival of the parent. Some organisms reproduce early in their life cycle while others delay long enough for growth and development before reaching their reproductive maturity. The probability of future survival of the adult can be reduced through the actions of:

- Finding a mate
- Defending a breeding territory
- Daily feeding and protecting of their young

The king penguin (*Aptenodytes patagonicus*) is an example of a species with a breeding cycle lasting more than 1 year. In a reproductive study on the king penguin's survival and its probability of subsequent breeding the research showed king penguin males are more prone than females to skip a breeding season once they engage in chick-rearing their offspring.

The energy burden of a lengthy chick-rearing period in king penguin males negatively impacts resource allocation toward replenishing their body stores. It is believed the males often skip participation in the subsequent breeding season for this reason. It is noted that similar behavior was not observed in the female king penguin, whose breeding success in the following year remained unaffected (**Fig. 11.12**).

The balancing of reproductive expenditure vs survivorship is also evident in pink lady's slipper (*Cypripedium acaule*). In a 1990 plant study on this rare endemic orchid species Primack and Hall focused on a population of pink lady's slipper growing in a northeastern woody region of Massachusetts. The study included two plots. One was hand pollinated to increase the lady slipper's reproduction rate. The other plot was not manipulated and represented the normal reproductive rate (control group). Plants that were hand pollinated showed an immediate rise in fecundity over those in the control, nonmanipulated group. Plant growth was then observed during subsequent years of the study. By the third and fourth year the cost expenditure for increased reproduction in the hand-pollinated group resulted in overall decreased

Fig. 11.12

Penguin colony in Salisbury Plain, South Georgia, sub-Antarctic.

Source: Image courtesy of Mike Reyfman

plant growth and flowering rate compared to the control group. It appears that an increase in allocation of energy to reproduction, rather than time and energy for growth, led to diminished fecundity (5%–16% drop in probability of flowering). The age of reproduction is tightly aligned with optimal age-specific patterns of mortality (survivorship) vs fecundity (birth rate). Attempts to force reproduction in the pink lady's slipper shifted its life history continuum further away from optimal reproductive fitness that this species had adapted to.

Reproductive effort is defined as the portion of its energy budget (total energy expenditure, [TEE]) that is devoted to reproductive success. Energetic costs can be intrinsic or extrinsic.

Intrinsic energetic costs:

- Gonad development
- Nutrient demands
- Production and storage of gametes

Extrinsic energetic costs:

- Movement to breeding area
- Competition for mates
- Nesting
- Parental care

The amount of energy invested in reproduction varies widely across organisms. Evolutionary adaptations are not intentional and do not involve conscious thought, which is true of any adaptations to improve chances of an individual's reproductive success. The goal of the different strategies selected through evolutionary processes is to maximize **biological fitness**. The measure of biological fitness is relative to other individuals in the population and is about the successful reproduction of the species. Thus all life histories must center around the topics we've discussed: fecundity, parity, and their energy costs to adult survivorship. In animal species in particular the strategies to maximize biological fitness must take into consideration not just parental investment but also male competition.

Parental investment is inclusive of the energy costs associated with a successful birth and support for the young until they are capable of doing so for themselves. This includes acquisition of food resources, protection from predators or competitors, and any parental care designed to improve the offspring's chances of survival, even when at the cost of the parent's ability to invest in other offspring. Parental investment includes supplementing the egg with energy resources or other nutrition to the developing embryo (both internal fertilization and external). From the standpoint of TEE, species with high parental investment may experience a sharper drop in their adult survivorship when resources are scarce.

The time and resources allocated to adult maintenance (intrinsic) and avoidance of predation (extrinsic) are offset by what is allocated to producing its offspring. A higher degree of extrinsic factors arguably selects for evolution of shorter life spans and earlier peak fertility. The energy allocated to reproduction will always involve trade-offs. In an optimized life history conflicts between the demands for survival and reproduction are resolved to an individual's advantage for fitness. If an organism produces many offspring then each individual offspring will receive a smaller investment.

Alpine plants thrive in a harsh environment of low precipitation and temperatures and a shortened growing season. Any energy invested in flowers and seed production would be costly in this harsh habitat. And the risks of unsuccessful flowering are high due to early freezing in the warm season resulting in loss of flowers or poor pollination, delayed embryogenesis, and lack of time for seed maturation, dispersal, and its germination. Alpine plants therefore implement a lower risk, low energy form of asexual reproduction called clonal propagation (e.g., modular growth).

Animals that live in unpredictable environments, or areas where parental care is difficult to achieve, have lower probabilities of offspring survival. The strategy of having sizable offspring litters is justified by the increased chance that a few will survive and reproduce. For comparison the ocean sunfish releasing millions of eggs. In contrast the shark invests considerable parental time to nurture just one offspring. This is the case for most long-living animals, that is, more energy investment allocated for parental care. The amount of energy allocation varies with number, size, and maturity at birth. Species that are born relatively helpless required less energy allocation before birth but require additional parental care after birth. Species young that are born at a more advanced stage of development require more energy *before* birth but less parental care will be needed after birth. In genera, birds and mammals provide some parental care. And invertebrates (octopi, crustaceans, millipedes) often care for their eggs by brooding and defending them.

In the primate world paternal care is common and particularly noteworthy in the New World monkeys. The *Callitrichids* are small arboreal primates ranging from Brazil and Bolivia to Costa Rica. This small-bodied monkey family includes the cotton-top and emperor tamarins, which regularly give birth to twins in the wild, weighing 15% to 20% of the mother's body weight. For most of the parental care period the male Tamarin carries the infant, which is unable to self-thermoregulate during its first 2 months of life (**Fig. 11.13**).

Parental investment is much less significant in marsupial young (kangaroos, koala bear, possums) because the still-developing pups won't leave their maternal pouch until reaching independent homeostasis (**Fig. 11.14**). Marsupial investment in the fetus before birth is small with a shorter gestation. But there is the added investment of nurturing pups while still within the pouch. This mode of reproduction is thought to have evolved as an adaptation to environmental instability in the biogeographically distinct continent of Australia. It is proposed that the intermediate parental investment strategy of marsupials was favored in harsh environments because mothers could abandon a reproductive attempt (her developing pups) more readily. This would increase the likelihood of her survival to another reproductive event compared to the mammalian reproductive strategy with lengthier gestation periods and longer parental investment after birth.

Fig. 11.13

Emperor tamarin of the Amazon basin.

Source: Brocken Inaglory under CC by S-A 3.0

Fig. 11.14

Kangaroo and her pup.

Source: Pixabay under CC 2.0

Mode of Parity

Species differ in the number of reproductive events they experience during their life spans. This is known as mode of **parity**. **Semelparous** (Latin derived *semel* = once; *pario* = beget) organisms reproduce only once. Semelparous species include annual plants, insects, and a few species that live several years, like the Pacific salmon and the Giant Pacific octopus. The energy investment in growth, development, and energy storage occurs with one large burst of reproductive effort. The organism then dies after its mission to reproduce is completed. This mode of reproduction is consistent with *r*-selected strategies. There will be a large burst of reproductive effort with low survivorship and low parental investment. But some semelparous plants are longer-lived, for example, the agave and bamboo plants (life spans up to 100 years). These longer-lived plants reproduce *vegetatively* until the "big bang" of sexual reproduction at the end of their life span (**Fig. 11.15**). For true semelparous organisms death occurs immediately after reproduction. The century plant agave and bamboo exhibit a modified semelparity. This is because they vegetatively die back after reproduction, and new root structures will sprout in a new location.

Fig. 11.15

Century plant, an agave that flowers once every 10 to 25 years, immediately dies off but not before its root suckers grow into new plants.

Source: Stacia Zachary, US Air Force, Public Domain

Fig. 11.16

Alaskan sockeye vs Atlantic salmon. These salmon are in different genus and have different life histories.

Source: wolfpix is licensed under CC BY-ND 2.0

Iteroparous (Latin derived *itero* = repeat; *pario* = beget) organisms reproduce multiple times. Iteroparous species include most vertebrate animals, shrubs, trees, and perennial plants. This strategy is favored when adult survival to the next breeding period is likely, that is, species have low to moderate mortality. Evolutionary adaptations that favor iteroparity include greater offspring protections, including *in utero* growth, nursing (in mammals), and greater survivorship through minimized predation risks. The Atlantic salmon (Genus *Salmo salar*) is iteroparous in contrast to the semelparous Pacific salmon (Genus *Oncorhynchus*) (**Fig. 11.16**). The difference in life histories that evolved between the two salmon genus is the lower mortality risks in Atlantic salmon. They do not face strong competition to acquire their nesting sites and will not expend significant energy allocation to complete a single, large reproduction event. Instead Atlantic salmon's energy allocation is spent on its own survival in a fluctuating environment (periodic predation, limited food resources). For the iteroparous species the ability to have more than one reproductive event maximizes their opportunities to overcome risks incurred in the young offspring.

Biological fitness is always maximized when trade-offs lead to successful reproduction—the overriding goal of a species. Varying environmental conditions may favor iteroparity over semelparity or *vice versa*. Not unlike other life-history traits covered in this chapter, parity is a continuous variable. Semelparity and iteroparity are the extremes along a continuum of possible strategies of reproductive effort through time. The semelparity strategy concentrates reproductive effort in discrete time and the iteroparity strategy distributes reproductive effort over longer timescales.

Sexual Selection

Mating systems describe the pattern of mating between males and females in a population. In sexually reproducing species the selection of mates operates within a social framework. Life history traits of males are usually strongly affected because of competition for mates.

Monogamy is the formation of a pair bond between one male and one female. Monogamy is common in birds but rare among mammals (3%–9%). Monogamy is favored when both parents must cooperate to successfully raise their offspring. A pattern of monogamous behavior can occur for the duration of a breeding seasonal or only when young are born underdeveloped, or **altricial**. The offspring will require both parents to provide food, warmth, and protection, as is the case with nesting hatchlings. In mammals monogamy occurs in species where a male provides resources for the female and their offspring, and he defends their territory. But in instances when lactating mothers contribute all parental care for the survival of the young the males are considered free, in terms of fitness, to seek as many new partners as possible. Many apparently monogamous birds "cheat" by having extrapair copulations. Both female and male birds have been observed engaging in this behavior. The "cheating trait" in monogamous birds increases their fitness by producing offspring with another mate and by increasing the overall genetic variation among their offspring.

Polygamy is the acquisition of two or more mates by one individual. A pair bond exists between the individual and each mate. The individual with multiple mates usually does not care for its offspring. A polygamous mating system is favored when resources needed to reproduce (i.e., food or habitat) are unevenly distributed. Environmental and behavioral conditions lead to various types of polygamy. **Polygyny** occurs when a single male mates with multiple females. This is the most common mating system of animals. **Polyandry** is when the female has a pair bond with two or more males. Polyandrous mating is the less common form of polygamy. In some bird species the females compete for and defend their males and needed resources, especially a favorable habitat. In polyandry the female will lay a clutch of eggs with each male, and the male remains to incubate the eggs and care for its young.

Sexual Dimorphism

Sexual dimorphism is the phenotypic condition in dioecious species where the opposite sexes of the same species have distinct characteristics beyond those differences in their sexual organs (gonads), for example, attention-gathering colorful plumes and vocal cords found in males but not present in the females. These sexual dimorphic traits and behaviors increase a male's chances of mating success. Significant time and energy allocation will be spent in the process of locating, attracting, and mating with the sex partner. The development of these traits is not present in the female, who instead devotes much of her energy to preparing for and developing embryos through the birthing process. Charles Darwin first developed the theory of sexual selection and proposed two processes to explain sexual dimorphism: intersexual selection and intrasexual selection. Mating selection processes pertain to physical adaptation in the male (male:male competition) and aesthetic sense in the female, who in many cases plays a crucial role in selecting the most fit males.

Intrasexual selection involves competition between members of the same sex, usually male–male competition through exaggerated secondary sexual characteristics, including larger body size, enhanced aggressiveness, and even weaponry. Typical sexually dimorphic male evolution consists of armament for fighting between males, like the development of antlers, spurs, or canines to battle for preferential access to females.

Mating competition occurs when females mate only with a single male, typically the "winner" of the competition. Direct male competition often includes physical aggression between males.

In the polygynous Northern elephant seal, where males can weigh 3300 to 5100 lbs and measure 13 to 16 ft in length, a high-ranking bull mates with up to 50 females in a season (**Fig. 11.17**). The male-on-male fights for dominance are brutal, with significant bloodshed, though not usually to the death. The younger, less-capable males are chased away. Only a small percentage of males in the Northern elephant seal population—the healthiest, strongest, and most genetically inclined in the herd—will have the opportunity to breed during their lifetime.

Intersexual selection involves an individual of one sex selecting an individual of the opposite sex based on "attractiveness." Usually it is the female's choice but there

Fig. 11.17

Northern elephant seal.

Source: lutramania under CC BY-SA 2.0

can be male choices. Intersexual selection may lead to exaggerated characteristics in males related to sexual displays, including:

- Bright or elaborate feathers, fur, scales, or skin
- Vocalization
- Secondary sexual characteristics, including massive antlers and horns

The peacock is one of the finest examples of intersexual selection (**Fig. 11.18**). Males compete for access to females by flamboyant displays of plumage, covered with dozens of blue and green "eye spots." When being chased by predators the long, flamboyant plumage of peacocks has an added feature: the long tail feathers fall off easily when predators latch onto them, allowing the peacock to fly away to safety.

Female selection of a mating partner through display of attention-grabbing traits does present a disadvantage in higher predation risks to the highly visible male. A leading hypothesis to explain why females prefer showy male traits is the good genes hypothesis. First proposed in the 1980s by evolutionary biologist W.D. Hamilton and behavioral ecologist Marlene Zuk, it is based on the concept that sexually selected, showy male traits are "honest indicators" of good genetic quality. If a male can carry off a big flashy appendage around its predators and survive, what better indication that he has strength, quickness, and overall genetic superiority? Thus female preference is the selective force behind the evolution of exaggerated features like bright colors and showy plumage. The same can be said for male birds, whose measure of fitness is the repertoire and complexity of song they possess. Since acquisition of food is related to fitness females sometimes choose a mate based on acquisition of this resource. Often the food is located in a high-quality territory that the male defends.

Fig. 11.18

In the wild a peafowl (male) displays tall tail feathers to attract a peahen (female) for mating.

The good genes hypothesis has been challenged in more recent studies. In Atlantic mollies (*Poecilia mexicana*) the female selection criteria have been shown to change during a strong predator presence, favoring the less-showy males. And in the large-horned beetle the female selection process actually reversed natural selection away from larger, more robust beetles. This is an interesting case where the most fit (e.g., males with massive mandibles [jaws], broad masculinized heads, and necks with streamlined abdomens) were not the favored mating partner. These brawnier beetles could easily win the sexual selection battle if the goal was to lend their "good genes" to the gene pool. But it turns out that this phenotype (big chests/small abdominal region) is a disadvantage in the opposite sex. Specifically, the streamlined abdomens would lead to a reduction in reproductive success as the female would not possess her optimal body form for producing offspring. When the big horned males were preferentially killed there was an increase of 20% reproductive success in a controlled study using high predation. This example demonstrates the potential for an evolutionary tug-of-war compromise between sexual selection and natural selection.

Plant Mating Systems

Plants have two main types of mating systems. The first is **outcrossing**, that is, cross-fertilization between two individuals. The pollen (male gametes) from one individual will come in contact with the ovum (female gamete) of another individual. The second mating system is called **autogamy**, that is, self-fertilization events on a single plant between its own pollen and ovum flower parts. Outcrossing increases genetic diversity while self-fertilization (autogamy) does not. Different adaptations favor either of the two mating systems in their environment.

Adaptations that favor autogamy include:

- Timing to maturity of male (pistil/pollen) and female (stigma/ovary) at same time
- Closed flowers that enclose and protect the sexual organs
- Bending pistils toward stigma to increase chance of self-pollination over random pollen sources

Adaptations that favor outcrossing include:

- Separate timing to maturity of male (pistil/pollen) and female (stigma/ovary) parts
- Separation of the male (stamen/pollen) and female (pistil/ovary) into two separate flowers
- Removal of ovary through development of pulpy fruit to attract animals for dispersal
- Means of attracting pollinators (showy flowers and nutrient-rich nectar)

Outcrossing often relies on abiotic factors (wind and water) or pollinators and passive animal dispersal for its success. The gain in increased genetic diversity and hybrid vigor offsets its greater energy investment needed to produce large, external flowering organs that attract dispersal agents (pollinators).

In summary, all species developed life histories that account for the environmental challenges of individual survival vs the risks and costs of reproductive success.

Glossary

Fast-slow life history continuum model
Holobiont
reproductive effort
Monogamy
Polygamy
Polygyny
Polyandry
Intersexual selection
intrasexual selection
biological fitness

Citations

1. T. Garland, Jr., Trade-offs, Curr Biol 24 (2) (2014) R60–R61. https://doi.org/10.1016/j.cub.2013.11.036
2. P.W. Hughes, Between semelparity and iteroparity: empirical evidence for a continuum of modes of parity, Ecol Evol 7 (20) (2017) 8232–8261. https://doi.org/10.1002/ece3.3341
3. S.C. Stearns, The Evolution of Life Histories, 1st ed., Oxford University Press, Oxford, 1992.
4. S.L. Bock, et al., Spatial and temporal variation in next temperatures forecasts sex ratio skews in crocodilian with environmental sex determination, Proc R Soc B Biol Sci 287 (1926) (2020) 20200210 https://doi.org/10.1098/rspb.2020.0210
5. R.J. Spencer, F.J. Janzen, A novel hypothesis for the adaptive maintenance of environmental sex determination in a turtle, Proc Biol Sci 281 (1789) (2014) 20140831. https://doi.org/10.1098/rspb.2014.0831
6. T. Enriquez, H. Colinet, Cold acclimation triggers major transcriptional changes in *Drosophila suzukii*, BMC Genomics 20 (1) (2019) 413. https://doi.org/10.1186/s12864-019-5745-7
7. C.J. Krediet, K.B. Ritchie, V.J. Paul, M. Teplitski, Coral-associated micro-organisms and their roles in promoting coral health and thwarting diseases, Proc Biol Sci 280 (1755) (2013) 20122328. https://doi.org/10.1098/rspb.2012.2328
8. Y. Sawall, et al., Extensive phenotypic plasticity of a Red Sea coral over a strong latitudinal temperature gradient suggests limited acclimatization potential to warming, Sci Rep 5 (8940) (2015). https://doi.org/10.1038/srep08940
9. W. Booth, Facultative parthenogenesis discovered in wild vertebrates, Evolutionary biology. Biol Lett 8 (6) (2012). https://doi.org/10.1098/rsbl.2012.0666
10. S. Milius, Odd head of seahorse cloaks its sneak attacks, Sci News 157 (11) (2000) 168.
11. F. Jiguet, P. Jouventin, Individual breeding decisions and long-term reproductive strategy in the king penguin *Aptenodytes patagonicus*, Ibis. 141 (2008) 428–433.
12. R.B. Primack, P. Hall, Costs of reproduction in the pink lady's slipper orchid: a four-year experimental study, Am Nat 136 (5) (1990) 638–656. http://www.jstor.org/stable/2462226
13. C. Körner, Concepts in Alpine plant ecology, Plants (Basel) 2 (14) (2023) 2666. https://doi.org/10.3390/plants12142666

14. M. Edwards, J. Deakin, The marsupial pouch: implications for reproductive success and mammalian evolution, Aust J Zool 61 (2013) 41–47. https://doi.org/10.1071/ZO12088
15. F.R. Bernthal, J.D. Armstrong, K.H. Nislow, N.B. Metcalfe, Nutrient limitation in Atlantic salmon rivers and streams: causes, consequences, and management strategies, Aquat Conserv 32 (6) (2022) 1073–1091. https://doi.org/10.1002/aqc.3811
16. National Academy of Sciences (US), Mate choice and sexual selection: what have we learned since Darwin?, in: J.C. Avise, F.J. Ayala, (Eds.), In the Light of Evolution: Volume III: Two Centuries of Darwin, National Academies Press (US), Washington, DC, 2009. https://www.ncbi.nlm.nih.gov/books/NBK219729/
17. Burney J. Le Boeuf, Male-male competition and reproductive success in elephant seals, Am Zool 14 (1) (1974) 163–176. https://doi.org/10.1093/icb/14.1.163
18. S.L. Balenger, M. Zuk, Testing the Hamilton–Zuk hypothesis: past, present, and future, Integr Comp Biol 54 (4) (2014) 601–613. https://doi.org/10.1093/icb/icu059
19. D. Bierbach, Predator-induced changes of female mating preferences: innate and experiential effects, BMC Evol Biol 11 (2011) 190. http://www.biomedcentral.com/1471-2148/11/190
20. K. Kiyose, Sexually selected traits and life history traits of larger and smaller males of the horned flour beetle *Gnatocerus cornutus*, Ecol Entomol 46 (2021) 807–815. https://doi.org/10.1111/een.13016

CHAPTER

Species Interactions

12

Chapter Outline

12.1 Species Interactions Alter Population Dynamics

At the end of this section students should be able to:

- Explain why resource allocation is a major factor in species interactions.
- Describe relationships between species as neutral, beneficial, or harmful.
- Provide an example of species interaction on population size.

Species interactions affect the population dynamics for each species involved. For example, close interactions can influence the processes of birth and death by deferentially influencing individual survival and reproduction. Close interactions can act as agents of natural selection. Any species interaction that impacts life histories will thereby take on a vital role in a species' evolution.

Because many resources are in limited supply competition among species can be intense (**Fig. 12.1**). As such, acquisition of resources by an individual in one species reduces its availability for individuals in other species. For example, phototrophs all require the same resources (water, light, essential nutrients). In a dense forest with

Ecology for the 21st Century. DOI: https://doi.org/10.1016/B978-0-443-24031-7.00012-X

Fig. 12.1

Species interaction: intraspecific male-combat to gain access to females.

Source: stevevoght under CC BY-SA 2.0

a large canopy only filtered light reaches the understory (lower strata). Light availability is even scarcer on the forest floor due to shading from the overhead canopy as well as any understory plants. Approximately 2% of solar radiation reaches the forest floor, where large-leafed shrubs and new tree saplings thrive in scattered patches of sunlight (climate.gov). Shade-loving species, for example mosses and low-growing ferns, grow ubiquitous in the shaded forest with few competitors for resources. For heterotrophic animals a broad range of potential species interactions is possible, including predator–prey relationships, direct competition for resources, and even use of another organism as its habitat (e.g., fungi living on a host). In every habitat species interactions are capable of driving adaptations.

In 1988 Snow & Snow published *Birds and Berries: a Study of an Ecological Interaction*, describing biological interactions as a natural world in which no organism exists in absolute isolation. Snow & Snow stated that for its survival and the functioning of an ecosystem as a whole, every organism must interact with its environment and any other organisms within it. Direct interactions may occur only once in a season, for example, pollination, or for an entire lifespan, for example, through symbiotic interactions. Indirect species interactions occur when one species affects another species through intermediaries. This may occur through shared resources or through common enemies.

Symbiosis

Symbiotic relationships are close interactions between individuals of different species over an extended time period, including one entire phase of its life cycle or throughout its life cycle. Symbiotic interactions may impact the abundance and distribution of the associating populations. Some relationships are beneficial for both species, with each protecting the other from predators by mutual symbiosis. Using positive, negative, and neutral designations—(+), (–), or (0), respectively—species are impacted by a symbiotic interaction, as follows:

- Mutualism— both species mutually benefit from the interaction (+/+)
- Commensalism—one species benefits, the other is unaffected (+/0)
- Neutral—neither species affects the other, relationship is (0/0)
- Competition—detrimental to the populations of both species (–/–)
- Parasitism and predator–prey—one species benefits, the other is harmed (+/–)

Many examples of **mutualism** exist in pollination biology. An example of mutualism is the relationship between termites and protozoan and bacterial symbionts that live within the termite's gut. This relationship benefits both participants and is designated (+/+). The microscopic symbionts aid in the digestion of cellulose, which benefits the termite. This is because woody cellulose fibers are indigestible to the termite, even though wood is its main food source! Protozoan and bacteria provide a digestive function to break down the cellulose and in return get to live in the termite's intestinal tract, which hosts an optimal environment with ample food and a warm, protected niche for temperate microbial symbionts.

Lichens, composed of fungi and photosynthetic land algae, represent another example of mutualism found in nature (**Fig. 12.2**). Here the algal species is the primary producer that supplies enough carbon source for both members of the symbiont. The physical lichen structure is formed by the fungal species and serves to protect both species from the elements. The lofty proximity of fungi on branches and tree trunks facilitates access to atmospheric components to meet the algal photosynthetic requirements. Through coevolution the lichen–algal relationship reveals loss of unnecessary genes based on their shared lifestyle.

- The mutualistic interaction is mutually beneficial as it provides a benefit for both species and is designated (+/+).

A **commensal** relationship occurs between many birds and trees. The tree receives no benefit but will remain unharmed when birds build nests among its branches. The lightweight nest will put little strain on the structural integrity of the branch. And leaves are positioned above and surrounding the nests so they will still be able to complete their primary productivity function without interference. The bird receives a major benefit in safe sanctuary high above the ground, where its eggs and hatchlings are less vulnerable to predators. Another example of commensalism is the growth of an orchid on a tree limb. Unlike the mutually beneficial lichen–algal relationship only the orchid receives a benefit. Orchids are **epiphytes** and while some of their nourishments are extracted from the tree in most cases the host tree neither benefits nor is harmed by the presence of the orchid.

- This commensal interaction causes no harm while benefitting one participant and is designated (+/0).

Neutralism occurs whenever two populations interact in a community with no effect on the evolutionary fitness of either. In neutralism interactions are incidental or indirect and are considered to be without effect on either species' population dynamics. An example of neutralism is the presence of humans and pigeons together in cities. The pigeons favor human cities for the chance of picking up food scraps. Humans are not significantly affected by the presence of pigeons in the city. The presence or absence of one or the other participant is of no consequence. However, due to the complexity of all interactions within ecosystems the complete lack of competition between or benefit to either species in a neutralistic relationship would be an extremely difficult task to prove. As such, true neutralism is considered very rare or nonexistent. Neutralism is mostly used to describe situations where interactions are merely insignificant or negligible. In the example given pigeons did not directly interact with or need humans and only received benefit as a by-product (food scraps) of human presence. Another example would be the relationship of white-tailed deer and snakes sharing space within a forest. There are no direct species interactions occurring between the two species. Nonetheless studies have shown that in areas of high deer density there was a correlated increase in biodiversity of snakes and other reptiles and amphibians. The indirect benefits were caused by deer waste that led to richer soils bringing in more insects, which in turn provided more food resources for snake populations.

Fig. 12.2

Green lichen in Veneto, Italy.

Source: Umberto Salvagnin under CC by 2.0

- The neutral interaction is indirect and coincidental without harming either participant, designated as (0/0).

In **parasitism** an organism is dependent on the resources of another living organism, its host, and derives nutrients from it. In the parasite–host interaction one species is feeding off the other species, reducing its fitness though not necessarily killing it. Typically the host is weakened as the parasite siphons resources that the host needs for its own maintenance. Helminths are parasitic worms that reside in the intestinal tract of humans and other animals. The species–species interactions between helminths are their hosts can be lengthy, as the parasite needs to complete its life cycles then deposit eggs in the host feces, which facilitates its dispersal to a new host. Tapeworms have been known to grow up to 50 ft long in some animals. Sometimes the parasite requires two separate host species to complete its life cycle. The unicellular protozoan parasite, *Plasmodium spp.*, is the infectious agent of malaria and requires both mosquitoes and a vertebrate host to complete its life cycle (**Fig. 12.3**).

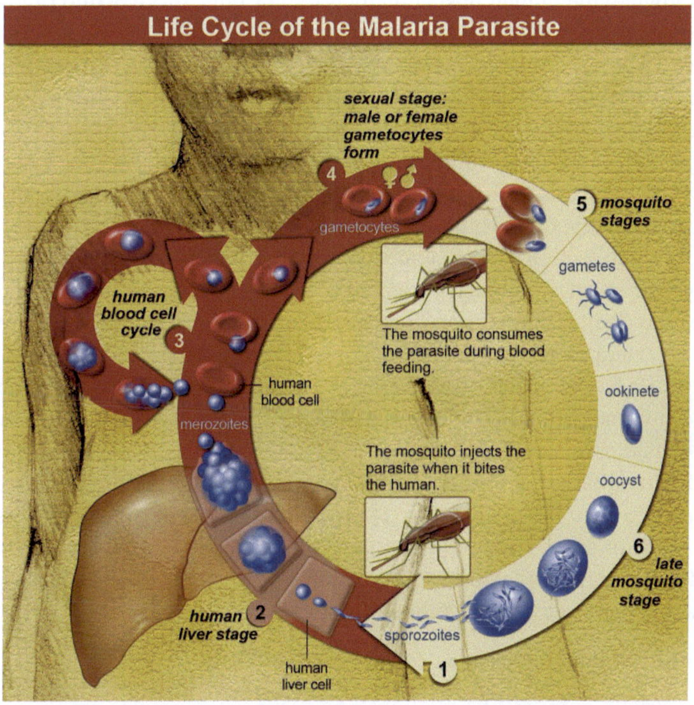

Fig. 12.3

Complex parasite–host relationship. The infectious agent of malarial disease, a protozoan *Plasmodium sp.*, requires both a specific mosquito species, *Anopheles*, for its juvenile stages and a vertebrate for its remaining life stages.

Source: US Center for Disease Control, Public Domain

A human contracts malaria after becoming infected with the parasite from a female mosquito of the *Anopheles spp*. The mosquito harbors the juvenile parasite, and its other life phases require a human host. Upon transmission of the adult parasite back to the mosquito during a "blood meal" asexual reproduction occurs in the mosquito's gut to complete its life cycle. Both the human (vertebrate) and *Anopheles* are required hosts and are harmed by their association with *Plasmodium*. Yet sometimes in parasite–host interactions some of the hosts species will not be harmed. Another protozoan, *Entamoeba coli*, exists in the human intestinal tract as a commensal.

- Parasite–Host interactions are designated as (+/–) if the host is harmed; designated as (+/0) if the host is not harmed.

One question that comes to mind in parasitic relationship is, "What is the ecological advantage for a life history strategy in which the essential host experiences mortality?," or "After the host's death how can the parasite remain viable?" In the case of malaria, a reemerging infectious disease with an annual mortality of half a million people worldwide, the human host actually does not limit survivorship of either the mosquito or parasite. As long as the mosquito survives, the protozoan parasite can hitch a ride to a new victim's bloodstream in this triad species interaction. Regarding the Anopheles mosquito, it has adapted to Plasmodium's presence with an enhanced immune response to avoid mortality.

The question on parasite viability in the ecosystem is an important consideration. As it turns out parasites have been shown to benefit control of dominant species and to increase competition and diversity among species. In Stephanie Godfrey's 2013 Wildlife publication it was demonstrated that parasites play an evolutionary role through the transfer of genetic material between species.

Besides the traditional parasite–host symbiotic relationship, there are numerous other species interactions designated as (+/–). The **predator–prey** dynamic occurs with one species feeding on the other. It is not a parasitic symbiont, however, because both species live independently of the other. If a predator kills its prey, then it becomes an agent of its mortality. As the number of predators ($N_{predator}$) rises the number of prey captured and killed will logically also rise. But then the increasing death rate and decline of its prey population (d_{prey}) potentially impact the predator's survival as well. The basis for decline in predator population is the reduction in its food supply, which is directly correlated with the population size of its prey. Reproductive fitness of a predator is therefore directly related to its consumption of prey.

- Predator–Prey interactions are designated as (+/–).

Finally, strictly **competitive** relationships occur when species are in rivalry for one or more resources. This can be food, mating, or territory. Competition includes interactions between different species (interspecific) or among the same species (intraspecific) competing for limited resources. For example, an *intraspecific* relationship exists between tree species when individual trees growing close together vie for the same sunlight and soil nutrients. Some tree species may adapt by growing taller or spreading their roots out farther. An *interspecific* relationship exists between lions and tigers competing for the same prey.

- In competitive interactions it is potentially harmful to both species (–/–).

Both predator–prey and competitive interactions can be modeled using the **L-V** equation. L-V is also known as the predator–prey equation. In this method the effects of the interaction on the species involved are predicted using a pair of linked first-order nonlinear differential equations. The model was developed independently in the 1920s by American mathematician and physicist Alfred J. Lotka and Italian physicist Vito Volterra. The L-V equation assesses potential benefits or death of a species involved in competition with another species after first determining each individual species' carrying capacity and population size. This is done separately and then the carrying capacity is calculated in association with the other species in the dynamic. As populations change through time according to the paired equations it is possible to assess a predator–prey relationship definitively. See Section 12.4 under Competitive Exclusion for more details on applications of Lotka-Volterra.

Figs. 12.4a–c illustrates how interspecific interactions between a predator and its prey (Canadian lynx and snowshoe hare, respectively) reflect cyclic population dynamics (Odum archives, 1953). The L-V model can be used to predict a range of outcomes whenever two species interact, for example, when a predator drives its prey to local extinction or if one species reduces the growth rate of a competing species over a specified time period. In the case of two species competing for the same resources, whenever *interspecific* competition is approximately equivalent between the two species a stabilization of both populations will result. In actuality one species gradually

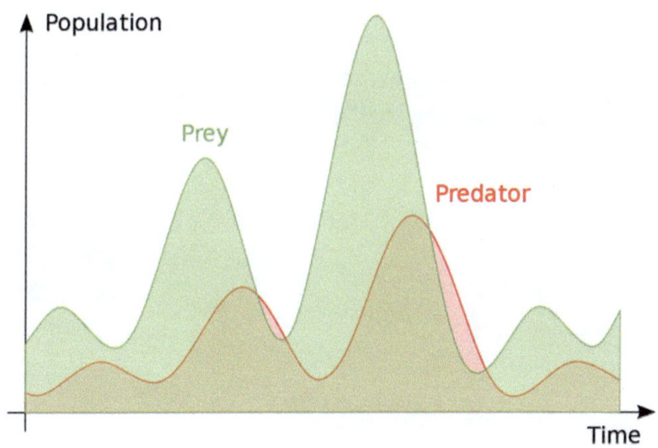

Fig. 12.4a

L-V model. L-V equations describe dynamics of species interactions in a predator–prey relationship.

Source: Curtis Newton under CC BY-SA 3.0

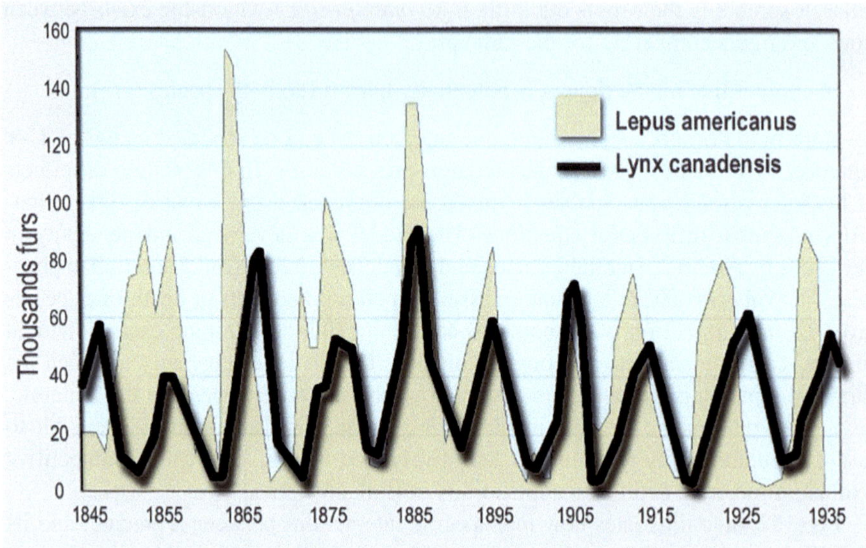

Fig. 12.4b

Lynx and the hare. As prey populations rise predator populations also thrive with their increased food resource that, in turn, diminishes prey populations, thereby decreasing predator population as its food supply becomes scarcer.

Source: Lamiot under CC BY-SA 4.0. https://en.wikipedia.org/wiki/Lotka%E2%80%93Volterra_equations#/media/File:Milliers_fourrures_vendues_en_environ_90_ans_odum_1953_en.jpgr

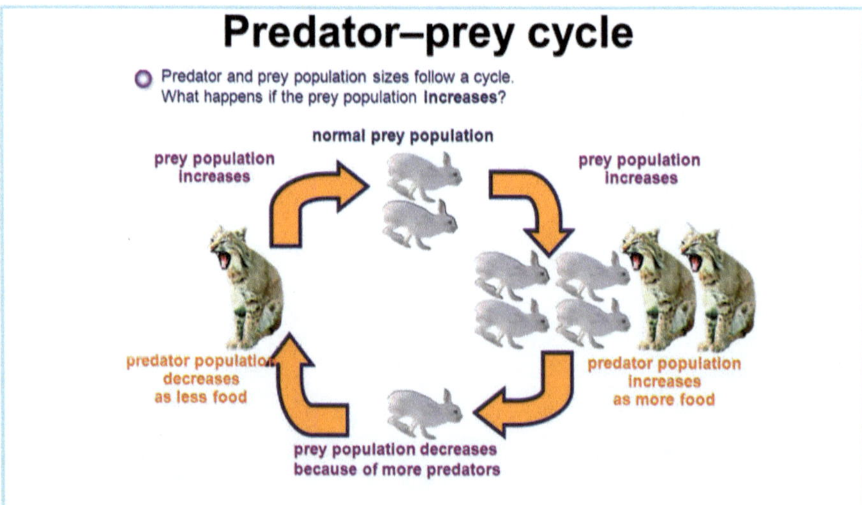

Fig. 12.4c

Predator-prey cycle.

Source: Lamiot under CC BY-SA 4.0. https://en.wikipedia.org/wiki/Lotka%E2%80%93Volterra_equations#/media/File:Milliers_fourrures_vendues_en_environ_90_ans_odum_1953_en.jpgr

outcompetes and excludes the other over time. In this scenario exclusion of one species will be dependent on the starting population sizes. That is, even with comparably equal interspecific effects the species with the higher initial population size is likely to drive the species with a lower initial population size to exclusion (local extinction).

12.2 Coevolution of Species

After reading this section students should be able to:

- Describe how species interactions are agents of natural selection.
- Relate the dynamics of species interactions to reciprocity in coevolution.
- Provide an example of both biological mimicry and camouflage.
- Explain how specialized pollination can increase rates of plant diversification.

The population dynamics of a single species will likely be influenced by a diverse set of species–species interactions. The individual interactions ultimately lead to different effects and include adaptations by natural selection. Because some characteristics and traits are expressed with variation among individuals within a population there will be differences in the types and degree of interactions. Unlike adaptation of an individual to its physical environment, species–species adaptations may result in reciprocal evolutionary responses designed to counter either species' advantage. In evolutionary biology there are many examples of mimicry and species coevolution to highlight type of reciprocity.

Camouflage and Biological Mimicry Developed as Survival Strategies

Organisms have a number of tools in their genetic toolbox that provide an added layer of protection from predators. An especially interesting tool is **mimicry**, whereby a species evades its predator because it has created an illusion that it is distasteful or threatening to the predator. Another prey strategy, **camouflage**, occurs by altering its phenotype to blend into the surrounding background, seeking to become invisible to its predator. Camouflage is limited by the ability of an animal to imitate morphological traits in the environment, including surface types, plants, or sessile animals like corals and sponges.

Of the two strategies, mimicry is more cost-effective since the organisms are not restricted to an environment that presents a safe background to disguise itself. The mimicry feature allows prey to move more freely without threat of attack. The camouflage strategy relies on phenotypic plasticity to meet the challenges from predators, which can include seasonal landscapes. Mimicry is more likely to be an established, permanent adaptation. Mimicry can also be an aspect of acquired behavioral traits.

Batesian mimicry is the term used when prey closely resembles an unpalatable species. For example, a fly that mimics a bee is less likely to be attacked by birds, which naturally avoid being stung.

Disruptive coloration is a form of camouflage common in butterflies. In this survival strategy the prey's identity or location is disguised or misidentified. The specialized adaptation is its coloration pattern. For example, circular patterns placed on the upper part of the wings, called eyespots, resemble the eyes of large owls. This is designed to confuse circling predators overhead.

In **aggressive lure mimicry** the predator (or parasite) develops characteristics that mimic a harmless species. The strategy is to avoid being detected by unsuspecting prey (or future hosts). One clever "wolf in sheep's clothing" strategy is a form of acquired behaviors. In the silver *Argiope* spider (*A. argentata*) elaborate patterns are woven in the middle of their webs. The patterns are designed to reflect UV light in such a way as to mimic actual nectar guide patterns used by flowers to attract bee pollinators. For bees in search of nectar, the Argiope spider anticipates fooling them with the illusion they created. Before the bees realize the light patterns are not nectar guidelines they will be entrapped in the spider's web (**Fig. 12.5**).

The coloration patterns of the devil's flower praying mantis (*Idolomantis diabolica*) are another example of aggressive lure mimicry. This exotic praying mantis has the capacity to impersonate a flower, not only in its morphology but behaviorally, as it will remain absolutely motionless in the presence of incoming prey (e.g., flies, moths, butterflies, and beetles). Most praying mantis are carnivores, considered friends of the garden, and have been known to feast on prey three times their size, including hummingbirds! *I. diabolica* is the largest praying mantis at 3 to 5 inches and considered one of the most beautiful.

Spider's Web: Argiope argentata/ Public Domain

Fig. 12.5

Argiope silver spider employs aggressive lure mimicry to recreate nectar guides for insect pollinators. UV-induced fluorescence of the yellow plains coreopsis reveals striking patterns in flowers used by bee pollinators.

Source: Craig Burrows, USFWS, Public Domain

Insects, initially fooled by the praying mantis's illusion, will already be in close proximity before becoming aware that the enticing flower is actually a predator. Too late, they are entrapped by the imposter flower's powerful legs and using the diabolical trickster's mandibles (jaw) it decapitates and feasts on its prey.

I. diabolica has more than one trick in its mimicry bag. When faced with an incoming predator the mantis deploys disruptive coloration mimicry. Praying mantis tip their body upward in the presence of predators. *I. diabolica* exposes an underbelly of brightly patterned red, blue, black, and white while raising its raptorial legs upward in a gesture that gives the illusion of being much larger. The tactic is to momentarily distract its predator long enough to make its escape. Called a **startle display**, intricate defense performances significantly enhance praying mantis biological fitness.

Both mimicry and camouflage are evolutionary engines in the perpetual struggle for existence between predator and prey. Since the predator dies if it does not get food, it must coevolve whatever is necessary to gain the selective advantage over its prey: speed, stealth, camouflage, a good sense of smell, sight, hearing, or immunity to poisons emitted by its prey, even developing characteristics like the right kind of mouth and digestive parts to capture and devour its prey.

Predator–Prey Dynamics

Coevolution is the process in which two species undergo reciprocal evolutionary change through natural selection. Coevolution is commonly seen in predator–prey dynamics because species are driven to adaptive changes in an attempt to thwart the other species' advantage. The apex predator snow leopard (*Panthera uncia*) has adapted through evolution and coevolution to successfully inhabit a wide range including some of the harshest climates in the mountains of Asia and the Indian subcontinent. As a predator it has acquired legendary stealth prowess and camouflage characteristics. For example, it is able to leap up to six times its length and climb steep mountain slopes that range from 6000 ft to 18,000 ft elevations (**Fig. 12.6**). Sometimes called the "ghosts of the mountains," camouflage is essential to survival for the snow leopard in a barren habitat with few prey options. The mountain ibex (wild goat) is prey for the lion and has coevolved to possess its own extreme prowess in the rocky terrains it cohabitates. The ibex escapes the snow leopard by having acquired the traits of extreme nimbleness with the ability to jump more than 6 ft *vertically* without a running start, easily traversing steep mountainous terrains ahead of its predator. Furthermore, to grip steep rocky cliffs, the ibex hooves have sharp edges with concave undersides that function like suction cups as it ascends cliffs and ledges (**Fig. 12.7**). The athleticism of each species is maximized through reciprocal coevolutionary adaptations.

Another example of coevolution between predator and prey is the complex relationship between a bat and moth. Bats use **echolocation** to navigate by ultrasonic sound waves rather than night vision to hunt nocturnal insects (**Fig. 12.8**). The bats are guided in the dark by emitting calls as high-frequency sound waves

Fig. 12.6

Snow leopard camouflaged in the upper valley of Phyang village of Ladakh.
Source: Tashi Lonchay Wiki Commons CC by SA-4.0

Fig. 12.7

Alpine ibex.
Source: Modification NPS, Public Domain

Fig. 12.8

Echolocation frequencies in bat–moth coevolutionary adaptations.

Source: Image by author

(12–210 kHz), which bounce off objects in their surrounding environment. Many bats can locate prey by listening for specific sound waves echoing when call signals are reflected off prey. In response to the predation by bats a number of adaptations have coevolved in insect species to improve their odds of survival. Moths are the preferred food source for many bats, and many moth species have developed an early warning system as a reciprocal coevolutionary adaptation. One adaptation is rapid fluttering of its wings to scatter the signal in an evasive maneuver tactic that confuses its predator. Another moth adaptation is the ability to hear an approaching bat's ultrasonic frequencies by enhanced sensory perception using auditory **tympanal organs**. Tympanal organs function similarly to human ears with sound waves triggering vibrations. The moth's highly enhanced tympanal organs are capable of picking up the ultrasound waves that send a stimulus to the insect's brain for perception of the incoming threat. The moth can then execute evasive flight maneuvers to avoid being killed.

In 2015 a report in *The Conversation* by ecologist D. Jacobs described a fascinating ongoing arms race between bats and moths, with some bats now equipped with a covert offensive weaponry (stealth echolocation) *inaudible* to moths' early warning system. Coevolution in some bats had countered the auditory defenses of moths by varying intensity or the duration of their calls so they were outside the detectable range of the moth's tympanal organs. For example, the North American spotted bat used ultra-low frequencies in the range of 12 kHz, and the African short-eared trident bat and the horseshoe bat used ultra-high frequencies in the range of 208 kHz.

Fig. 12.9

A barbastelle bat with its characteristic bumpy face.

Source: Image courtesy of Daniel Lewanzik, Max Planck Institute for Ornithology

Species coevolutionary interactions will be unique and specific between the two interacting species. For example, the Barbastelle bat in Europe uses echolocation at an ultra-low (10–100 times lower) intensity compared to calls emitted by other bat species that cohabitate in caves in the same region. As a result the Barbastelle bat feeds almost exclusively on the local moth population with a competitive advantage over the nonstealth mode bats (**Fig. 12.9**).

In the continuing saga of selective pressures and reciprocal adaptations in some cases the bat appears to be winning the arms race. At other times it is the moth's evasive armament that seems to have the edge. Coevolutionary processes act as agents of natural selection with genetic changes reinforced by beneficial (or less detrimental) interaction between the species.

Mutualism by Design

Many plants depend on animal pollinators to transport pollen between individuals. The relationship is mutually beneficial, with the pollinator receiving its main food resources in exchange for aiding a plant's reproductive success by distributing its seed. Characteristic attractants produced by the plant are rewards designed to attract pollinators, for example nectar (sucrose–fructose syrup) or nutrient-rich pollen. A striking example of magnified coevolution between a plant and its pollinator is the nectivore hummingbird and nectar-producing plants. Hummingbirds have very long bills and tongues specialized for extracting nectar from tubular flowers like the honeysuckle (**Fig. 12.10**) and passionflower plants.

Modified feeding structures developed over time through natural selection, with flower tubes and bill size variables being heritable characteristics. Nectar plants began to produce larger tube flowers with greater nectar production, which increased

Fig. 12.10

Ruby-throated hummingbird (*A. colubris*) and its extraordinary coevolutionary relationship with the honeysuckle plant.

Source: Dick Daniels under CC BY-SA 3.0

the number of visits by hummingbirds. The result is pollination occurring more often from this phenotype, resulting in enhanced fitness for the large nectar tube flowers with a corresponding shift in frequency in genotype within the population.

Concurrently any genetic mutation that increases bill length in the hummingbird would act similarly, selecting for a shift in the phenotype median. Over time coevolution between the hummingbird and the tubular flower gives rise to enhanced specialization (bill size, tubular length, and nectar quantity) that is preferred by both species. This raises a question, "Which came first the long-billed hummingbird or the long-tubed flower?"

Using the example of the 10 cm nectar tubes on the passionflower plant and its pollinator, the 11 cm sword-billed hummingbird, researchers at the Ludwig Maximilian University in Germany sought to solve the question. Through molecular analysis of 43 passionflower species, they first determined that long nectar tubes in passionflowers developed over 10 Mya, right around the time that sword-billed hummingbirds appeared. This indicates that the flower evolved to meet the needs of its pollinator. The study also revealed that around 4 Mya, during a period of very low sword-billed hummingbird population size, existing passionflower species responded by turning on an evolutionary switch to produce shorter flowers more easily assessable by other pollinators (birds, insects), in what the authors called "escape from extreme specialization." However, when the population size of sword-billed hummingbirds was restored the passionflower switched back to its earlier phenotype.

The species interaction between the hummingbirds and nectar-producing plants is clearly more favorable to each participant, even if not deterministically confined by its coevolution. This makes sense because other birds and insects are not nectivores and would not benefit from large quantities of nectar, thus would be less likely to visit nectar plants when pollen-producing plants are nearby. Recall the bioenergetics of nectar consumption in the hummingbird (Chapter 8) vs dietary preferences for specific pollen types by bumble bees (Chapter 10).

A delightful mutualistic interaction occurs between the predatory Peruvian tarantula *Xenesthis immanis* (up to 8 inches in length) and a tiny, narrow-mouthed microhylid frog, *Chiasmocleis ventrimaculata* (0.5 inches in length). The microhylid frog receives protector status by the spider in return for its caretaking services surrounding the spider's nest eggs. First recognized in Peru in 1989 by Crocraft & Hambler the coevolutionary strategy between the tarantula and the microhylid frog was also shown to have evolved independently between the two genera in other regions, including India and Sri Lanka. Normally the tarantula feeds on small spiders but microhylid frogs are unpalatable due to their skin toxins. But the microhylid genera specialize in eating endoparasites like *Dipteran* flies, which are harmful to the spider, and ants, which target spider eggs and juveniles and are too tiny for the spider to capture. The species dynamic is mutual as the frog benefits by sharing a burrow or tree hole with the spider and whenever food scraps from the spider's diet attract small invertebrates the frogs can feed on them. The formidable tarantula assumes a protective role against any likely predators of the frog, which include arthropods, geckos, and snakes.

In 1898 William Trelease, a prolific author and early president of the National Botanical Society, studied species of Joshua tree (*genus Yucca*), native to the arid southwest United States and its main pollinator yucca moths (*genus Tegeticula*) (**Fig. 12.11**). He noted the uniquely suited yucca floral shapes, which the moths entered to access its pollen and to complete their reproductive cycle. The female yucca moth **oviposits** in Joshua trees cut through the ovary wall of Joshua tree flowers to lay their eggs on its ovules. The ovules harbor the moth eggs in its protected, nourishing environment and the yucca moth reciprocates by pollinating the tree flower. Genotypic variations must have evolved to facilitate the precise alignment for reproduction to be successful. The moth's ovipositor organ must be long enough to reach the ovules without injuring it, which would be a detriment to providing intact seed nourishment for her new offspring. Coevolution favors matching the length of the ovipositor to the flower's **style**. If too many eggs were hatched, however, it would be detrimental to the tree, which has the ability to actively abort ovaries. This relationship is considered to be **obligate** mutualism because the interaction is required for both species to complete their lifecycle.

Coevolution in Character Displacement

Interactions that negatively affect both species in the relationship (–/–) can result in divergence of phenotypic characteristics. Divergence serves to *reduce* the intensity of the interaction and is often seen in competitive interactions of species seeking the same limited resources. Ecologists have speculated that species' interactions

Fig. 12.11

Female *Tegeticula antithetica* moth oviposits of a Joshua tree (*Yucca jaegeriana*). In the closeup view of the floral parts four petals have been removed to view the moth's tentacles, which are visible on either side of her proboscis and coiled around a ball of pollen.

Source: Image courtesy of William (Tad) Cole

between flowering plants (angiosperm) and pollinators have driven divergence and speciation ever since Darwin first hypothesized their coevolutionary relationship. Character displacement occurs when similar species overlap in geographic ranges and begin to accentuate their differences. It is biological competition for limited resources that drives character displacement.

In the eastern and western ranges of the North American Mohave Desert, there are two different populations of Joshua trees, which illustrate coevolutionary divergence. The species possess distinct phenotypes in height, breadth, and branching patterns. The trees are pollinated exclusively by two species of yucca moths. One moth species, the *Tegeticula antithetica*, pollinates primarily in the eastern range and the other moth species, *Tegeticula synthetica*, pollinates primarily in the western range. The moths differ significantly in body size (30%) and ovipositor (50%). The western range Joshua trees are pollinated by *T. synthetica* and their floral parts have significantly longer stylar canals than the Joshua trees in the eastern range, which are pollinated solely by *T. antithetica.*

Many studies focusing on the obligate relationship between the desert Joshua tree and *Tegeticula spp*. (Yucca moths) have led to a greater understanding of the role of pollinator coevolution as an evolutionary mechanism for *species divergence*. Floral features associated with pollinator mutualism are under strong selection and evolve more rapidly than vegetative parts. We can anticipate a fitness advantage when adaptations for reproduction success evolve more rapidly than other characteristics adapting to the environment. Recall that surviving long enough to reproduce is a driving force in an organism's life history. Furthermore, in the pollinator species dynamics demonstrated in this section note that it is the plant expressing a greater plasticity toward adapting to the needs of its pollinators.

In a 2008 study by Godsoe in the *American Naturalist* the Joshua tree was identified as the most likely host of the *Tegeticula* genus common ancestor. Using an analytical tool, Ancestral State Reconstruction, Godsoe demonstrated that the differences in morphology between two yucca moth species arose in conjunction with morphological changes in the Joshua tree's evolution. Godsoe's work emphasizes

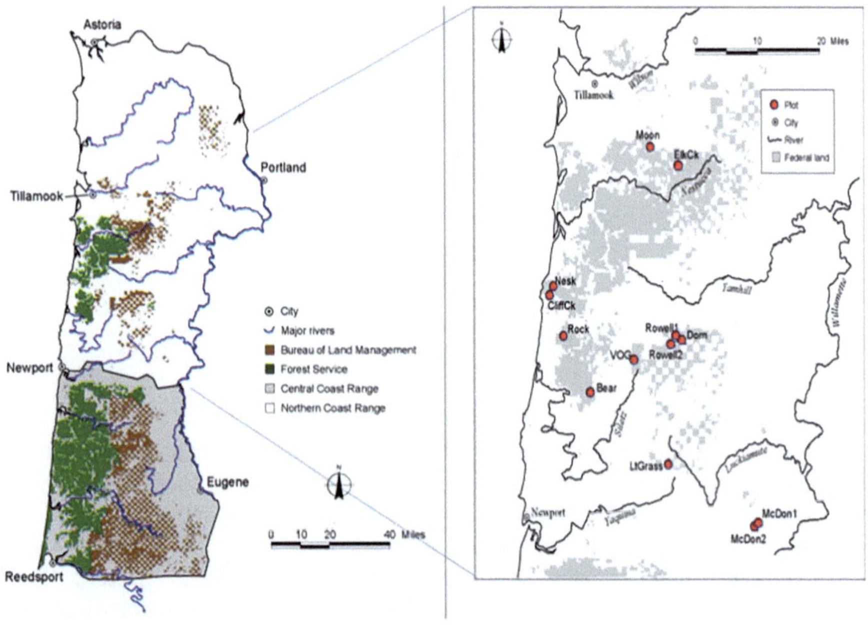

Fig. 12.12

Three hundred years of diminishing US virgin and old growth forests. In 2022 President Joe Biden called on the US Forest Service to prioritize the country's old-growth trees by amending the Northwest Forest Plan, which was enacted in 2024.

Source: Modification of A. Gray and National Park Service, Public Domain

the powerful analytical tools available for modern evolutionary studies. The current hypothesis is that pollinator selection *drove* divergence in the Joshua tree in eastern and western ranges of the Mojave Desert to match its pollinator moth morphology. As competition drove character displacement and divergence of the moth into two species the Joshua tree then coevolved to match its pollinator.

12.3 Interactions Influence Species' Niche

After reading this section students should be able to:

- Provide an example of species interactions influencing a species niche.
- Discuss advantages and disadvantages in generalist and specialist modalities.
- Describe fundamental and realized niches and give an example of each.

The locations where organisms are found define their habitat. The term ecological niche is more descriptive and explains the role an organism maintains within its habitat (i.e., how the species interacts). Species thrive where there are many usable resources compared to areas where there are many challenges, predators, or parasites. The niche of a species depends on both biotic and abiotic factors, which will determine whether the species will adapt and survive.

Generalist vs Specialist

The nature of species interactions can vary across geographic landscapes. Generalists are species that reside in a wider geographic distribution and experience greater variety of physical environmental constraints. Specialists have more restricted ranges and less adaptive pressures. The generalist has broader niches whereas the specialist has narrow niches. Generalists include the common cockroach but also coyotes, dandelions, and humans. Any changes in the environmental conditions will favor the adaptable generalist species for survival over specialist species. The generalist retains the ability to tolerate a wide variety of food resources and a wide range of living conditions. The opposite tends to be true of the specialist species. Specialists include endemic species like the northern spotted owl, which requires old-growth forests, as well as the giant panda of China, whose food source is primarily the bamboo plant. Specialists have an advantage when environmental conditions are more constrained. However, specialists are more prone to extinction during limiting resources or changing environments.

The nonmigratory northern spotted owls are majestic specialists with wingspans up to 4 feet. These birds suffer when their habitat is disturbed and fragmented (**Fig. 12.12**). Over the past 20 years their population size has declined by 40% due to loss of habitat (**Fig. 12.13**). The old growth forests do not quickly return after logging, and in the interim the northern spotted owl is left homeless. This is because older forests develop multiple layers from fallen trees, opening up the canopy and creating spaces for the owls to spread their wings to their full extent. Older forests also

Fig. 12.13

Northern Spotted Owls live in old-growth forests of the Pacific northwest, still largely untouched by human encroachment.

Source: Image courtesy of Francois-Xavier De Ruydts

contain fungus and other food sources for their favorite prey, the northern flying squirrel. And the monogamous northern spotted owls have established nesting sites that are not easy to rebuild in young woodlands that lack the complex structures and nesting cavities of the old growth forests.

Efforts to move the owls to newer forests have failed. Currently most remaining northern spotted owls are on federally protected lands. Besides being confined to 10% of their original habitat the northern spotted owl faces a new threat from competition. The barred owl leapfrogged from its native eastern US forests to the Pacific northwest forests and is now considered an invasive predator in the Pacific northwest. Barred owls are generalists, of the same genus and sometimes even mate with northern spotted owls. According to USGS the barred owls are quickly overtaking the old growth forests and suburbia, even as conservation efforts are under way to remove the barred owl from the forests. Along with recent pressures to open up federal lands, perhaps the best future for the threatened northern spotted owl will be realized through mating with its generalist competitor.

Genetic drift may occur when a species' range covers a wide geographic area and localized populations of the species face different extrinsic and intrinsic dynamics. Species that experience a broader range of biotic interactions and greater diversity from competitors, predators, or pathogens will face distinct selective pressures and adaptations to their local environment. For example, the common juniper, *Juniperus communis*, is found throughout the entirety of the northern hemisphere's cool temperate climate (Arctic South to 30°N) (**Fig. 12.14**). Based on known variation in selective pressures across this swath of Earth one might deduce there are many

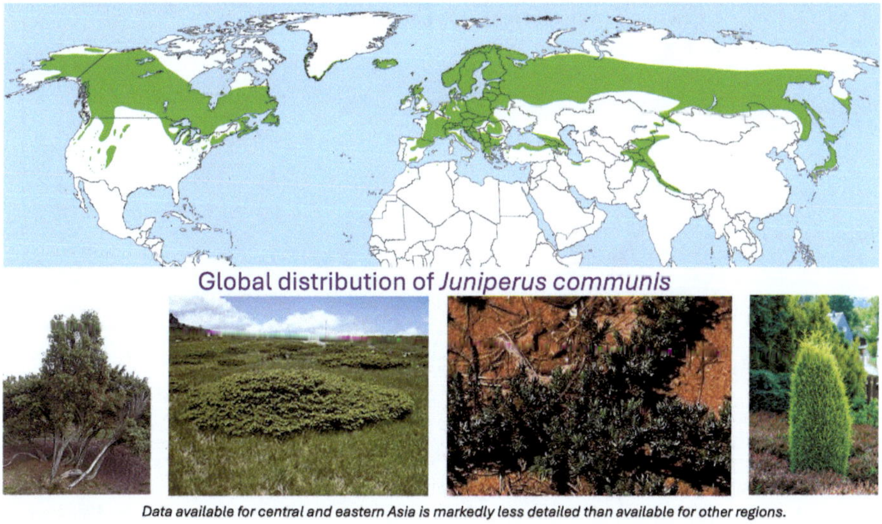

Fig. 12.14

Distribution and diverse morphologies of *J. communis*, inclusive of all subspecies and varieties (limited data available for central and eastern Asia).

Source: Modification of Caudullo, G. et al., 2017 under CC BY 4.0

subspecies of the juniper with extensive species morphological diversity. Indeed this is the case. *J. communis* grows either as a tree or shrub and can be found thriving in a wide range of soil types (sandy, loamy, clay), pH ranges, sunlight, and water. *J. communis* is the epitome of a generalist species.

Biodiversity Hot Spots

In contrast, specialist species are fastidious and limited in nature. They are often found in hot spots of global biodiversity. **Biodiversity hot spots** are classified as any region that has a high degree of biodiversity that is being threatened with destruction. These regions often contain endemic, specialist species that do not grow successfully under changing climatic conditions. Examination of hot spots on biodiversity maps illustrates that the trend toward higher biodiversity is clustered in the equatorial latitudes (**Fig. 12.15**).

To be designated as a hot spot the region must meet two strict criteria:

1. Contain at least 0.5% or 1500 endemic vascular plant species and
2. Have lost 70% or more of its primary vegetation.

Across the globe 36 areas have been identified that support 60% of the world's plant, bird, mammal, reptile, and amphibians. Biodiversity hot spots are environments with optimal growth conditions for many specialized, endemic species. An

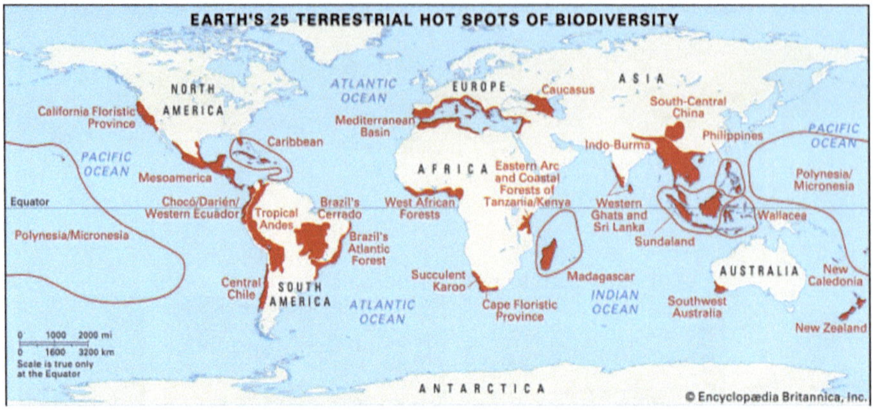

Fig. 12.15

Myers biodiversity hot spot map.

Source: After Myers

example of a biodiversity threat is the destruction of moist tropical rainforests for insertion of palm tree plantations. The endemic species of the southeast Asian tropical forest have small ranges, making those species now at risk for extinction. As the tropical forests are cleared out for conversion of land into plantations to produce palm oil or rubber many vulnerable endemic species will be irreparably lost.

In a 2006 study by Duke University 28 bird species, 147 amphibian species, and 42 mammal species were identified as "at risk" yet were not categorically listed as threatened or endangered by **IUCN**. Ongoing studies funded by Duke University reported a decade later (2016) that 42% of all species endemic to the region's forests faced higher risk of extinction from habitat loss than previously understood.

Habitat loss severely impacts fauna and animal diversity indiscriminately. A good example in the animal world is the solitary and territorial leopard (*Panthera*). The *Panthera* genus (genera includes the snow leopard) have an estimated loss of 75% of their historic geographic range, according to survey analysis by three conservation partner organizations: the National Geographic Society Big Cats Initiative, the Zoological Society of London (ZSL), and *Panthera* (**Fig. 12.16**). This is even more shocking because the *Panthera* species are highly adaptable generalists. Habitat loss is largely due to human encroachment and climate change but also trophy hunting and high demand for body parts (pelts, ivory, or bones used in traditional Chinese medicine). These combined factors are driving leopards into new and more-constricted ranges. The once-massive historic range of the leopard (13.5 million square miles) meant they could hunt a wide range of prey across parched deserts to the jungles of Java. Remaining individuals of the snow leopard (*Panthera uncia*) appear to be slowly adjusting to lower oxygen levels in

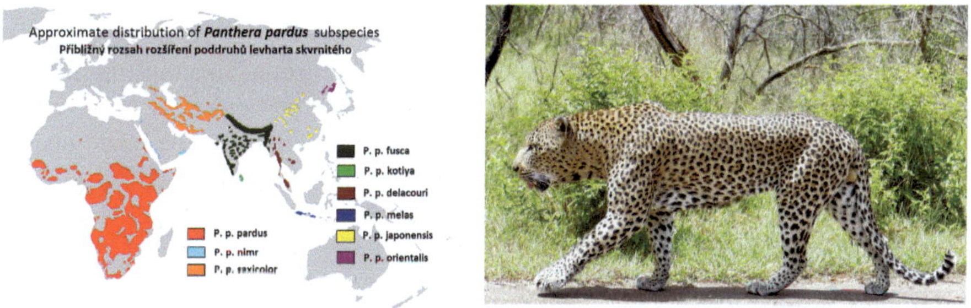

Fig. 12.16

Leopards (*Panthus sp.*) are now restricted to 25% to 37% of its historic range. Three of its nine subspecies (*P. p. pardus, fusca, and saxicolor*) reside in 97% of its remaining extant range. The other six subspecies are threatened with extinction, subsiding in isolation on small patch sizes of fewer than 100,000 km^2.

Source: Whitesachem under CC BY-SA 4.0

the extreme altitudes of the Asian mountain ranges of its forced migration due to habitat loss.

New susceptibilities to parasitic infections are emerging in immuno-compromised leopard species. The cause of immune depression is attributed to mounting environmental stressors while adjusting to their new life in new territories. One infectious agent, the microscopic mite *Notoedres cati*, is a bloodborne pathogen that infects snow leopards, causing fever, broad inflammation, and diminished energy. The apex consumer slowly dies as it is deprived of its selective advantage to hunt successfully for food resources in its sparsely populated rugged terrain.

Loss of Symbiosis

Even a mutually beneficial relationship can be altered when there is a change in the environmental background. Plants and fungi have existed in symbiotic relationships throughout a long history of evolutionary changes. Fungal mycelial threads will attach to the plant's roots and aid in nutrient extraction from the soil and water uptake for the plant's use. This supplemental boost is particularly useful in times of limited nutrients. In return, the plant provides the fungi with a ready carbon source in the form of sugars or starches. However, in lengthy periods of ample nutrients in the soil the plant would no longer need to rely on the fungi interaction. In this new environment the plant receives no benefit from the interaction. But under its coevolved symbiosis the plant would still be obligated to continue to feed the attached fungi. This would now be deemed an energetic loss of the plant's net carbon balance and its growth potential. In this instance the mutualistic relationship (+/+) changes to one that benefits one party but potentially harms the other, which trends toward a more parasitic (+/–) relationship.

The Ecological Niche

The characteristics that distinguish each species often reflect adaptations that allow individuals to survive, grow, and reproduce under a very particular set of environmental conditions. The **ecological niche** of a species is defined as the range under which a species is able to survive and reproduce and includes both chemical and physical attributes. There are two types of ecological niches:

- fundamental niche
- realized niche

The niche concept was independently developed by two ecologists, Joseph Grinnell (1917) and Charles Elton (1927). Grinnell, a field ecologist, emphasized the role of habitat and limitations imposed by the physical environment on species distribution. In **Grinnellian niche** theory the niche is distinguished by the different species occupying the same space but that only one species could hold a *particular* niche within that space. The British animal ecologist, Elton, emphasized in his classic book, *Animal Ecology*, that species interactions not abiotic factors are the primary determinants. In the **Eltonian niche** it is the trophic position of species in community interactions that drives species distribution. Elton placed less emphasis on competition between species. A good argument could be made for either driver of species' distribution.

In 1957 zoologist G. Evelyn Hutchinson developed a new niche concept, Fundamental vs Realized, as a compromise between Grinnell's and Elton's approaches to species distribution. An ideal niche would exist in absence of all competition. This is the **Fundamental Niche** concept and considers the environmental conditions under which a species can survive and reproduce, sometimes called its physiological niche. The fundamental niche includes the entire set of environmental conditions under which a species can persist. For example, there are three dimensions in an aquatic environment: temperature, salinity, and pH. Alteration in any of those parameters increases or decreases the size of a species' fundamental niche. Hutchinson's **Realized Niche** considers the portion of the fundamental niche that a species is actually able to use as a result of interactions with all other species in the area. A species-realized niche is typically a much smaller range than its fundamental niche.

A classic example of the difference between the two niche types (fundamental vs realized) is found in the barnacle types that reside in the deep and shallow intertidal zones along the rocky shores of the Atlantic northwest. The rocky shoreline is the fundamental niche of barnacle species. But competition from one barnacle species, *Balanus balanoides*, forces another barnacle species *Chthalamalus stellatus* to occupy a smaller realized niche on higher, drier surfaces. The difference between fundamental and realized niche introduces a central concept of the ecological niche, which is the **competitive exclusion principle**.

In 1961 field work conducted by Joseph Connell in the Shetland Islands off Scotland demonstrated the species interactions between barnacle species. Intertidal zones often show a vertical zonation pattern based on which species can survive in the open air at low tide or submerged under water at high tide. Adult individuals of the species *Chthalamalus* existed at the intertidal zone in higher locations than adult individuals of the larger species, *Balanus*, of another genus. But a completely different scenario was observed by Connell in young individuals of the *Chthalamalus*, which were distributed over a much larger range, including overlapping with *Balanus* in the lower intertidal zone (**Fig. 12.17**).

This led Connell to question whether each species was in its fundamental niche or whether they were being displaced due to competition. Connell monitored survival of both species by placing them in various vertical zones, with or without the presence of the other species. By mapping individuals over a 1-year period, including some sites where only young *Chthalamalus* were permitted to settle (by manual removal of *Balanus*), Connell's team made the discovery that adult *Chthalamalus* that did not come in contact with *Balanus* could thrive in the lower intertidal zones. His discovery demonstrated that submergence times of the lower zones had no impact on *Chthalamalus* growth but rather it was competition from *Balanus* that prevented its occupation. The opposite was true of *Balanus*, which could not survive in the upper intertidal zones with or without the presence of *Chthalamalus*, presumably due to exposure to the air during low tide.

The fundamental niche of *Chthalamalus* can be said to include the entire intertidal zone. In his direct observations Connell confirmed that *Balanus* smothered, undercut,

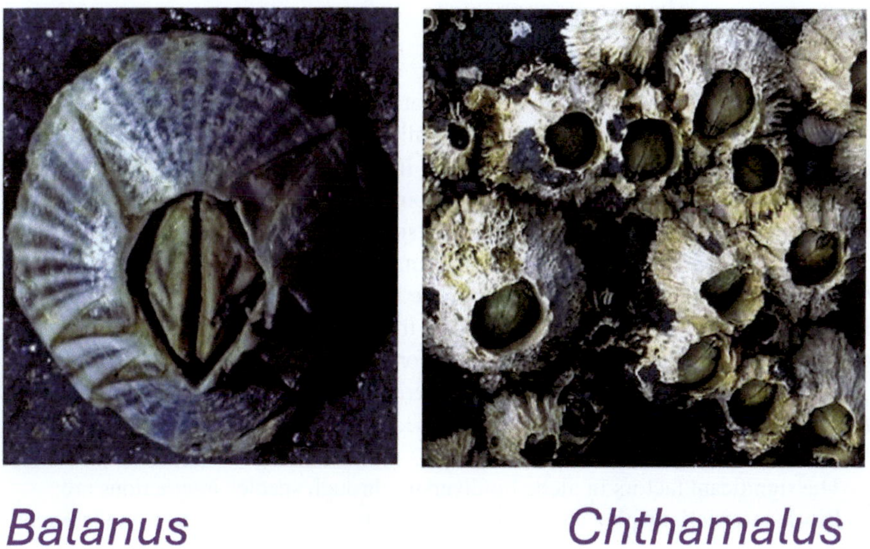

Fig. 12.17

Intertidal zone distribution of *Balanus balanoides* and *Chthalamalus stellatus*.
Source: After Connell, 1961

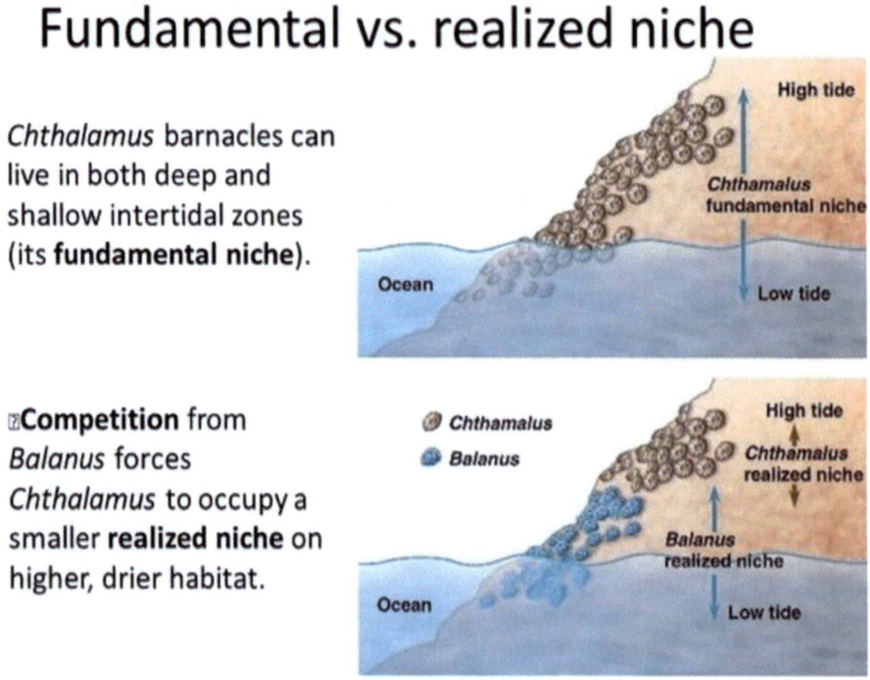

Fig. 12.18

Species competition in the intertidal zone.

Source: After Connell, 1961

or crushed *Chthalamalus* individuals and that the seasons of rapid *Balanus* growth experienced the highest *Chthalamalus* mortality rates (**Fig. 12.18**). The few surviving *Chthalamalus* individuals were smaller in size and produced fewer offspring. Connell demonstrated that interspecies competition was occurring and led to reduced fecundity for the *Chthalamalus* outside of its realized niche. *Balanus* and *Chthalamalus* compete and coexist on the same shore but their distributions overlap very little. *Balanus* outcompeted and excluded *Chthalamalus* from the lower zones. However, in the upper intertidal zones where the changing tides expose the barnacles to the atmosphere *Chthalamalus* is adapted for survival whereas *Balanus* is not.

Over evolutionary timescales species interactions play a major role in determining the range and conditions under which species can survive and reproduce within their ecological niche.

The significant factors in niche biodiversity through species interactions are:

Biotic interactions play an important role in determining a species' fundamental niche when the species interactions serve as selective agents for adaptations.

Adaptive radiation is a term introduced under evolution and speciation (Chapter 6). Adaptive radiation occurs when evolution of an animal or plant group expands into a wide variety of types adapted to very specialized niches and life histories. Through adaptive radiation, one species gives rise to a number of species that use different parts of the environment, such as habitats or resources.

Reproductive isolation was also covered under evolution and speciation. Reproductive isolation occurs when there are biological factors (barriers) preventing two species from producing viable, fertile offspring together. Reproductive isolation may result as a by-product of changes in morphology, behavior, physiology, and habitat preference. The changes lead to defining organisms as now-distinct species.

12.4 Competition and Predation

After reading this section students should be able to:

- Define interspecific or intraspecific relationships.
- Identify when species dynamics invoke the Competitive Exclusion Principle.
- List biotic and abiotic factors that influence species' relationships.
- Explain how species' interactions change along environmental gradients.

Interspecific competition occurs between two or more species and is a cornerstone of evolutionary ecology, affecting the populations adversely (–/–). Ecologists investigate interspecies competition to answer the following types of questions:

- How does competition affect the species involved?
- What type of interactions are involved?
- When does competition lead to species divergence and specialization?

It is possible for *intra*specific and *inter*specific competition to occur simultaneously, with intraspecific competition among members of the same population.

The Competitive Exclusion Principle

This principle describes why two species can't share the same niche in a habitat and coexist stably. Competitive exclusion was in effect in the vertical placements of barnacles along intertidal zones. It was also the driver for character displacement of Joshua trees to meet pollinator specifications. Darwin's theory of natural selection accounted for the competitive exclusion principle. And in 1917 it was the basis of the Grinnellian niche. But in 1934 Gregory Gause performed a series of classical experiments that demonstrated that complete competitors can never coexist at constant population values. Gause proposed that species with identical niches understandably have identical needs, which means they would compete for the same resources. Ultimately the weaker competitor will go extinct or make an evolutionary shift in its ecological niche. His landmark experiments were performed on the eukaryotic single-celled protozoan, *Paramecium* (**Fig. 12.19**).

In these studies two species, *Paramecium aurelia* and *Paramecium caudatum*, thrived when grown individually in the laboratory setting under the same conditions. But when grown in a mixed habitat (same test tube), under limited nutrients,

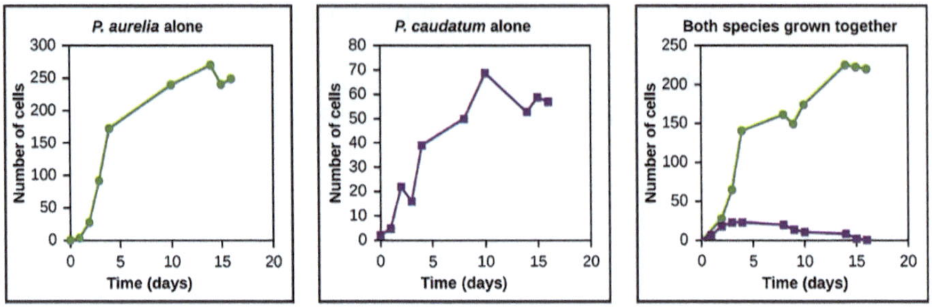

Fig. 12.19

Paramecium competition model.

Source: CNX OpenStax under CC BY 4.0

both species' growth habits decreased. Over time *P. aurelia* outcompeted *P. caudatum* for their shared resources, which led to the eventual extinction of *P. caudatum*. Under natural conditions it isn't realistic to find two species occupying identical niches in nature at the same time and in the same manner. Nonetheless it is reasonable to accept that the greater the extent in which two species' niches do overlap the stronger the competition between them will be. Competitive Exclusion Principle is commonly referred to as **Gause's Law**.

From research on competition exclusion, multiple factors are known to affect the outcome of an interspecific competition:

1. Environmental influence (survival, growth, and reproductivity)
2. Spatial and temporal gradients (variation) in resource availability
3. Competition exclusion with limiting resources
4. Resource partitioning

Environmental influence. Nonresource environmental factors have the potential to influence the outcome of competition between species. For example, water temperature affects the competitive ability of native fish species (brook trout, brown trout, and creek chub) in the Rocky Mountain Streams. The brook trout are found in higher elevations (cooler temperatures), the brown trout in median temperature ranges, and the creek chub is found in the lowest elevations and warmest climate. Taniguch et al. reported that under ambient temperatures of 20°C (68°F) the two trout species were competitively equal to each other but competitively superior to the creek chub (which spawns at 13°C to 18°C [55.4°F–64.4°F]) (**Fig. 12.20**).

However, when temperatures rose to 22°C to 24°C (71.6°F–75.2°F) creek chub began to display competitive success against both trout species. And at even higher temperatures Creek chub competitively dominated, with increasing lethality, first the brook trout and then brown trout species.

Temporal variation. Another nonresource factor that can influence competitive interactions occurs when environmental conditions are variable. The competitive

Fig. 12.20

Rocky Mountain River fish.

Source: Ellen Edmonson, Duane Raver, public domain

advantages of one species may change over time, favoring a different competitor. The influence might be due to greater adaptability in its phenotype, including behavioral plasticity. For example, ants have the ability to quickly respond to temperature changes by altering their metabolism and running speed. Another example is birds that quickly adapt to pitch or intensity changes in the environment and alter their vocalizations. Temporal and temperature variation in the environment make it possible for competitors to coexist whereas one species might exclude all of its competitor under constant conditions.

Environment gradient. Chipmunk species that coexist in the eastern slopes of the Sierra Nevada mountains reveal their competitive abilities under environmental gradients. Four species of chipmunks (*Eutamias*) that inhabit the Sierra Nevada have overlapping food requirements but occupy distinct altitudinal zones, determined in part by aggression between the species (**Fig. 12.21**). The interacting Chipmunk species and corresponding altitudinal zones are:

- Alpine chipmunk—alpine range
- Lodgepole chipmunk—pine-shaded forest range
- Yellow-pine chipmunk—shrub range
- Least chipmunk—sagebrush range

The yellow-pine chipmunk is dominant over the least chipmunk, which can live in all of the habitats on the slope but ends up restricted to sagebrush of the lower foothills. The least chipmunk tolerates heat stress better than all the other species. If the yellow-pine chipmunk is removed the least chipmunk will move higher, into the shrubby habitat of the Chaparral. But if the least chipmunk is removed the yellow-pine chipmunk will not move into the lower habitat.

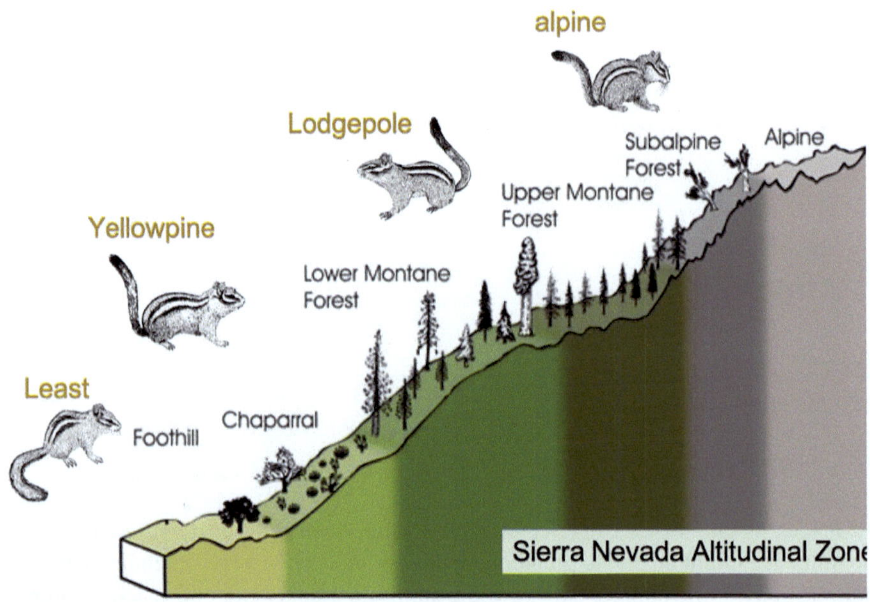

Fig. 12.21

Sierra Nevada altitudinal zones and chipmunk-realized niches.

Source: Modification of US National Park Service, Public Domain

The lodgepole chipmunk is dominant over the yellow-pine chipmunk, which determines the yellow-pine's upper range. But the lodgepole chipmunk is susceptible to heat stress and is restricted to the pine-shaded habitat in the Montane forests. Lodgepole chipmunks are the most aggressive, which likely determines the lower range for the high-altitude Alpine chipmunk. The range of each species is established by a combination of aggressive competitive exclusion and a realized niche based on ability to survive and reproduce in the different environments.

Competitive exclusion. This describes a species being excluded from a local community through its competitive interactions with other species. **Competitive release** occurs when a species is able to expand within its species' niche because its competitor has been removed. Competitive release can happen in several ways, for example, when a species enters a new habitat currently free of potential competitors. Competitive release might also happen if the competitor in its niche had been removed by another means. Connell demonstrated the ecological niche, fundamental vs realized, in the growth habits of barnacles in the intertidal zones of Shetland. The process of niche exclusion was shown to occur by species competition, with the successful competitor *Belanus* excluding another species, *Chthalamalus*, from a community, and a **releasing factor** (exposure level) restricting *Belanus* dominance to the lower intertidal zone and freeing *Chthalamalus* to inhabit the upper level.

Ecological systems are inevitably complex structures with a whole gamut of intraspecies and interspecies interactions and a large number of focal species as competitors. But competitive release is not simply the result of a trophic cascade. The distinction between a trophic cascade and competitive release is that competitive release involves no more than two trophic levels, and the "release" is in the form of competition. In a trophic cascade there will be at least three trophic levels involved and the release will be due to predation.

Birds in temperate and tropical communities exhibit interspecific territoriality. Birds will even defend their territories from members of closely related species (e.g., the gray and dusky flycatchers in the western United States). And other bird species defend territories against a broader range of competitors (acorn woodpeckers defend against other woodpeckers, blue jays, and squirrels). In plants and **bryozoan** another series of experiments demonstrated that competition for an essential resource can influence the ability of an individual to exploit its other resources, affecting survival and growth.

Resource partitioning is an evolutionary change in species' interactions in response to selective pressures from interspecific competition. If you have ever spent time bird watching you may have noticed great morphological diversity in New World warblers (**Fig. 12.22**) even when occupying the same breeding grounds (coniferous trees) and resources (insects and caterpillars). Robert Macarthur, an influential population ecologist whose 1958 PhD thesis *Ecology* documented seven variations in five warbler species, the Cape May, yellow-rumped, black-throated green, Blackburnian, and the Bay-breasted warblers.

Fig. 12.22

Cape May warbler.

Source: After Menge, 1976

The variations are as follows:

- habitat use
- nesting site
- territoriality
- time of activity
- foraging direction
- style
- tree zone

If Gause's law holds true one of the warbler species should be outcompeting all the others. But this doesn't appear to be the case in the five warbler species occupying the same niche and appearing to assume identical roles in the bird community. Macarthur argued that while there is some overlap in usage warblers established a zonation of realized niches within a single tree. Some warblers occupied the top portions and outermost branches where they easily captured prey, some used the interior branches, and others used the lowest, or basal regions (**Fig. 12.23**). The Cape May warbler occupied spaces on the outer periphery of the uppermost reaches of the tree while the Bay-breasted restricted itself to the mid-interior. Niche zonation appeared with all warblers present—with the exception of the yellow-rumped, which did not stay in any specific territory—moving around the tree more freely.

Fig. 12.23

MacArthur's warblers.

Source: After MacArthur 1958 under CC By-SA

Niche partitioning. On closer inspection of feeding habits the Cape May warbler chased flying insects more often than the Blackburnian warbler, through use of vertical movements in the outer periphery. The black-throated green hovered more extensively than the Bay-breasted warbler. And the yellow-rumped bird exhibited a varied feeding habit. Furthermore, MacArthur noted food shortage limited the size of the warbler populations and concluded bird behavior permitted access to distinctly different food resources. The warblers even appeared to compromise on nesting times so that corresponding peak food requirements required for parenting did not overlap. Thus warbler interspecies interactions were based on communal partitioning of their limiting resource and occupation of different niches. MacArthur's detailed observation of the variation in life history of warblers introduced a framework concept for niche partitioning. Niche partitioning is important to the study of species biodiversity and explains how multiple species can become more efficient at performing their ecosystem functions and services in relation to how a solitary species can perform. MacArthur's thesis is a contemporary of the Eltonian niche, proposing that the five warblers were sympatric species coevolved to occupy separate, behavioral niches.

Modeling Interspecies Dynamics

The L-V model (predator–prey equation) is frequently used to chart interspecies dynamics through time. For this application the L-V function requires the adoption of specific assumptions about the environment and the predator–prey populations:

- Prey population is not limited in food resources.
- Predator population food supply is dependent entirely on the size of the prey population.
- Rate of change of population is proportional to its size.
- No environmental change occurs that would favor one species.
- Predators have limitless appetite.

The conditions of these L-V assumptions cannot always be met in nature and interpretation of data using the L-V model requires an intuitive understanding of species' dynamics by the ecologist.

Note: In this course the L-V model is defined and conceptualized in theory and by examples. Modeling exercises to mathematically solve predator–prey dynamic problems will not be covered.

Competition Field Studies

Competition is clearly a complex dynamic, involving many abiotic and biotic factors. Demonstrating competition in the field is understandably more challenging than demonstrating competition in a captive laboratory setting. Under natural conditions in the field researchers have little control over the environment, which can make it difficult to determine if a population is at or below carrying capacity. There may be inadequate knowledge of the life history requirements for a particular species. And it can at times be difficult to distinguish between subtleties that distinguish different species.

Removal experiments are sometimes used as a straightforward approach when working in the field. With this method removal of a potential competitor is followed up by monitoring the resulting impacts on the focal species. However, removing individuals of a species may have other direct and indirect environmental effects that unintentionally alter the outcome. For example, if the focal competitors are plants then removing one plant species might increase the light energy reaching the soil surface thereby increasing soil temperatures. Warming the soil has the consequence of increasing evaporation rates, which now reduces soil moisture as well as alters the decomposition rate. Altered nutrient cycling will, in turn, affect belowground resource availability.

In summary, competition involves a variety of environmental factors that vary in time and space and the outcome under one set of environmental conditions may very well be different than the outcome under a totally different set of conditions. Applying the scientific method with rigorous repetition and variation on the experimental method is a necessity for interpretation of good ecological field research.

Glossary

Bryozoan
startle display
mandibles
competitive release
releasing factor
echolocation
tympanal organs
parasitism
neutralism
obligate
epiphyte
ovivposits & flower's style

Citations

1. B. Snow, W. David, Birds and berries: a study of an ecological interaction, Poyser (1988) 268.
2. A. Brune, Symbiotic digestion of lignocellulose in termite guts, Nat Rev Microbiol 12 (2014) 168–180. https://doi.org/10.1038/nrmicro3182
3. H. Song, et al., A comparative genomic analysis of lichen-forming fungi reveals new insights into fungal lifestyles, Sci Rep 12 (2022) 10724. https://doi.org/10.1038/s41598-022-14340-5
4. Ohio State University, "Snakes, Salamanders And Other Creatures Thrive In Areas With Higher Deer Populations.", ScienceDaily (28 October 2008). www.sciencedaily.com/releases/2008/10/081020120129.htm
5. S. S. Godfrey, Networks and the ecology of parasite transmission: a framework for wildlife parasitology, Wildlife 2 (2013) 235–245. https://doi.org/10.1016/j.ijppaw.2013.09.001
6. A. Åkesson, et al., The importance of species interactions in eco-evolutionary community dynamics under climate change, Nat Commun 12 (2021) 4759. https://doi.org/10.1038/s41467-021-24977-x
7. G.A. Jamie, Signals, cues and the nature of mimicry, Proc Biol Sci 284 (1849) (2017) 20162080. https://doi.org/10.1098/rspb.2016.2080
8. C.L. Craig, Signal polymorphism in the web-decorating spider Argiope Argentata is correlated with reduced survivorship and the presence of stingless bees, its primary prey, Evolution 55 (5) (2001) 986–993 JSTOR http://www.jstor.org/stable/2680311 Accessed 11 Mar. 2024
9. Z. Mirzaee, Biology, ecology, and biogeography of eremic praying mantis Blepharopsis mendica (Insecta: Mantodea), Peer J 12 (2024) e16, 814. https://doi.org/10.7717/peerj.16814

10. E. Font, Mimicry, camouflage and perceptual exploitation: the evolution of deception in nature, Biosemiotics 12 (2019) 7–24. https://doi.org/10.1007/s12304-018-9339-6
11. O. Johansson, E. Nyam, P. Lkhagvajav, J. Alexander, G. Samelius, Predation patterns and hunting behavior of snow leopards: insights from an Ibex hunt, Snow Leopard Reports (2023) 2. https://doi.org/10.56510/slr.v2.14323
12. T. Neil, Moths and bats have been in an evolutionary battle for millions of years – and we're still uncovering their tricks, The Conversation (2022). https://theconversation.com/moths-and-bats-have-been-in-an-evolutionary-battle-for-millions-of-years-and-were-still-uncovering-their-tricks-175673.
13. D. Jacobs, Explainer: the evolutionary arms race between bats and moths, The Conversation (2015). https://theconversation.com/explainer-the-evolutionary-arms-race-between-bats-and-moths-43675.
14. E.J. Temeles, A new dimension to hummingbird-flower relationships, Oecologia 105 (4) (1996) 517–523. https://doi.org/10.1007/BF00330015
15. S. Abrahamczyk, D. Souto-Vilarós, S.S. Renner, Escape from extreme specialization: passionflowers, bats and the sword-billed hummingbird, Proc Biol Sci 281 (1795) (2014) 20140888. https://doi.org/10.1098/rspb.2014.0888
16. R.B. Crocraft, K. Hambler, Observations of a commensal relationship of the microhylid frog Chiasmocleis ventrimaculata and the burrowing theraphosid spider Xenesthis immanis in southeastern Peru, Biotropica. 21 (1989) 2–8.
17. W. Trelease, Further studies of yuccas and their pollinators, Annu Rep Mo Bot Gard 4 (1893) 181–226.
18. Gucker, C.L. (2009). "Yucca brevifolia." *Fire Effects Information System*. US Department of Agriculture, Forest Service, Rocky Mountain Research Station, Fire Sciences Laboratory. Archived from the original on 2009-04-12.
19. C. J. van der Kooi, et al., Mutualisms and (a)symmetry in plant–pollinator interactions, Curr Biol 31 (2) (2021) R91–R99. https://doi.org/10.1016/j.cub.2020.11.020
20. T. Van der Niet, R. Peakall, S.D. Johnson, Pollinator-driven ecological speciation in plants: new evidence and future perspectives, Ann Bot 113 (2) (2014) 199–212. https://doi.org/10.1093/aob/mct290
21. W. Godsoe, Coevolution and divergence in the Joshua tree/Yucca Moth Mutualism, Am Nat 171 (6) (2008) 816–823.
22. https://www.nwf.org/Educational-Resources/Wildlife-Guide/Birds/Northern-Spotted-Owl.
23. J.M. Jenkins, et al., Social status, forest disturbance, and Barred Owls shape long-term trends in breeding dispersal distance of Northern Spotted Owls, Am Ornithol 121 (2019) 1–17. https://doi.org/10.1093/condor/duz055
24. N. Myers, et al., Biodiversity hotspots for conservation priorities, Nature 403 (2000) 853–858. https://doi.org/10.1038/35002501
25. Duke University. "Many more species at risk from Southeast Asia tree plantations, study finds." ScienceDaily. ScienceDaily, 9 August 2016. www.sciencedaily.com/releases/2016/08/160809170457.htm
26. A.P. Jacobson, Leopard (Panthera pardus) status, distribution, and the research efforts across its range, Peer J 4 (2016) e1974. https://doi.org/10.7717/peerj.1974
27. S. Ostrowski, M. Gilbert, Diseases of free-ranging snow leopards and primary prey species, Snow Leopards (2016) 97–112. https://doi.org/10.1016/B978-0-12-802213-9.00009-2

28. A. Pocheville, The ecological niche: history and recent controversies, in: T. Heams, P. Huneman, G. Lecointre, M. Silberstein, (Eds.), Handbook of Evolutionary Thinking in the Sciences, Springer, Dordrecht (2015), https://doi.org/10.1007/978-94-017-9014-7_26
29. G.E. Hutchinson, Population studies: Animal ecology and demography, Bulletin of Mathematical Biolog 53 (1–2) (1991) 193–213. https://doi.org/10.1016/S0092-8240(05)80046-1
30. J.H. Connell, The influence of interspecific competition and other factors on the distribution of the barnacle Chthamalus Stellatus, Ecology 42 (4) (1961) 710–723. https://doi.org/10.2307/1933500 JSTORAccessed 12 Mar. 2024.
31. G.F. Gause, The Struggle for Existence: A Classic of Mathematical Biology and Ecology, Dover Publications, Inc, New York, 2019. ISBN 9780486838298a. Retrieved Feb 2025, 20.
32. Y. Taniguchi, et al., Temperature mediation of competitive interactions among three fish species that replace each other along longitudinal stream gradients, Can J Fish Aquat Sci 55 (1998) 1894–1901.
33. L.W. Buss, Habitat selection, directional growth and spatial refuges: why colonial animals have more hiding places, in: G. Larwood, B.R. Rosen, (Eds.), Biology and Systematics of Colonial Organisms, Academic Press, London and New York, 1979, pp. 459–497.
34. S.D. Wilson, D. Tilman, Plant competition and resource availability in response to disturbance and fertilization, Ecology 74 (2) (1993) 599–611. https://doi.org/10.2307/1939319. Accessed 14 Mar. 2024.
35. T.W. Sherry M.K. Extensions and limitations of MacArthur (1958): A review of ecological and evolutionary approaches to competition and diet in the New World wood warblers (Parulidae), Ornithology 139 (2) (2022). https://doi.org/10.1093/ornithology/ukac010

CHAPTER

Community Structure and Dynamics 13

Chapter Outline

13.1 Community Structure

After reading this section students should be able to:

- Describe a community defined by its species composition.
- Interpret a chart of overlapping species abundance under environmental gradients.
- List measurable indices to monitor community structure.
- Explain why keystone species disproportionately impact community structure.

The Biotic Community

The biotic community is the living part of an ecosystem. Biotic communities bring about energy transformation and the accumulation of biomass. The species of plants and animals of a biotic community will exhibit varying degrees of species diversity. A few species will be in great abundance along with a percentage that is rarer but contribute more diversity to the community (**Fig. 13.1**). Perhaps not surprising,

Ecology for the 21st Century. DOI: https://doi.org/10.1016/B978-0-443-24031-7.00013-1

Fig. 13.1

Blue starfish (*Linckia laevigata*) lying across *Acropora* coral in the Ribbon Reefs of the Great Barrier Reef.

Source: larrywkoester under CC BY 2.0

species diversity is high in natural communities and much lower in managed ecosystems, such as croplands. As an ecosystem experiences extremes of environmental change communities that are rich in a variety of living organisms will become more resistant. The overall biodiversity in communities is quantifiable through applying a **biodiversity index**. Ecologist or conservationists use a biodiversity index to develop a baseline to compare the ecological health of different areas. Since regions with few species may be at risk if something in the environment changes, regions with higher biodiversity are more likely to contain species capable of adapting to a changing landscape. This is one of the concerns with growing monocrops as the bulk of the world's food supply.

A basic biodiversity index:

- # species in area/total # individuals in area = biodiversity index

Species Composition

The biological structure of a biotic community is formed by its living components and the total of all the different species is known as its **species composition**. Species' composition can be measured by two biodiversity indices: species richness and species biodiversity. Both indices may vary seasonally or due to other disturbances. **Species richness** is the number of species living in an ecological community, landscape, or region. **Species biodiversity** is the more descriptive index because it includes not just species richness but also **species evenness**, which considers every species' concentration. Species' composition is expressed as a percentage and accounts for the species' biodiversity and its contribution to the community structure. During conservation studies monitoring species' composition helps to gain an understanding of the influence and stability of its members. For example, carrying capacity will be influenced by the species composition in an area.

Relative species abundance describes the total number of individuals for a specific species within an ecological community, relative to a total count of all the community's individuals. Oftentimes the species identified as having the highest relative abundance within an area has a role in forming the community habitat, helping to control its biodiversity, and is known as its foundation species. Examples of foundation species include seagrass meadows, kelp forests, and coral reefs.

Foundation species are the "base" or "bedrock" of a community. These species have the greatest influence on a community's overall structure and factor high in species' composition studies. Often the foundation species are primary producers and contribute most of the energy for the community. For example, in the kelp forest off the coast of California kelp and brown algae are foundation species in that aquatic environment. Some foundation species physically modify their environment by producing and maintaining habitats to benefit the entire community.

A **species' ecological niche** results through development of specific adaptations designed to enhance biological fitness under a certain set of environmental conditions. By the same token these special adaptations will restrict or prevent success of the same species under a new set of environmental conditions. For example, plants that thrive in warmer climatic environments are unlikely to be equally successful in cooler conditions. In the case of ectotherms (animals adapted to relying on external sources of heat) their species' ecological niche will have restricted distribution ranges and possibly limitations on its daily and seasonal activity. It is within their species' ecological niche that species have the best chances for survival and population growth. The overall community structure can be thought of as an expression of the species' ecological niche.

Coral polyps of a coral reef, though not photosynthetic, are hosts to photosynthetic symbionts (dinoflagellates) that inhabit their body tissues; together they make up the holobiont. The structural components of the coral reef community include their calcified exoskeleton ($CaCO_3$) of living and dead coral polyps. The reef structure serves as a protective home from oceanic waves, currents, and predators for many endemic species of the coral reef community (**Fig. 13.2**).

Fig. 13.2

Corals reefs are made of calcium carbonate, formed over thousands of years. Polyps are colorless but attract algae, called zooxanthellae, which live inside the cells of the polyp and give corals their vibrant color.

Source: US National Park Service (NPS), Public Domain

In most terrestrial habitats vegetation provides the main structure of the community. Plant structures support many important ecosystem services. This is because vegetation can mitigate the impact of erosion and delay precipitation run-off when the tree canopy captures rainfall, which also helps prevent flooding. The complexity of forest ecosystems resists the impact of climate extremes to a greater extent than open pastures. And the greater leaf index in the understory adds evaporative cooling layers while serving as proxies for net productivity of terrestrial carbon stocks.

Extrinsic Factors Create Environmental Gradients

A number of extrinsic factors, including temperature and precipitation, impact species composition, richness, and community structure. Charting the presence and intensity of abiotic factors along an environmental gradient illustrates the dependence of species abundance on extrinsic factors. In **Fig. 13.3** the overall distribution pattern of multiple species' fundamental niches shows an overlap wherever conditions are most compatible to species' livelihood (shaded area). Areas are considered primary constraints if species do not occupy them. Along the entire environmental gradient (plotted on X-axis) overlapping environmental niches identify areas of shared available resources. Note that overlapping niches also have smaller population sizes (plotted on Y-axis). Recall Gause's law, which states that two species competing for the same limiting resource cannot coexist at constant population values. At least one or both will decrease in population size.

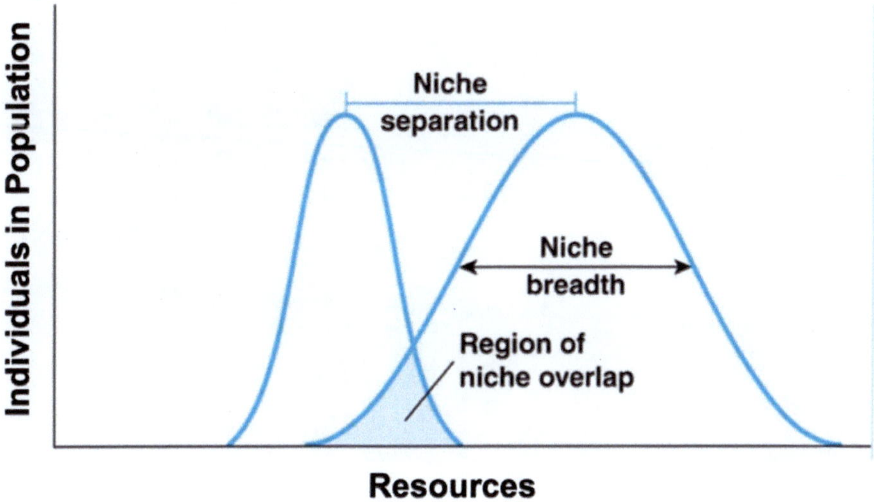

Fig. 13.3

Niche overlap in species competing for resources (seed/nuts/berries and space, etc.).

Source: Image by author

The actual community structure at any point along the environmental gradient is a function of the species' realized niche. Another level of complexity in community structure exists because of interspecies interactions. For example, one species' influence on the presence and abundance of another species impacts the fundamental niches and species' relative abundance distributions along niche gradients. Competition and predation may reduce abundance or exclude species altogether while mutualism may increase abundance or extend the distribution of a species. Across the landscape there are always small variations in the abiotic environment, and the differences in environmental tolerances among species at different locations will change the community structure.

Walker Branch Watershed is located in Oakridge, Tennessee, and is part of the US Department of Energy's Oak Ridge Reservation. The watershed has been the site of ongoing environmental research since the 1960s. For example, the USGS studied the biological structure and geographic range of 28 North American birds. In the relatively small breeding region of the Walker Branch watershed the distribution range of four bird species overlapped at some point during a single year, shown in **Figs. 13.4a, b**. These birds were identified as the pine warbler, cerulean warbler, ovenbird, and warbling vireo. The pine warbler has its highest abundances

Fig. 13.4a

Pinewarblers are rarely found away from pine trees, except during migration.

Source: NA Breeding Bird Survey, USGS, Public Domain

Species	Number of individuals	Relative abundance
Carolina Chickadee	244	24.6
Red-eyed Vireo	139	14.0
Eastern Wood Pewee	90	9.1
Tufted Titmouse	64	6.5
Blue Jay	55	5.5
Yellow-billed Cuckoo	41	4.1
Hooded Warbler	39	3.9
Downy Woodpecker	36	3.6
Red-bellied Woodpecker	27	2.7
Blue-gray Gnatcatcher	25	2.5
White-breasted Nuthatch	24	2.4
Pine Warbler	22	2.2
Scarlet Tanager	21	2.1
Cardinal	20	2.0
Carolina Wren	16	1.6
Wood Thrush	15	1.5
Summer Tanager	15	1.5
Cerulean Warbler	9	0.9
Yellow-shafted Flicker	8	0.8
Acadian Flycatcher	7	0.7
Indigo Bunting	7	0.7
White-eyed Vireo	6	0.6
Yellow-breasted Chat	6	0.6
Ruby-throated Hummingbird	5	0.5
Hairy Woodpecker	5	0.5
Ovenbird	5	0.5
Kentucky Warbler	5	0.5
Yellow-throated Vireo	4	0.4
Worm-eating Warbler	4	0.4
Brown-headed Cowbird	4	0.4
Chipping Sparrow	4	0.4
Prairie Warbler	3	0.3
Rufous-sided Towhee	3	0.3
Pileated Woodpecker	2	0.2
Great Crested Flycatcher	2	0.2
Brown Thrasher	2	0.2
Ruby-crowned Kinglet	2	0.2
Warbling Vireo	2	0.2
Black-throated Green Warbler	2	0.2
Yellowthroat	2	0.2
Total	**992**	**100.0**

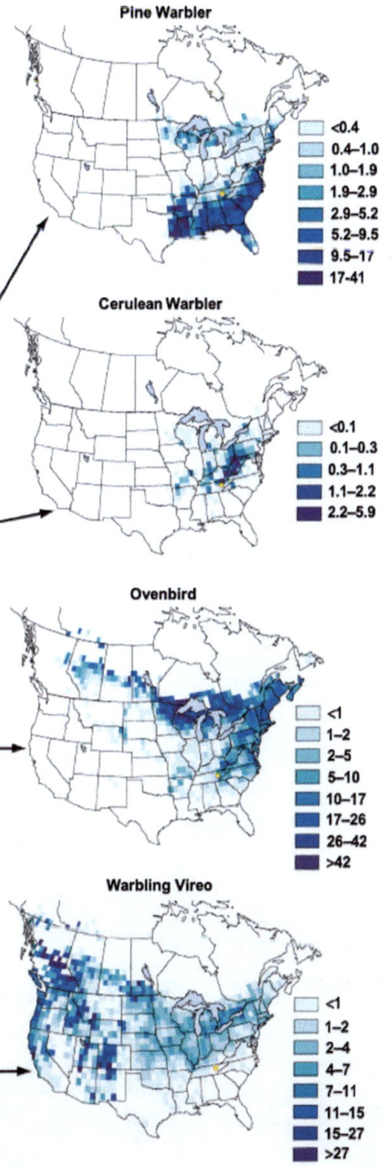

Fig. 13.4b

Species relative abundance (percent individuals) in Walker Branch Watershed in Oak Ridge, Tennessee. Arrows show distinct geographic range and abundance of four warbler species with one range overlap occurring at the watershed.

Source: NA Breeding Bird Survey, USGS, Public Domain

in the southeastern sector of the United States. The ovenbird has its highest abundances in the northeastern United States and parts of Canada.

In the figures the Oak Ridge Reservation is the ecological niche overlap, represented by the shaded area, of an environmental gradient for the four birds. The region has high biodiversity throughout its diverse habitats spanning 33,500 acres and is home to many hundreds more birds, including many threatened species. This is possible because of large expanses of mature and pristine hardwood forests. Notably, the region has never been fragmented by development or farmland. There exists an abundance of bird-friendly brush corridors, abandoned pastures, grasslands, cedar glades, barrens, and wetlands (Tennessee Wildlife Resources Agency).

The Amazon River Basin's aquatic fauna is another example of high species' diversity. Distribution of fish species falls into several fundamental niches along an environmental gradient. The downstream regions of the basin exhibit entirely different habitat conditions than its upstream portions. The environmental heterogeneity is due to its biogeographic history, variability in soil type, topography, plant physiology, and hydrology. It is here that the Peruvian Fitzcarraldo (P-F) region, situated in the Western Amazonian rain forest (200–500 m above sea level), is home to headwaters of four large Amazonian tributaries (**Fig. 13.5**). This P-F region (purple dots on

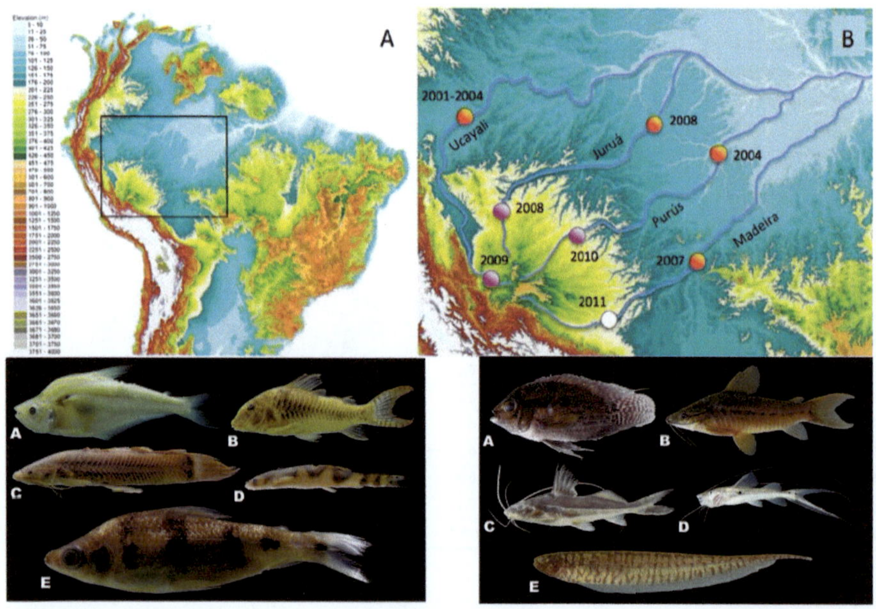

Fig. 13.5

(a) Topography of Fitzcarrald Arch using Digital Elevation Model (DEM), 25 m intervals; purple circles between 212 m and 310 m of Fitzcarrald Arch. (b) Examples of species collected.

Source: Image courtesy of Roberto Esser Reis

map) is noteworthy for having among the highest species richness and density of any vertebrate fauna on Earth. There are more than 5600 aquatic species, with many specialists and all inhabiting less than 0.002% of the world's total liquid water supply. Why so many species can coexist in such a small habitat has long been of interest to conservation ecologists. Each of the four tributaries draining into the P-F region are isolated from one another. Investigations supported the hypothesis that fish species in the eco region pre-date a Late Miocene-Pliocene, c. 4 Mya geological uplift. Both habitat specialization and geographic isolation were shown to have contributed to the high species richness in the P-F region of the Amazon Basin.

Diffuse Species Interactions

Species interactions are inclusive if there are more than two species. Most terrestrial and aquatic environments are inhabited by a diverse population of species, including transient occupants, with potential competition among any number of species competing for a limited supply of resources. Interactions involve any number of species in association within its habitat, and the results are *diffuse* rather than exclusive. Community dynamics are often diffuse, involving a greater number of species than those under investigation. Repeated studies using different species and time points within the community help to improve the validity of research findings in community ecology.

Diffuse competition describes the total competitive effects of a number of interspecific competitors. A 2002 study of competitive interactions between plant species within an old field community was done by selectively removing species and assessing the effects on the growth of the remaining species. At the Institute of Ecology in Jena, Germany, studies on species richness was gathered data over six years which demonstrated that removing a single species had little effect, whereas removing groups of species tended to have a much stronger effect. The results of the study suggested that competitive interactions within a community are weak and diffuse rather than direct and specific.

If relative abundance of a species is affected by multiple competitive interactions then designing an experiment that identifies all of the interactions would prove difficult to assess because weak and diffuse competition would likely take a few years to detect (**Fig. 13.6**). And removing only one or even a few of the competitors should have little effect on the abundance of that species. During an extensive 15-year follow-up of Jena's initial species' richness experiments Weisser et al. monitored and assessed 60 species, including running analysis on belowground resource partitioning and trophic cascades. The long-term study demonstrated that greater biodiversity is a positive driver in ecosystem function. With more diversity several mechanisms can act simultaneously to reduce pathogen attack and enhance the traits for resource partitioning.

In many food webs predation can affect the outcome of other competitive interactions occurring among the prey species. In the Rocky intertidal zone of the Pacific northwest diverse invertebrate herbivores, including mussels, barnacles, limpets, and

Fig. 13.6

Community interactions can be weak and diffuse with multiple groups feeding on different resources.

Source: Lumen Learning under CC BY 4.0

chitons, are able to coexist even though they are direct competitors for space and resources (**Fig. 13.7**). A predator, the purple starfish (*Pisaster*), preys on the invertebrate species. If *Pisaster* starfish is removed from the community dynamic the structure would be significantly altered. Zoologist Robert T. Paine (University of Washington) was one of the first scientists to demonstrate the effect of predation on food webs using the Pacific northwest's Rocky intertidal zone food web. Paine removed *Pisaster* from some areas (experimental) but left other areas undisturbed (control). Without a predator one might predict that species' abundance of all 15 prey species in the study would increase. Surprisingly, with no *Pisaster* the number of prey species *decreased* from 15 to 8. The removal of *Pisaster* and concurrent reduction in species diversity reflects an *increase* of competition among the remaining species.

Apparent competition is possible when a predator species feeds on two or more different species. If the predator species is removed the populations of each prey species will switch to being a population regulated by purely intraspecific, density-dependent mechanisms.

Once the starfish was removed the strongest competitors among the species began to exclude the other species, which caused the decrease in species diversity

Fig. 13.7

Rocky intertidal zone of the Pacific northwest houses many coexisting invertebrate herbivores, mussels, barnacles, limpets, and chitons.

Source: NOAA, Public Domain

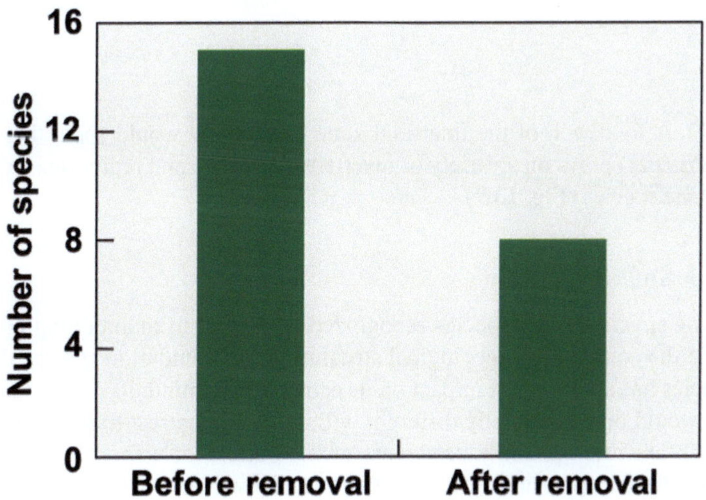

Fig. 13.8

Removal of *Pisaster* indirectly led to reduction in species diversity in the intertidal community.

Source: After Paine, 1974

Fig. 13.9

Representative food web illustrating the importance of *Pisaster* species to other community members.

Source: After Paine, 1974

(**Fig. 13.8**). A food web of the intertidal zone community would show that purple starfish (*Pisaster*) preys on a variety of invertebrate species and represents an important keystone species (**Fig. 13.9**).

Keystone Species

A **keystone species** is any species recognized as critical to maintaining biodiversity within the community's ecological structure. Like foundation species the keystone species has a significant impact on its ecosystem community. The community structure would be dramatically different without it. In contrast to foundation species, which are often primary producers at the bottom of a food web, keystone species are found at or near the *top* of the food web. Importantly, there are no replacement species that could fill their ecological niche. While not necessarily the largest or most abundant species, keystone species will always have a huge influence on the food web.

The concept of a keystone species was first introduced during Paine's research on *Pisaster ochraceus* in the Pacific. With the removal of *Pisaster* from its community structure one of its natural prey, mussels, experienced a population increase. This in turn completely altered the species composition and reduced biodiversity as the aggressive mussels began to crowd out other species. These other species included sea snails, limpets, and bivalves supported by benthic algae. By removing its keystone species Paine demonstrated that the intertidal plain's biodiversity decreased by 50% within 1 year after its keystone species, *Pisaster*, was removed.

Canis lupus, commonly known as the gray wolf, of the Greater Yellowstone Ecosystem (GYE) is an important North American keystone species. The expansive GYE spreads across the western states of Montana, Wyoming, and Idaho and has a diverse temperate climate range. The natural resources and ecological niches of the GYE include active geothermal basins, mountains, forests, meadows, and freshwater habitats. The gray wolf is a top predator with a diverse range of prey, including elk, bison, rabbit, and bird species, and scavenger species like the vulture. At least in part, these species are controlled in the GYE by the presence of gray wolves. Their feeding and nesting behaviors have been shown to be largely dependent and reactionary to gray wolf activity throughout the ecosystem. The gray wolf was removed from this ecosystem by active predator control measures. The unforeseen consequences provided strong evidence for the importance of keystone species in their ecological niche.

Livestock became vulnerable prey to gray wolves beginning with the 1800s westward expansion of American settlers. The wolves targeted livestock partly because their native prey base was being decimated through agricultural encroachment and habitat loss. With the rise of the gray wolf threat to their livestock settlers advocated for a planned elimination (active predator control) of gray wolves along with other large predators (bears and cougars). The homesteads and farming activities were quickly spreading across the majority of the large predators' historical range.

Predator control measures continued until the first half of the 20th century, with all three predators (wolves, bears, and cougars) systematically killed to protect settlers' livestock as well as deer and elk. The deer and elk were considered more desirable by the human inhabitants and presented no threat. In 1872 an occasional gray wolf was still found prowling Yellowstone. Around this time the GYE was established under the Yellowstone National Park Act. The new federal legislation's stated purpose was that the Secretary of the Interior "shall provide against wanton destruction of fish and game found within said Park." Legislators who wrote the national park law had the large predators in mind but lacked a comprehensive understanding of the ecosystem and interconnections of species within their community. As such its authors singled out the gray wolves' habit of killing prey species to be a "wanton destruction" of other animals. This led in hindsight to the ill-guided policy of eliminating gray wolves. It indirectly led to costly and dire ecological consequences with decades of poor land management.

In the Southern Greater Yellowstone ecosystem (**Fig. 13.10**) the grizzly bear and the gray wolf became locally extinct. The 1980s marked the last sighting of grey

Fig. 13.10

Yellowstone National Park territory.

Source: US National Park Service (NPS), Public Domain

wolves in the park. By the mid-1900s wolves had been almost entirely eliminated from all 48 US continental states. As might be expected removal of the wolves led to their prey, for example, elk populations, to rise significantly.

Soon after the departure of its keystone species (wolves), the park began experiencing land erosion and plant die-off. This was attributed to elk overpopulation, which reduced the habitat availability for other species. For example, bird populations that forage vegetation along the rivers were in decline while some local bird populations went extinct as their nesting sites were destroyed. While elk continued to multiply inside the GYE deciduous trees suffered from overgrazing. The entire community structure was altered. Without the keystone gray wolves another prey, the coyote, dramatically increased in population size. This had the effect of reducing the population of its prey, pronghorn antelope. Another species impacted was the moose, as a competitor for resources that feed on willow and other woody plant species along the shores of the rivers were negatively impacted through resource competition with the now-rising populations of elk. Field observations have since shown that when wolves kill elk they leave partially consumed carcasses, which indirectly creates a year-round abundant food source for other carnivores, including the coyote, eagles, bears, raven magpie, red fox, and approximately 20 other species (**Fig. 13.11**). It soon became evident that the deplorable park conditions caused by an increase in elk population were due to elimination of its natural predator, the gray wolf.

In the 1960s and 1970s national awareness that there were environmental consequences from removal of keystone species led to the passage of several laws designed to correct park management strategies. But without including funding to reintroduce the gray wolf by 1974 they were officially listed as endangered. Gray wolf recovery

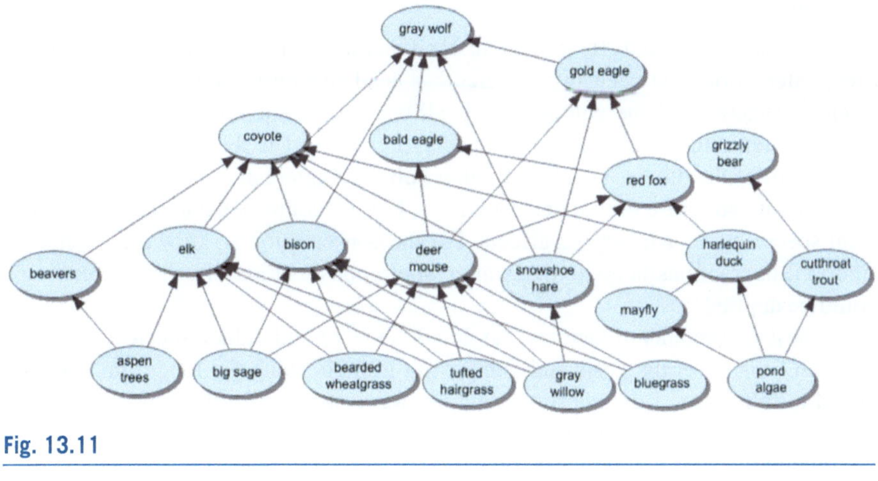

Fig. 13.11

Hayden Valley in GYE has an extensive array of producers, primary consumers, secondary consumers, and tertiary consumers (only some pictured here).

Source: Image by author

was then mandated under the new Endangered Species Act with federal funds for wolf reintroduction allocated in 1995. After a 70-year absence, 31 gray wolves were relocated to the United States from Canada. Within the next few years wolf management became decentralized to state control. Between 2008 and 2011 the gray wolf was officially removed from the endangered species list in the states of Montana and Idaho and continued to improve elsewhere.

Reintroduction of the gray wolf to Yellowstone National Park has had positive ecological impact, as the regrowth of shrubs restored an important food source (berries) for the threatened grizzly bear. The park also witnessed a resurgence of aspen, cottonwood, and willows in some parts of the park, which restored habitats and populations of birds and beavers in what appears to be an ecological cascade. In 2017 the USFWS also removed the restricted Yellowstone population of grizzly bears from the threatened species list. Note in **Fig. 13.11** that it is the gray wolf that holds the keystone species position in this ecosystem.

13.2 Community Dynamics

After reading this section students should be able to:

- Explain how food webs illustrate indirect interactions.
- Conceptualize how community equilibrium can change through time.
- Describe the levels of succession in a community and its impact on species diversity.
- Discuss the long-term impact on the community from invasive species.

Community Hierarchy

Food webs are useful tools to help understand hierarchical control of communities. Direct interactions between two species can lead to indirect interactions across the entire spectrum of the community. If all species interactions are important to community structure then when one species is removed it should have an equal effect on all of the others. If only a smaller subset of interactions is of importance to community structure then species are more loosely connected and removing different species will have varying effects. For the conservationist, knowing the most important interactions helps to determine where conservation efforts and funding should be applied.

In most community dynamics the importance of individual species remains unknown because of the costs and complexity in performing the long-term studies needed to fully answer the question. And extensive studies like those performed in Jena on grassland species interactions take a long time to achieve meaningful results. It is possible to derive some useful insight on indirect community interactions by examining their food web. For example, an indirect interaction occurs when species

Fig. 13.12

Canadian Boreal Forest.

Source: Mark Baldwin Smith and peupleloup under CC BY-SA 3.0; CC BY-SA 2.0

A does not interact directly with species B but instead influences species C, which in turn directly interacts with species B.

In the boreal forests of North America (**Fig. 13.12**) the Canadian lynx does not directly interact with white spruce but they do prey on snowshoe hare. But the snowshoe hare—named for its snowshoe-shaped hind feet lined with stiff hair to support its weight in the snow—does directly interact with the white spruce. The hare feeds heavily on its twigs, buds, and bark, which is a negative interaction for the white spruce. Under heavy predation of snowshoe hare by the lynx a positive affect is seen in white spruce populations.

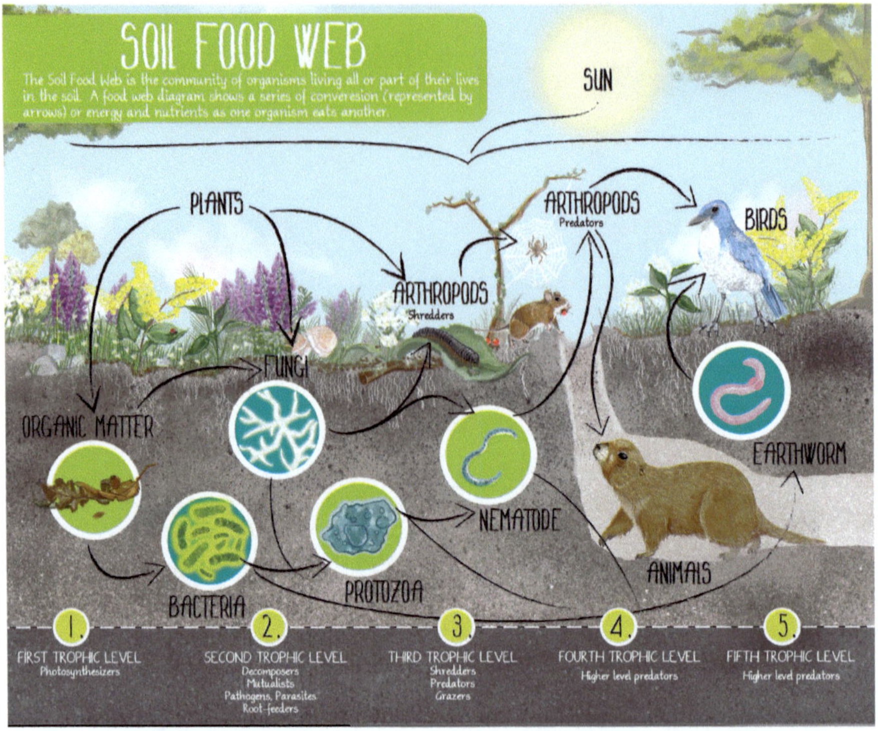

Fig. 13.13

Aggregation of species forming the food web for Canadian boreal forests in trophic levels. Arrows point from prey to predator.

Source: USDA, Public Domain

One approach to understanding the influence of species diversity on community structure and dynamics is to group species according to functional categories (e.g., feeding types). Groupings allow researchers to look at the community in more generalized terms. For example, in soil ecology (**Fig. 13.13**) species were aggregated by trophic levels to simplify species interactions. The food web doesn't just delineate energy flow through the trophic levels (**Fig. 13.14**); it can be used to highlight the direct or indirect interactions between species.

Food webs can highlight two types of control:

- **Bottom-up control**—population abundance is controlled by the degree of productivity and abundance in the populations found in the trophic level below.
- **Top-down control**—predator populations control the abundance of prey species, which in turn influences abundance of food sources by its consumption.

Resource competition among plants can be complex with variability in their limiting resources. Resource variability occurs in specific nutrients, water, light, and the

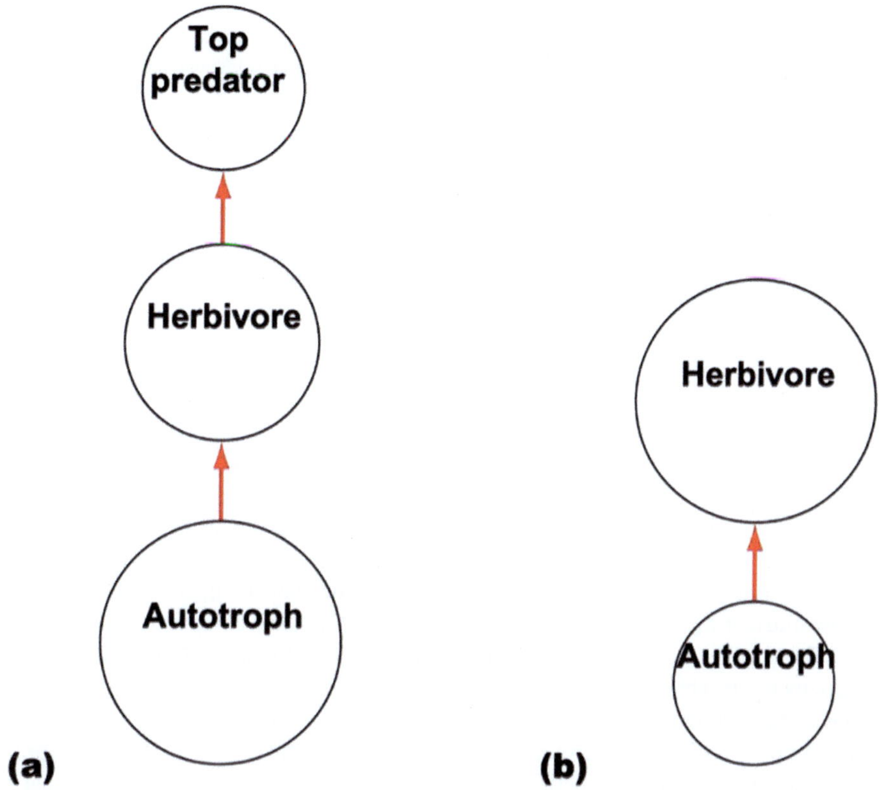

Fig. 13.14

(a) Food web consisting of a single basal species (autotroph), an intermediate herbivore that feeds on the autotroph, and a top predator. (b) Removal of the top predator results in an increase in the intermediate species, which in turn results in a decrease in the basal species on which it preys.

Source: Image by author

ways in which plants compete for them (**Fig. 13.15**). There is also a soil/air interface that creates a natural division between the aboveground and belowground intra- and interspecies dynamics. These interactions involve **competitive displacement**, also known as exploitative competition. In general, high availability of nutrients supports increased rates of photosynthesis, plant growth, and greater plant density, with an inverse relationship to species richness and diversity. Belowground competition is generally strong regarding mineral uptake and the varying mobility of more than 20 essential nutrients. The belowground competition is generally stronger than aboveground interactions. Soil ecology is therefore an important consideration for any plant community dynamics study.

Fig. 13.15

Species diversity decreases as dominant upland species flourish with greater access to resources (water, light, nutrients).

Source: South Carolina Department of Natural Resources, Public Domain

In high-nutrient soils dominant species out-compete and the aboveground plants quickly increase in size. Competition for light leads to taller plants with deeper, broader canopies creating shade optimized to exclude their competition. But during low nutrient availability the dynamics convert to weaker competitive displacement. The species that normally dominate under high nutrient availability are now less competitive. Their ability to displace less-competitive species is partly reduced because aboveground growth rates and biomass have been reduced.

In competition for belowground resources nutrient uptake is proportional to the size of the root system. Competition for nutrients leads to selection for plants that have longer overall root biomass. Enhanced root masses reduce soil nitrogen drastically, which eventually kills off neighboring plants. In summary, competition for nutrients leads to a decrease in the plant growth, survival, or fecundity.

Regarding the overall structure of community we can conclude that competition and predation are among the strongest influencers. Their affects can reduce abundance or exclude species. In contrast mutualism increases abundance or extends species distribution. The actual community structure at any point along an environmental gradient is a function of the species' realized niches. Their responses are modified by interactions with other species in the community and these interactions may be strong or diffuse.

Ecological Succession

Community dynamics consider the incremental environmental gradients in community structure as they change over time. But community dynamics can experience abrupt environmental disturbances and disasters. All ecosystems experience some fluctuation in their environmental conditions; however, sometimes these changes are drastic responses to disturbances by natural and unnatural causes, such as volcanoes, earthquakes, storms, fires, and climate change (**Fig. 13.16**). When communities have

Fig. 13.16

Rangitoto Island, a volcanic island near Auckland, New Zealand, with a lava field beginning primary succession vegetation.

Source: Nevilley, under CC BY-SA 3.0

a stable structure they are considered at equilibrium. After an environmental disturbance the community may or may not restore its equilibrium state. The introduction of invasive species and the structural components they ultimately threaten reveal just how delicate equilibrium is within communities.

The process of change that occurs in an ecological community on newly exposed or disturbed sites is an ecological phenomenon known as **ecological succession**. The time scale for all stages of ecological succession isn't defined and can take decades (e.g., a wildfire in Yellowstone National Park) or can occur over the period of millions of years (after a mass extinction event). Ecological succession is a unifying concept because it integrates many ideas already introduced: species life history strategies, population behavior, community dynamics, and ecosystem processes. Succession progresses over time through a sequence of appearing and disappearing species entering the scene as the conditions of the ecological niche change. The first colonization of species occupying the newly exposed or newly formed land occurs in the primary succession stage. Recall the Grime Model and its criteria for plant species survival based on degree of disturbance or stress.

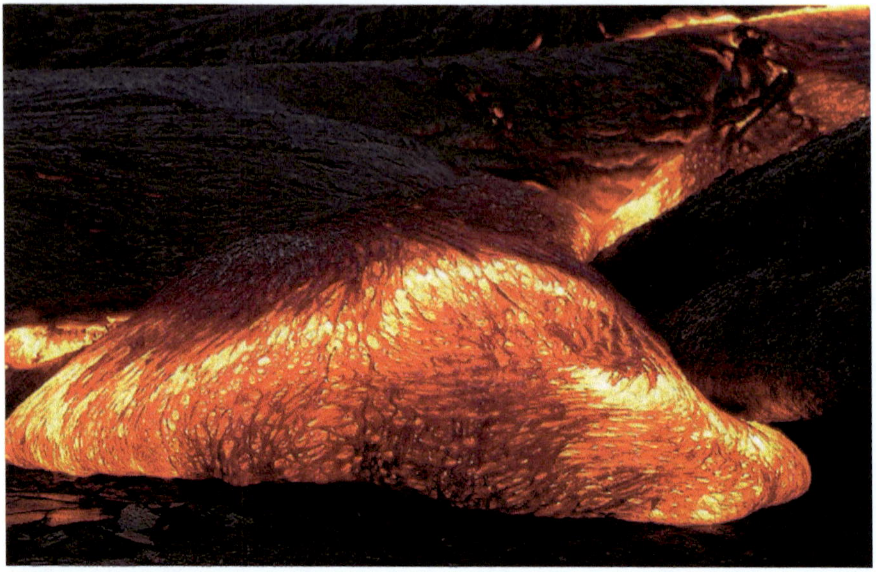

Fig. 13.17

Lava from volcano forms new rock.

Source: Hawaii Volcano Observatory (DAS), Public Domain

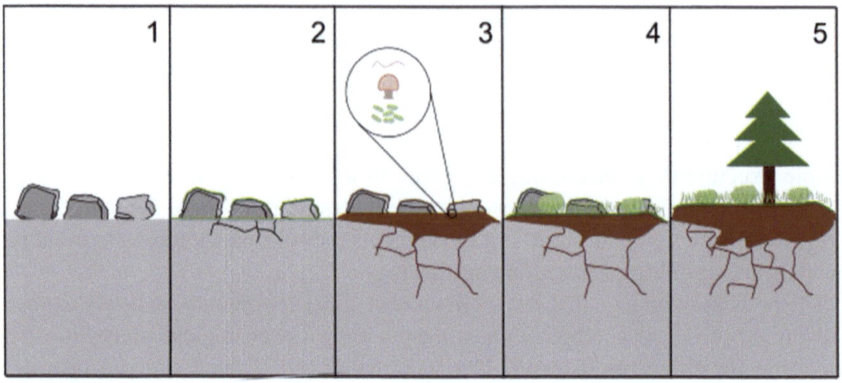

Fig. 13.18

Primary succession on bare rock.

Source: Wikiwand, under CC BY 4.0

Primary succession occurs after a major environmental disturbance (**Fig. 13.17**), in which communities need to develop on entirely new land without the benefit of established topsoil, for example, at river deltas, glaciers, sand dunes, or on exposed rock. The establishment of **pioneer species** occurs first, usually hearty plants, mosses, and lichens that have few soil requirements (**Fig. 13.18**). Species that grew during the earliest succession events will begin to die off and form a layer of decomposing organic material that aids new soil formation.

A primary succession event can be identified in the field by examining distribution of plants. For example, glacial retreat exposes bare rock for colonization. The regions further from the glacier will have had more time to develop new life forms and larger trees will be present. Closer to the disturbance will be evidence of pioneer species and grasses that ecologist can easily identify.

In the **secondary succession** stage the ecosystem is still partly disturbed with remnants from the previous community still optimized for the terrain, even as new species begin to emerge. This will include grasses, then shrubs, then smaller trees with relative stability within the ecosystem. Not all new plant species will thrive as the new, larger plants will eventually outcompete smaller shade-intolerant plants. Eventually the area will reach an established community with a set of organisms considerably different from its pioneer species. The composition is now in its intermediate species stage.

Fig. 13.19

Secondary succession in a boreal forest after a forest fire. Primary succession is succeeded by grasses and perennials. During the forest's intermediate succession shrubs and young saplings will begin emerging. The climax community is dominated by mature evergreen and deciduous trees that are stabilized until the next succession.

Source: Modification of Hannu, Public Domain

The secondary succession stage is much more common than the primary succession, for example, whenever forests are cleared by wildfire (**Fig. 13.19**). Wildfires will burn most aboveground vegetation and kill or disperse resident animals. In this case the topsoil is retained, and the ecosystem is still partly intact. Since the burned debris released organic and inorganic nutrients in the form of ash the soil will be enriched from the fire. Even areas damaged by severe fires are capable of quickly sustaining new life. In a forest prior to a fire the vegetation is dominated by tall trees shading the undergrowth and low-lying species. Once the trees are destroyed by fire early plants are typically sun-loving annuals from seed dispersal or seeds not destroyed by the fire. In just a few years these are replaced by spreading grasses and then the process of woody shrubs followed by small deciduous and evergreen trees (pine, oak, and hickory).

It generally takes another 150 years to go from the intermediate stage to an ecosystem's original community equilibrium stage. It is only at this point that the species composition will no longer be in flux. Once at the equilibrium state the ecological succession has reached its **climax community** and will be relatively stable, dependent on weather and other fluctuations, until the next major disturbance.

In forest management the forest fire is considered an essential and natural force that positively influences an ecological community. The natural cycle of fire functions

Fig. 13.20

Bandipur National Park, India, forest fire in 2019 destroyed 10,000acres of forest and wildlife.

Source: NaveenNkadalaveni, under CC BY 4.0

to enhance plant regeneration, nutrient cycling, habitat revitalization, disease control, and biological diversity by altering predator–prey dynamics. In the forests of the Sierra Nevada of the western United States, one of the most conifer-diverse ecoregions on the planet, years of logging and deliberate suppression of natural fires (occurs every 5 to 35 years) have altered the forest dynamics, including its behavior toward fire. By suppressing natural fire cycles fuels (woody plants) build up, leading to larger and more devastating fires (**Fig. 13.20**).

Ongoing efforts by the US Forestry Department to restore the Sierra Nevada include letting natural wildfires run their course, which ensures restoration of thriving forest plant communities well adapted and resilient to fire effects (**Fig. 13.21**).

Perusal of old US Forest Service records reveals that up until the 1930s on average 30 to 40 million acres across the American west had burned. The practice of fire rejuvenation wasn't lost in commercial agriculture, where the benefits of fire and plant succession have a rich and long history. One example is the traditional ritual of burning blueberry fields along the rocky coastal communities of Maine.

To nourish and protect the blueberry fields of Maine, a commercialized burner is used in early spring to kill all vegetation, pests, and weeds. The blueberry bushes themselves are fire resilient. The field will then be allowed to remain fallow for one season, giving it a chance to regenerate with less weeds and pests. Field burning has been practiced in this region since before chemical farming and continues partly

Fig. 13.21

Sierra Nevada wildfire.

Source: NaveenNkadalaveni, under CC BY 4.0

due to the geologic substrates, which are sands and gravel. Chemical weed control is potentially hazardous as it would readily infiltrate into the subsurface and enter groundwater.

Successional pathways are influenced by a number of factors, including substrates and soil type. An initial colonist species diversity may be restricted due to soils that are water saturated or serpentine, or volcanic lava rock. For example, plant communities that establish in serpentine soils must be able to tolerate moist bogs or rocky barrens and the distinctly harsh environmental condition compared to surrounding nonserpentine soils. The species profile of the initial community can also affect the success of compatible emerging colonists. Any subsequent disturbance when the community is in flux may alter the successional pathway. Animals, including insects, can inhibit or reverse succession by tracking over soils, matting, or feeding on fauna. Animals can also facilitate succession by loosening soils to germinate seeds and dispersal. Community stability can be described by its resistance to change and resilience following disturbances, the latter being expressed via successional pathways.

Invasive Species Threaten Community Dynamics

In Chapter 10 the impact of invasive species on population dynamics was introduced. Here we explore the consequences of invasive dominant species on the community structure and dynamics.

Fig. 13.22

Kudzu, a Japanese vine species invasive in the southeast United States.

Source: Scott Ehardt, Public Domain

An **invasive species**, like the Kudzu vine, is a species that is not native to a specific location (**Fig. 13.22**). According to the IUCN invasive species are introduced species that tend to take over the ecological community's space and resources and spread to a degree that causes damage to the environment, human economy, or human health. Exotic species can be introduced to new regions through trade, travel, or tourism. Exotic species include pathogenic organisms, such as epidemic microbes, insects, or fungal pests capable of decimating forests. Invasive species include overgrowth of aquatic weeds that suffocate lakes and streams, leading to low photosynthetic productivity and eutrophication.

The Bluegill Invasion

During the 1960s an innocent wildlife souvenir gifted from one nation to another had the unintentional consequence of causing a decade-long environmental crises. On his visit to the Shedd Aquarium in Chicago the Crown Prince Akihito of Japan, an amateur ichthyologist, was gifted 15 bluegills (Illinois's state fish), which he stocked in the moats surrounding his palace. Within a few years bluegills became an ecological nightmare for Japanese freshwater lakes and rivers, destroying its fish biodiversity. In Lake Biwa, Japan's older freshwater body of water, bluegills decimated the native Crucian carp population, a prized culinary delicacy. By 2005 bluegills had reached an estimated 25 million and were added to Japan's Invasive Alien Species Act. More than 60 years later a promising solution is being attempted with new molecular technologies using clustered interspaced short palindromic repeats (CRISPR) bioengineering that target's bluegill reproductive success. But there is ambivalence in the public's view about human intervention and any downstream impacts of a genetically modified organism (GMO) added to their native ecosystems. Public awareness of the environmental crises resulting from out-of-control invasive species highlights socio–political ramifications and an urgency for human response.

The Asian Carp Story

In the United States another environment cris is also began in the late 1960s with the introduction of the Asian grass carp. This was part of a management plan by the Arkansas US Wildlife and Fisheries. Initially fisheries and sewage treatment facilities sought the fish's excellent filter-feeding capabilities to clean weed-choked ponds of excess plankton. By the mid-1970s Arkansas and Mississippi fish farmers were stocking Asian carp as water filters, importing not just grass carp (*Ctenopharyngodon idella*) but three other Asian carp: bighead, silver, and black carp: *Mylopharyngodon, Hypophthalmichthys noblis*, and *H. molitrix*, respectively. When the carp species escaped during flooding they quickly out competed native species for food. The first record of a bighead carp caught in natural waters (Ohio River, Kentucky) was in 1981 but by the 1990s flooding of Mississippi fish farms led to establishment of several carp species in the Mississippi River and migration to the Missouri and Illinois Rivers (USGS; Ohio DNR). Migrations now threaten the Great Lakes. According

Fig. 13.23

Silver Asian carp can grow to 100 kg.

Source: West Virginia Fish and Wildlife Services, Public Domain

to USGS, as late as December 2019 650 grass carps were captured within the Great Lakes Basin.

Black carp are voracious predators of native mussels and snails, outcompeting the native fish predators for these limiting food resources. Silver carp consumes both primary producers and consumers, eating even the plankton that native mussels and snails feed on (**Fig. 13.23**). Between the two, major fluxes in the food web threaten aquatic communities. Along the Illinois River there are regions where Asian carp are estimated to comprise 95% of the river's aquatic biomass! Lake and stream specialists estimate that when carp populations rise to densities over 80 to 90 lbs per acre they contribute to algae blooms and water pollution. In the Illinois and Mississippi Rivers Asian carp are a threat to native fish populations. The ecological implications extend to local economies and recreational anglers. The State of Illinois DNR oversees the annual removal of 750 tons of carp along its Illinois River Basin. In the Great Lakes Asian carp are a threat to lake trout and salmon fisheries.

The Chicago Sanitary and Ship Canal, built in 1900, is the major waterway linking the Great Lakes to the Mississippi River via entry to Lake Michigan. To prevent Asian carp from leaving the canal (where they are now amassing) and entering the Great Lakes electric barriers were installed to barricade their migration (US Department of Interior). However, this may not be sufficient to curb the threat, given the invasiveness of Asian carp. Several US states and the Canadian government sued the US Federal government to permanently barricade the locks of the Chicago Sanitary and Ship Canal from Lake Michigan. Lawsuits filed by five states (Michigan, Wisconsin, Minnesota, Ohio, and Pennsylvania) were rejected at the Supreme Court level as federal law maintained that the US Army Corps of Engineers must retain navigational access through the water bodies. As of 2014 the US Army Corps submitted plans to close the canal, at a cost of nearly $15 billion. The closure of the canal and restoration of its natural barrier between Lake Michigan and the Mississippi River Basin will take many years to complete. In the meantime the Asian carp reportedly escaping the electric barriers is worrisome. Should the barricades fail Asian carp could become the destroyer of one of the largest freshwater fisheries in the entire world.

Scientists and wildlife management teams look for solutions to the carp crises from previous successes in minimizing ecological invasions from threat to nuisance status. In the later 20th century an exotic eel-like invasive species of the Great Lakes, the sea lamprey, became uncontrollable. Early awareness of the threat to the Great Lake's trout population and prompt action to eradicate the parasitic sea lamprey was key to its successful resolution. Once biologists assessed their vulnerabilities during spawning season in small tributaries they were able to strategize how to control the sea lamprey's growth.

Population and community ecologists also team up to address the Asian carp threat. As far back as 2002 the Minnesota Clean Water Legacy funded $1.8 million to start up a research center at the University of Minnesota. The MAISRC (Minnesota Aquatic Invasive Species Research Center) has been at the forefront of important research on Asian carp eradication strategies. In 2010 the president of the United States pledged $78 million to prevent Asian carp in the Mississippi River and Chicago Waterway System from invading the Great Lakes. And by 2012 an international study between the US Department of Fisheries and Oceans Canadian confirmed that as few as 10 female and 10 male Asian carp of reproductive age were sufficient for the nonnative fish to disrupt the native fishery, alter the ecosystem, and create another food web in the Great Lakes.

Asian carp behavioral studies to identify vulnerabilities have revealed that they congregate in large pools under the ice in winter and that up to 90% of carp populations migrate together (**Fig. 13.24**). The MAISRC now operates electric fences to guide invasive carp to specific entrapment sites. In 2018 MAISRC collaborated with a Washington state company to fight the carp invasion with a flexible pneumatic tube. The tube, called the Salmon Cannon, was originally designed to launch migrating salmon past damns in the northwest to reach their spawning sites. Its application in the Asian carp will be the reverse, to suck up thousands of congregated carp from the state's lakes and marshes. But for the cannon to work the migrating carp must jump into the apparatus. The research center is at work on a means of elevating the carp en masse. Meanwhile the Illinois Army Corps of Engineers is installing a series of deterrents, including the electric fence, a bubble barrier, and a series of underwater speakers at Joliet Lock and Dam in Joliet, in anticipation to ward off adult fish and eggs from migrating. The Army Corp's projected cost to deter the Asian carp at the Great Lakes is $778 million. Is it time to consider the GMO strategy to tackle Asian carp in the United States?

Fig. 13.24

Jumping Asian Carp.

Source: USFWS, Public Domain

Invasive species like the Asian carp provides an excellent example on the importance of species evolving together in their ecological niche. In the carp's natural habitat in Asia specific predators eat a greater percentage of the juveniles than endemic predators of North America are capable of consuming. Since native carp infant mortality is traditionally quite high (prey for many birds, fish, and invertebrates) the Asian carp life history reveals they have a prolific reproductive strategy to produce hundreds of thousands of eggs/spawn. The only significant predator of adult carps has been humans. As a staple diet item in most Asian countries, as the older carp are consumed it reduces fecundity in the wild population, leading to the Chinese overfishing crises.

Promising news on the Asian carp crisis in North America comes from a surprising twist on the narrative; Chinese conservationists have traveled to the Great Lakes region to study why carp are thriving in the US Midwest. This interest is because in China the endemic silver and bighead carp are a common food source, high in omega-3 fatty acids and lower in toxins since they are vegetarian feeders. They are so popular that their populations have become threatened due to overfishing, water pollution, and construction of hydroelectric dams.

Zoologists at the St. Louis Zoo are testing Asian carp as the protein source to feed their penguins, sea lions, and pelicans. Analysis of the stomachs of large aquatic predators (alligator car, blue catfish, white pelican, and eagles) has revealed the Asian carp are indeed their prey, at least up to juvenile size. Since Asian carp are edible they could theoretically supply the American diet with ample, inexpensive fish. But the carp are bony and not palatable in western cultures. Nonetheless American ingenuity hasn't overlooked another way to tackle the Asian carp problem, with highly nutritious filleted carp on the rise in the US specialty markets.

The Humongous Fungus

An invasive species dominating community dynamics in the plant world, at the expense of community structure, is the giant honey mushroom, *Armillaria ostoyae*. This invasive fungal species was introduced in Chapter 10. Initially brought to the Pacific northwest from its native subalpine and boreal regions of northern America and Europe, *Armillaria spp*. propagates as a microscopic fungal spore. Infected tree roots are filled with a latex-type white mass of fungal mycelium, which extracts water and carbohydrates from the tree while making it harder for the tree roots to uptake water and nutrients, as thin, black fungal filaments (the rhizomorphs) invade neighboring tree roots.

When a pathogenic parasite is dependent upon the resources of its host for survival mortality rates logically should be balanced in both predator and prey (recall L-V model). But in the northwest US's wide expanse of hardwood and conifer trees the Armillaria fungi has unlimited resources and no known natural predator. Today

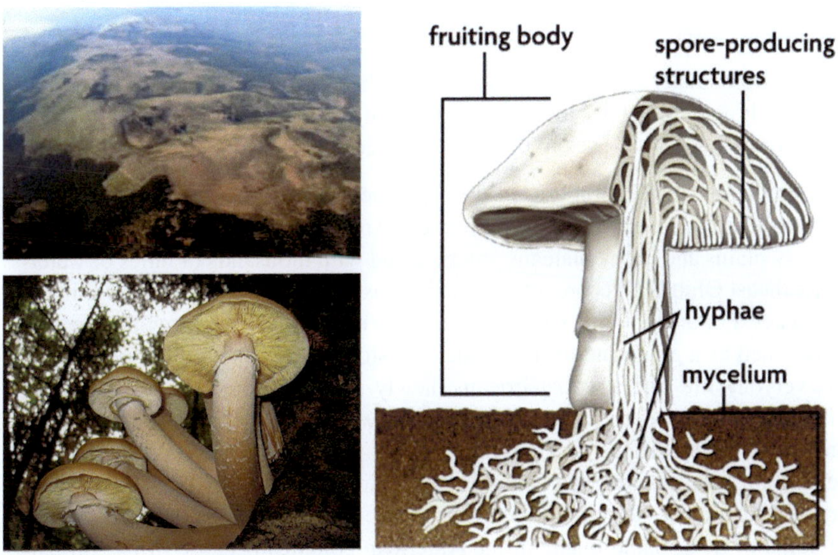

Fig. 13.25

Malheur National Forest. In 1998, using aerial photographs to mark the area of dead trees (2200acres), forestry scientists eventually identified the invasive culprit as a single individual of the species, *A. ostoyae*.

Source: Modification of WikiVisually, US Forest Service, Public Domain

880 hectares of the Eastern Oregon Malheur National Forest are infected with the pathogenic mushrooms, which has permanently altered the forest community structure (**Fig. 13.25**). Any species adapted for survival as an interdependent member in its native community structure, when transplanted unintentionally by humans, has the potential to be altered. For aggressive invasives the species' new role is that of an alien terminator. This can happen whenever flora or fauna is dropped into the midst of an entirely different ecosystem and has a marked advantage in its interspecies dynamics. A balance will eventually be restored in the community dynamics, whether by depletion of resources or new adaptations occurring in other species to combat the survival threat.

Community ecologists, wildlife fisheries management, and the interests of governing policies all coincide around the threat of invasive species on economic stability. In the next chapter we will highlight the importance of ecosystem services to the human food supply and local economy. According to Wildlife and Fisheries professor Peter Sorensen, "It's time to declare war on aquatic exotic species," which include not only the silver and black Asian carp but other invaders without natural predators, like the zebra mussel.

13.3 Exploring Ethnobotany

After reading this section students should be able to:

- Define ethnobotany and make connections to the Gaia hypothesis.
- Describe how biodiversity loss impacts the livelihood of indigenous cultures.
- List several indigenous uses of the sassafras tree.

The southeast Ohio ecoregion, in the Appalachian foothills, contains some of the richest herbal treasures in North America. An eastern temperate forest biome with grassy plains and rocky plateaus, interspersed with thousands of miles of waterways, southeast Ohio is an ecoregion recognized as one of the most biodiverse in the United States. For an ecologist studying the area's **ethnobotany** this region has long been identified as a goldmine for medicinals. Historically the local flora was essential to everyday life of Native Americans, newly settled colonials during the 15th to 17th centuries, and even the enslaved peoples who were escaping from southern regions of the United States. During their journey to freedom this population relied on the land for food, shelter, and healing while being hunted along the Underground Railroad! Remarkably, slaves captured from western African countries were able to locate some of those same plants in the American colonies because of the similarity in latitude and biogeography. For several centuries prior to the desperate trail of slaves into the free lands of the north explorers from Europe had been frequenting the colonial lands seeking flora specifically for its healing properties. Treatments for syphilis, malaria, and many other ailments brought Europeans to study the culture and ethnobotanical practices of the native civilizations.

The Gaia Hypothesis

Perhaps in this time of planetary crises it is prudent to protect and conserve the multitude of therapeutic remedies and botanical treasures that were discovered in the Ohio Northwest Territory (pre-statehood). It is well known that the tropical rain forest hot spots contain potential cures for diseases, although those are now disappearing at an alarming rate. Often biodiversity loss is concurrent with loss of native cultures and empirical knowledge of the practical uses of fauna and flora. The causes of biodiversity loss are multifold, including habitat fragmentation, urbanization, and climate change events, all of which have impacted the southeast Ohio ecoregion. In this section we will discuss a few of the important species but first we'll introduce the concept of **Gaia hypothesis**, a blending of science and lore that elevates ethnobotany's importance in the 21st century.

Gaia hypothesis states that our world blesses us with herbs. To heal ourselves we must know ourselves and our ecological surroundings, using intuition and common sense that tells us we are one and all. **Dr. James Lovelock** (born 1919, now aged 102) developed the Gaia hypothesis to bridge the worlds of science and spirituality (**Fig. 13.26**). Gaia is the mythical Greek goddess of Earth he was drawn to as he researched Earth's biogeochemistry and recognized Earth was a living entity or Earth spirit. Using this conceptual framework we can draw upon our evolutionary connections to all living things, plant and animal with needs met by

Fig. 13.26

Engraving of Hildegard Von Bingan (ca 1150) and the futurist Dr. James Lovelock (2005).
Source: W. Marshall under CC BY 4.0

the environment. The relationship between healing plants and humans is "Gaia in action" and, indeed, by living directly from the land a person would be at one with nature in a spiritual sense. That a healing is transformative, beginning with a simple act of taking herbal medicine, goes beyond treating pathologies or revitalizing the body. It is very much place-based, as one is "home, healed, and whole." The famous visionary of ancient times **Hildegard Von Bingan** (1098–1179) spoke of the *viriditas* or greening power. She described a bridge between the natural and human worlds connected throughout time.

1. The **sassafras tree** (*Sassafras albidum*) is an example of local flora that was used by local inhabitants of southeast Ohio for medicine and so much more (**Fig. 13.27**). This tree (three species extant, one now extinct) has a distinctive root beer-flavored bark used in teas by many native Indian tribes, especially as a blood cleanser (spring tonic). The leaves were often applied as a poultice for various skin conditions including scurvy. Sassafras is now known to have antiseptic properties. The ethnobotanical practices of sassafras were adopted by the backcountry Appalachian Scotch–Irish who emigrated into the region and revered its unique properties, even to believing carrying a piece of its wood could ward off witches. Because the wood was decay resistant it proved useful in the making of canoes. Chickasaw Indians used the dried sassafras leaves as

Fig. 13.27

Sassafras tree.

Source: Randy Everette under CC by SA 3.0 and USGS, Public Domain

a spice, a practice that flourished in the deep south in Cajun or creole gumbos as filé. European explorers were most interested in the sassafras tree bark for its ability to treat syphilis and gonorrhea and the bark was an important trade commodity with Europe.

Throughout Northeastern America tribes found more than 80 ethnobotanical uses for sassafras. Among those applications, the root bark was used to reduce fevers and alleviate chronic rheumatism and gout; it was also used as an antiinflammatory. Women included the bark in teas prepared for menstrual pain. The volatile oil of sassafras, the efficacy of which is due to the chemical constituent safrole, was used in high dosages to cause vomiting, spontaneous abortion, or death. Used in minuscule dosages it was a popular flavoring to mask unpleasant flavors. For example, children in the 19th century were given dosages of opium for its calming effect and the scent of safrole oil was added to mask the taste of opium!

Whether regional native tribe, incoming colonist, or explorer, there has always been a plethora of biota and empirical knowledge that crossed cultures and Appalachian generations. Today herbalists have experienced somewhat of a revival, perhaps due to the high costs of modern medicine or a desire to escape the high-tech world to connect with the spirit of Gaia. With fewer untouched landscapes overharvesting brings new threats to some of the most potent plants as they are driven to extinction. For example, the prized goldenseal and Ginseng (*G. panax*) are indigenous to an ecoregion that includes southeast Ohio and remains in high demand for Chinese Traditional Medicine.

Overharvesting in the remaining wild habitats causes the most ecosystem damage. But the impact from footpaths and trampling plants or unintentional entry of disease and invasives into pristine areas are also a concern. In an attempt to counteract loss of species wild harvesting of medicinal roots in the United States is prohibited on state and federal lands. The private foundation **United Plant Savers (UPS)** was founded specifically to identify species "At-Risk" of being overharvested. UPS published an "**At-Risk Assessment Tool**" in the *Journal of Ethnobiology* with guidance on harvesting substitutes for the rarest plants. Given there are many plant selections with similar medicinal properties it makes good sense to conserve and protect the most threatened species. Fortunately it is often observed that the most medicinally relevant species are also the hardiest. Ironically, to the gardener some of the best medicinals are unwanted, very persistent weeds (plantain, red clover, pokeweed, etc.). Here are a few more native species with applications as therapeutic medicinals or other ethnobotanical uses:

2. **White trillium** (*Trillium grandiflorum*)—native to Ohio and its state wildflower, the leaves are edible, used in salads, and taste like sunflower seeds. Native Americans found uses for white trillium, which was called Birthroot, for female reproductive problems and as an antiseptic, diuretic, and antispasmodic.
3. **Red clover** (*Trifolium pratense*)—native of Europe and transplanted to North America, this little gem is a weedy invasive that is packed full of nutrients (high in calcium, chromium, magnesium, niacin, phosphorus, vitamin C,

potassium, and potassium). It also acts as a **phytoestrogen** and was once prized by Native Americans as an antiinflammatory and expectorant. It is often part of spring tonics for its function as a blood cleanser.

4. **Dandelion** (*Taraxacum officinale*)—many colonials developed a spring ritual of gathering dandelion spring greens similar to ramps. Another weedy gem, dandelion is favored by the bumble bee and is great for the environment. The dandelion was first identified in ancient literature as a medicinal in Arab culture in the 10th and 11th centuries (era of Hildegard Von Bingan). All parts of the plant were used—roots, leaves, and flowers—to make a spring tonic high in nutrients (vitamins A, C, E, K, B, copper, magnesium, potassium, and manganese). The lowly dandelion was even listed in the US Pharmacopoeia until 1926, which was around the time herbal medicine went into decline. Dandelion properties include use as a digestive bitter, detoxifier (blood and liver cleanser), mild laxative, and to alleviate headaches.
5. **Black haw** (*Viburnum prunifolium*)—native eastern North American understory shrub, with berries used to feed the local fauna and its bark used by the Cherokee for muscle spasms. Black haw was known as a female herb, to strengthen the uterus and pelvis during pregnancy and childbirth.
6. **Blue cohosh** (*Caulophyllum thalictroides*)—known to the Cherokee, Chippewa, Iroquois, and many more Native American tribes for the treatment of pain, tension, stagnation, rheumatism, colic, and excessive bleeding from the lungs. Like black haw (*Viburnum prunifolium*) blue cohosh was commonly associated as a female herb to aid menstruation and childbirth.
7. **Bloodroot** (*Sanguinaria canadensis*)—used by shaman as potent medicine for wounds that didn't heal and culturally for body painting by many Native American tribes. Ingestion of the dried root often led to vomiting (emetic), acting as a stimulant for sluggish mucosal and secretory organs. In high dosages is deadly. High in the alkaloid sanguinarine, now prohibited from sale in the United States, this chemical was once used commercially in toothpaste as an antimicrobial agent for gingivitis and periodontal diseases. Topical applications of bloodroot include treatment of skin cancers.
8. **Lily of the valley** (*Convallaria majalis*)—heart medicinal similar to foxglove (*Digitalis purpurea*), it is high in cardiac glycosides but more easily eliminated from the body than foxglove. It is considered a favorite heart medicinal by UK medical herbalists, especially in cases of edema or nervous disorders affecting the heart.
9. **Polkweed** (*Phytolacca americana*)—versatile native plant used for its dyes for body paint (berries), as a poison to stun fish, and medicinally as a potent immune system stimulant (antiviral). The early spring leaves are edible but late summer berries and root extracts were carefully prescribed by shamans and wise women after dried (can be deadly to the unexperienced), "medicine or poison is in the dosage."

Ethnobotany reveals a new purpose in protecting biodiversity through ecosystem services. The Gaia hypothesis highlights the role of human coevolution with plants, as evidenced by the large quantity of medicinal products from plants that benefit human conditions.

Further references on medicinals in the wild can be found at:

- UPS, Goldenseal Botanical Sanctuary, Rutland, Ohio
- *Journal of Ethnobiology*
- William Cook's Physiomedical Dispensary, 1869
- Native American Ethnobotany Database
- The Medicine Gardeners Blog

Glossary

Species biodiversity
Species ecological niche
Competitive displacement
At-Risk Assessment Tool
United Plant Savers (UPS)
Phytoestrogen

Citations

1. G.B. Bonan, Forests and climate change: forcings, feedbacks and the climate benefits of forests, Science 320 (2008) 1444. https://doi.org/10.1126/science.1155121
2. E. Martínez-García, et al., Drought response of the boreal forest carbon sink is driven by understorey–tree composition, Nat Geosci 17 (2024) 197–204. https://doi.org/10.1038/s41561-024-01374-9
3. J. Albert, T. Carvalho, J. Chuctaya, P. Petry, R. Reis, B. Rengifo, H. Ortega. Fishes of the Fitzcarrald, Peruvian Amazon. In: J. Albert, ed. Lulu Press, Raleigh NC. 2012 1–9. ISBN: 9781300185840.
4. W. Weisser, Biodiversity effects on ecosystem functioning in a 15-year grassland experiment: patterns, mechanisms, and open questions, Basic Appl Ecol 23 (2017) 1–73.
5. K.D. Lafferty, T.H. Suchanek, Revisiting Paine's 1966 sea star removal experiment, the most-cited empirical article in the American naturalist, Am Nat 188 (4) (2016) 365–378. https://doi.org/10.1086/688045
6. Wolf Wars: America's Campaign to Eliminate the Wolf. 27 Ep4: The Wolf That Changed America. Sept. 14, 2008. Accessed 16 Mar 2024. https://www.pbs.org/wnet/nature/the-wolf-that-changed-america-wolf-wars-americas-campaign-to-eradicate-the-wolf/4312/
7. Yellowstone Wolf Project. 2017. Yellowstone Wolf Project [online]. Yellowstone National Park: Yellowstone Forever. https://www.yellowstone.org/wolf-project/
8. E.T. Aschehoug, The mechanisms and consequences of interspecific competition among plants, Ann Rev Ecol Evol Syst 47 (1) (2016) 263–281.
9. https://www.atlasobscura.com/articles/why-do-people-burn-blueberry-fields Accessed 16 Mar 2024.
10. C. Elliott, The prince, the mayor, and the U.S. fish that ate Japan: an innocent gift from Chicago to Prince Akihito in 1960 caused a decades-long ecological crisis that Japanese scientists are now close to solving, National Geographic: Environment

(2022). https://www.nationalgeographic.com/environment/article/the-prince-the-mayor-and-the-us-fish-that-ate-japan
11. J.L. Teem, Genetic biocontrol for invasive species, Front Bioeng Biotechnol 8 (2020) 452. https://doi.org/10.3389/fbioe.2020.00452
12. J.V. Adams, Trade-offs between suppression and eradication of sea lampreys from the Great Lakes, Journal of Great Lakes Research 47 (1) (2021) S782–S795. https://doi.org/10.1016/j.jglr.2021.04.005
13. D. Li, W. Prinyawiwatkul, Y. Tan, Y. Luo, H. Hong, Asian carp: a threat to American lakes, a feast on Chinese tables, Compr Rev Food Sci Food Saf 20 (2021) 1–23. https://doi.org/10.1111/1541-4337.12747
14. J. Castrodale, These Invasive Carp Are Getting a New Name so Americans Will Eat Them, *Food and Wine*. 9 Feb (2021). https://www.foodandwine.com/news/asian-carp-seafood-new-name Accessed 16 Mar 2024.
15. A. Melotto, R. Manenti, G.F. Ficetola, Rapid adaptation to invasive predators overwhelms natural gradients of intraspecific variation, Nat Commun 11 (2020) 3608. https://doi.org/10.1038/s41467-020-17406-y
16. G. Visconti, "The Gaia hypothesis, evolution and ecology", in: G. Visconti, (Ed.), Climate, Planetary and Evolutionary Sciences, Springer, Cham, 2021. https://doi.org/10.1007/978-3-030-74713-8_11
17. L. Jaeger, Hildegard of Bingen (ca. 1098–1179): Building Bridges Between Mysticism and Science: Women of Genius in Science, Springer, Cham, 2023. https://doi.org/10.1007/978-3-031-23926-7_2
18. A. Weigelt, et al., The Jena Experiment: six years of data from a grassland biodiversity experiment, Ecology 91 (3) (2010) 930–931. https://doi.org/10.1890/09-0863.1

CHAPTER

Tracking Biodiversity

14

Chapter Outline

14.1 Ecosystem Loss

After reading this section students should be able to:

- Highlight the key points of the 2022 IPBES Report.
- Explain the benefits of ecosystems as a service.
- Relate current threats to biodiversity as human loss on a grand scale.
- Provide examples of ecosystem loss and successful restoration efforts.

In July of 2022 the **Intergovernmental Science-Policy Platform on Biodiversity and Ecosystem Services (IPBES)** released its report on the irreparable harm by overexploitation of wildlife and ecosystems occurring through human activities. The list of human activities is extensive, including commercial farming practices, pollution, and altering habitats to the extent that species are being driven to extinction. Currently one million species are at risk across the globe at a rate faster than at any time during Earth's history, short of a mass extinction event. With loss of biodiversity an ecosystem's community structure and diffuse interactions are altered, leaving species more vulnerable and less resilient to challenges. There are increased risks to entire populations when healthy ecosystems are destabilized.

Any materials extracted from Earth and used for human needs are called natural resources. Examples of natural resources are wood, water, fossil fuels, soil, metals,

Ecology for the 21st Century. DOI: https://doi.org/10.1016/B978-0-443-24031-7.00014-3

and stone. Given that all human life depends on services derived from ecosystems it isn't an exaggeration to say that, first and foremost, human self-interest is linked to protecting biodiversity.

Ecosystem Services

Ecosystem services include anything to do with a benefit to humans, including jobs, economic growth, health, and any well-being tied to the natural world. Ecosystems provide valuable resources in both rural and urban environments and help to reduce the impacts that extreme events may have on infrastructure and human communities. For example, forests, barrier beaches, or wetlands act as buffers against fires, floods, and severe storms. Billions of people rely upon wild foods and natural medicines for their health and well-being. Perturbations of ecosystems, whether driven by climate change or other disturbances, have direct human impacts through loss of ecosystem services. The many detrimental impacts that occur when ecosystems are destroyed include reduced water supply and quality, loss of wildlife habitats, and forest carbon sinks. Urban forests have a role to help clean air, cool buildings, and provide aesthetics and recreation parks. In rural areas commercial forestry and agriculture provide economic value and recreational opportunities and cultural resources add important societal benefit. Forested land in the United States absorbs and stores the carbon equivalent of approximately 17% of the nation's fossil fuel emissions. According to the US Marine Fisheries Services, the world's oceans absorb 25% of all human-caused CO_2 emissions.

Oceanic ecosystems provide many services besides regulation of Earth's climate, including providing seafood, recreation and tourism, cultural heritage, and transportation of goods. In 1982 the **UN Convention on the Law of the Sea** designated that the area of ocean surrounding nations that they could claim exclusive rights to extends 200 nautical miles (370.4 km) from their coastlines. The coastal states have jurisdiction over only their natural resources (maritime food, offshore oil, energy from water and wind, etc.), which fall within their **exclusive economic zone (EEZ)**. Recall this area is also known as the continental shelf and is rich in biodiversity of marine plants and animals spanning from tropical coral reefs to Arctic waters' ecosystem.

Wetlands contribute a number of functions that benefit people, most notably as a means of floodwater control and as reservoirs of high biodiversity. Wetlands also have a vital role in water filtering, groundwater replenishment, protection from storms, and shoreline erosion. Just as important, wetlands contribute to fixing and sequestering carbon and other nutrients. For these reasons wetlands help mitigate the impact of climate change.

Antarctica also imparts a rich and diverse range of ecosystem services. This includes climate modulation, recreation, cultural heritage, science education, and high primary productivity. Ecosystem services can be found in every environment and act as a bridge between humans, their well-being, and the natural world.

Ecosystem services can be categorized into four essential types:

- **Provisioning services**—benefits extracted from nature include items like foods (vegetables, fish, livestock), drinking water, timber, wood fuel, fossil fuels, clothing, housing, and medicines.
- **Regulating services**—basic services that make life possible for humans. Includes processes like fauna purification of the air and water, detritivore recycling of wastes, pollination of flowers and crops, and erosion control by tree roots. It also includes carbon storage and climate regulation.
- **Cultural services**—include nonmaterial benefit for the cultural advancement of people. Includes a role in regional or national cultures, recreation, or creativity inspired by interactions with nature.
- **Supporting services**—fundamental processes that sustain life forms. Includes photosynthesis, biogeochemical cycling, and soil formation.

Conservation and Environmental Sustainability

A conservationist is someone who works to preserve or restore natural resources, biodiversity, and ecosystems. Their main goal is for the preservation of healthy ecosystems for the benefit of future generations and the planet. Conservation biologists measure biodiversity, focusing on preservation of biodiversity, and will implement management plans to diminish harmful activities, including social, economic, and environmental factors. The work of a conservationist might be instrumental in the construction of a wetland to help divert stormwater runoff while treating industrial wastewater.

Lake Victoria Cichlid Diversity

The highly diverse fish family *Cichlidae* (cichlid) are freshwater fish found in tropic zones across the Americas, mainland Africa, and southern Asia (**Fig. 14.1**). An ichthyologist studying cichlids discovered hundreds of variations in morphological and behavioral specializations, adapted to specific habitats within the Victoria Lake ecosystem. The documented feeding strategies alone included cichlids that consumed floating plankton, scraped algae from rocky surfaces, foraged for insect larvae in the benthic layer, crushed snails, bit tails, and even consumed the eggs of other cichlid species.

The wide diversity of cichlids in African Great Lake Victoria resulted from the process of adaptive radiation, the same process that resulted in the 15 species of Darwin's Galápagos archipelago finches. But the expansive 500+ species of cichlids in Lake Victoria makes adaptive radiation in Darwin's finches seem to be a modest event. Lake Victoria is nestled in the **East African Great Rift Valley (GRV)**, bounded by Uganda, Kenya, and Tanzania. GRV is historically known for its high diversity and speciation events. It is even the most likely site for the cradle of the human species. Evidence for *Homo sapiens* originating in the GRV is that the oldest humanoid specimens were found in Kanapoi, Ethiopia, and the Omo River remains

Fig. 14.1

Lake Victoria African cichlids.

of Kenya (**Fig. 14.2a**). The GRV is also known for its volcanoes, extreme climatic and geographic variety, and biodiversity. It's not surprising this region houses a site of immense lake biodiversity and speciation events.

During the 1980s when biologists were documenting the remarkable cichlid variety, some species were already disappearing. The invasive Nile perch (*Lates niloticus)* had been introduced to Lake Victoria by fisheries in the 1960s to help feed the African people who lived around the lake. The goal was to improve the commercial fishing industry (**Fig. 14.2b**). By the 1980s Nile perch were already consuming and

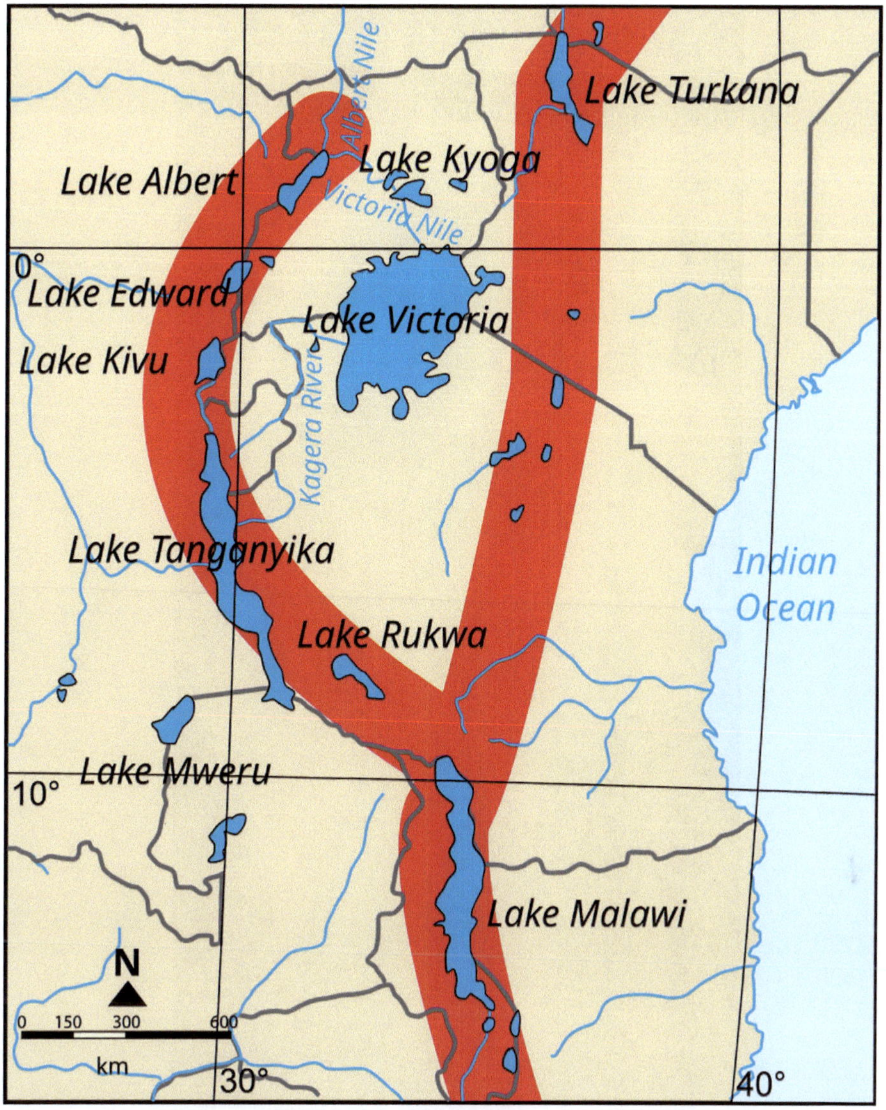

Fig. 14.2a

East African GRV (bold red lines) encircles Lake Victoria. The cradle of human civilization resides in the GRV, between Lakes Turkana and Victoria (Kanapoi) and the Omo River.

Source: smugger888 under CC by 2.0

driving numerous cichlid species to the point of extinction. Approximately 200 different cichlids species went extinct during this short period of time. The accelerated extinction was due to predation by the Nile perch and the contributing extrinsic factors of land clearing and agricultural runoff into the lake. In the race to catalog the hundreds of newly discovered cichlid species scientists were finding many species

Fig. 14.2b

Nile Perch.

Source: Reinhard Kraasch under CC by-SA 3.0

going extinct before they were even named. Today cichlid diversity in Lake Victoria is only a fraction of what it once was.

Can you think of another species that was introduced into a new region to provide an ecosystem service only to result in negatively impacting the entire ecosystem?

Prairies: An Ecosystem in Danger

Prairies are among the most endangered ecosystems in the world, particularly tall grass prairies, of which only 1% to 4% of their range remain intact. Prairie ecosystem health is vital, with many conservationists citing its unique biodiversity and role globally as a climate regulator. Because prairies serve as sinks to store carbon they provide a valuable ecological service. Atmospheric carbon, emitted through natural processes and by human activities (e.g., fossil fuel burning), is the primary component of greenhouse gas. Reports from the IPCC have consistently reported that increased greenhouse gases are exasperating the conditions for climate change. Present levels of greenhouse gases have led to sea level rise, higher temperatures, and altered rain patterns. Seeking ways to diminish the rise in atmospheric carbon, scientists investigate the potential of natural carbon sinks, especially the remaining large expanses of rain forests and prairie grasses.

Prairie grass carbon reserves are the result of carbon sequestration, which occurs below ground. Prairie roots can spread up to depths over 15 ft (**Fig. 14.3**). Indeed the estimated underground carbon stored in prairie lands is greater than the carbon stored aboveground in forests. The expansive grass prairies were once the largest carbon sink in the world, far greater even than the Amazon rainforest. This gives context to the devastating global effects occurring with the destruction of many prairie grasslands.

Root systems of prairie grasses are massive, with depths capable of extending further below the soil than their vegetative parts rise above the soil (up to 15 ft!) The densely tangled root structures provide several valuable ecosystem services. In addition to their vital role in carbon storage prairie plants in the landscape accomplish the following:

- Slow rainwater runoff
- Reduce soil erosion
- Mitigate severity of flooding
- Rebuild damaged soil structure

Ecosystem in Recovery: Barrier Reef of the Bay Islands

The **Mesoamerican Reef** (Great Mayan Reef) spans 620 m, from Mexico's Yucatan Peninsula to Hondura, and is the second-largest coral reef in the world. The Honduras coastline is stabilized from oceanic storms in the south by coral reef barriers of the bay islands, Roatan, Guanaja, and Utila that provide sheltered waters for marine life (**Fig. 14.4**). There are more than 30 species of whales, dolphins and porpoises,

Fig. 14.3

Prairie Roots Project's mission, Iowa, is to increase public awareness of the importance of prairie roots.

Source: Image courtesy of Jim Richardson, The Land Institute, CC BY NC ND

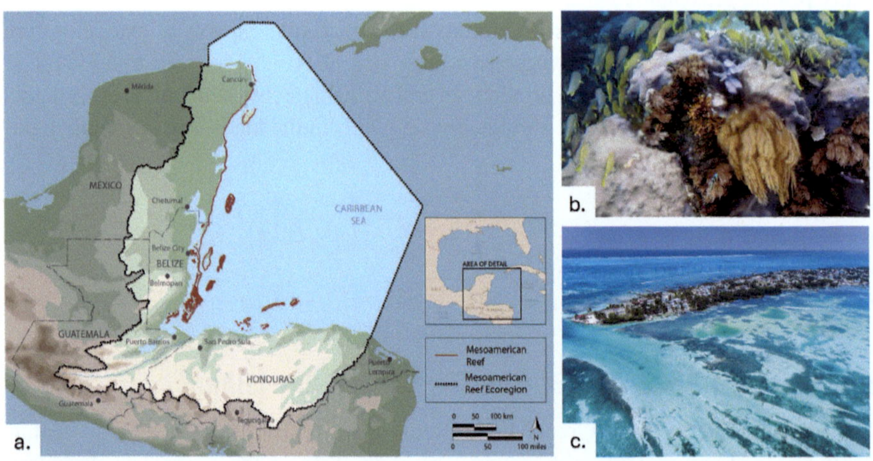

Fig. 14.4

(a) Barrier Reef of the Bay Islands located in the Mesoamerican Barrier Reef System.
(b) Bay Islands, off Honduras coastline.

Source: BlogMama under CC BY-SA 2.0
Map credit: Source: Google Earth

sea turtles, mollusk, pipefish, sea horses, and sea snakes; more than 1500 species of fish reside in the Mesoamerican Great Barrier Reef, estimated to be more than one billion years old.

In the 1980 s, according to the **UN's International Coral Reef Initiative (ICRI)**, the reefs "experienced its first large-scale impact to coral reef when pathogenic diseases became widespread throughout the Caribbean coral. The first mass coral bleaching from elevated sea temperatures began to destroy coral beds all across the Caribbean." Already under duress the bay Island barrier reef incurred critical physical damage by Hurricane Mitch in 1998. Another major storm, Hurricane Iris of 2001, exacerbated the already adverse impacts with river runoff and sedimentation spillage from coastlines. By 2006 an ICRI survey reported an overall decline in live coral in some regions by as much as 50%. Increasing severity of storms was not the only threat to the barrier reef system, as overharvesting and pollution runoff contributed to decline of the coral ecosystem.

Coral restoration efforts require a multipronged approach at local and global levels, including:

- Removal of nuisance and invasive species
- Providing emergency responses to physical damage in high-risk areas
- Implementing innovative technologies to develop more resilient coral species
- Controlling the spread of coral diseases to reduce coral larvae mortality

In follow-up studies between 2007 and 2008 the ICRI reported the Mesoamerican Reef was improving. The live coral biomass had increased from 11% to 13% in some areas. And while fish biomass was still on the decline the recovery of coral was a promising story on joint conservation efforts, given the ecosystem's complexity. Unfortunately by 2020 the Mesoamerican reef appeared to be once again on the decline.

Sea Urchins: Recovery of a Keystone Species

The exceptionally vibrant coral reefs of the Caribbean Sea are an important habitat for many species of marine fishes and invertebrates. Faced with emerging environmental challenges (increasing storms, rising temperatures, and overexploitation), the ecosystem has been under constant stress. The devastating consequence of the decline of the coral ecosystem is the loss of biodiversity and mass die-offs of surviving populations.

The long-spined sea urchin (*Diadema antillarum*) is an important grazer in the Caribbean reef system, clearing macroalgae so coral larvae can grow. In the mid-80s an unknown pathogen nearly wiped out the entire population. The mass die-off left many coral reefs smothered and choking with overgrowth of algae, further damaging the already fragile reef's architecture. Restoration of the long-spined sea urchins off the coast of Honduras has been slow but may provide the key to bringing back the coral reefs of the Mesoamerican region. Recovery efforts of this key species have involved building resilience through translocation of sea urchins into protected low-density areas to promote a natural recovery process of corals.

14.2 Tracking Biodiversity

At the end of this sections students should be able to:

- Describe the metrics used to prioritize conservation targets.
- Identify several major databases and organizations tracking biodiversity.

Measuring Diversity

Biodiversity is measured as the number and kinds of species existing in a specific location or across the entire planet. Biologists employ various methods to classify biodiversity and look at its various levels, including genetic composition and ecosystem diversity. The information gained is useful toward planning strategies to preserve biodiversity.

Genetic diversity provides the raw material for the adaptive strategies employed during species evolution. A species' future potential in a changing environment depends upon the genetic diversity within a species' population. For example, if a plant population is exposed to a new fungal disease selection will act on genes for resistance to the fungus as long as those gene variants exist within the genomes of some members of the species' population. But if resistance genes do not exist in the gene pool the plant population will not have the potential to evolve resistance and could face extinction. In the cases of staple food crops this would be detrimental to the local economy. Genetic diversity is just as important in taxonomic families. A genus that contains many different types of species will have more genetic diversity in its toolkit than a genus containing only species that look and respond alike and have similar ecological niches. Realistically, if a conservationist had to choose between saving only one genera of species the genus with the most genetic diversity has the greatest potential for subsequent evolution would be the more favorable candidate (**Fig. 14.5**).

A case study performed on cheetah felines at an Oregon breeding colony for large cats makes a strong case for this point. The cheetah is the single surviving species of the genus *Acinonyx*. At the Oregon facility captives were inadvertently exposed to a deadly virus (feline infectious peritonitis [FIP]), which swept through the cheetah population, killing 50% of the population. During this viral episode none of the resident captive lions developed symptoms. While the cheetah populations have famously low genetic variation lion populations are known for their high genetic diversity. If an epidemic were to sweep through vulnerable species, their chance of extinction is high. Given the choice to save only one of the two species, the lion would have the highest odds of adapting and surviving. The cheetah would require more stringent efforts and possibly molecular engineering or mixed-breeding technologies to rescue the species.

Chemical diversity. Most genes code for cellular proteins, the macromolecules that carry out metabolic processes needed for an organisms' continued existence. Biologists can define genetic diversity as *chemical diversity* since species produce a

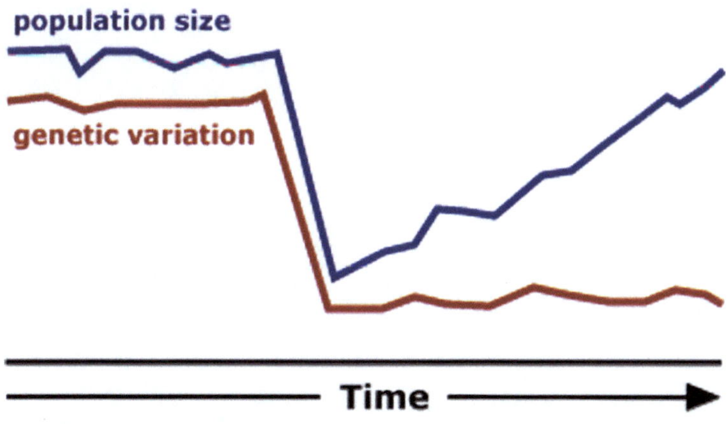

Fig. 14.5

Genetic variation does not rebound as quickly as population size.

Source: Image courtesy of UC Berkley EvoLibrary

diverse array of chemical molecules to alter cellular function and to export. And any single mutation in the genetic code of a protein can entirely alter its function or effect the organism's survival and selective advantage in changing environments.

For example, cephalopods (squids, octopuses, cuttlefish, and nautiluses) have shown a remarkable propensity to edit their own genome. Research collaborators report that octopuses have the capacity to alter the coded messages written in their DNA into customized commands suited to meet the demands of a changing environment (**Fig. 14.6**). Custom DNA synthesizes novel proteins to enhance the octopus's ability to survive under new constraints. The flexible strategy uses a different evolutionary trajectory than used by vertebrates. The cephalopod's genome gets rearranged or expanded to produce new biochemical products.

The array of chemicals produced by plants is astounding. **Phytochemical** diversity provides a potential benefit for humans as they represent a primary source of pharmaceuticals, with an estimated 25% of all drugs sourced from plant biochemistry. More than 7000 medical compounds derived from plants are prescribed by physicians. Some experts estimate that there are more than 21,000 medicinal plants in the world's remaining tropical forest ecosystems yet to be discovered. Measuring biodiversity through its chemical diversity provides an important function in maintaining human health and welfare (**Fig. 14.7**).

Humans consume a specialized diet harvested, or cultivated, from animals, plants, and fungi based on obtaining a greater yield of food energy and adopting globalized agricultural practices. But the many modern advances in food production come at the expense of diversity. For example, only a handful of varieties of banana are widely cultivated, despite substantial diversity in the banana species gene pool. There are an estimated 30 crops providing 95% of the world's

Fig. 14.6

White-spotted octopus *Callistoctopus macropus*, Mediterranean Sea.
Source: Yoruno under CC-BY-SA-3.0

food-energy needs. In the past 100 years the United States and China have lost 90% of their fruit/vegetables and rice varieties, respectively. Loss of agricultural diversity presents a major threat to global food security. An increasing homogeneity in global food supplies results from human migration, economy, and "big ag" **monocrop agriculture**. The severest threats of monocropping are concentrated in the heaviest populated regions, specifically China and India. It is logical to surmise that survival of the human population depends upon sustaining a diverse, stable food source. And that the decline in plant crop biodiversity is troubling to biologists and agricultural scientists.

Monocropping has increased the number of global seed bank depositories, where seeds of agricultural diversity are gathered and stored. One of the first seed banks was created by Nikolai Vavilov in St. Petersburg, Russia, during World War II. The world's largest collection, covering 13,000 years of agricultural history, is stored in a seed bank known as the **Doomsday Vault**. This vault is located between Norway and the North Pole, in the Norwegian Svalbard archipelago and managed by the Crop Trust. In summary, monocrop practices lead to limited *genetic diversity* and diminished *chemical diversity* as a food source.

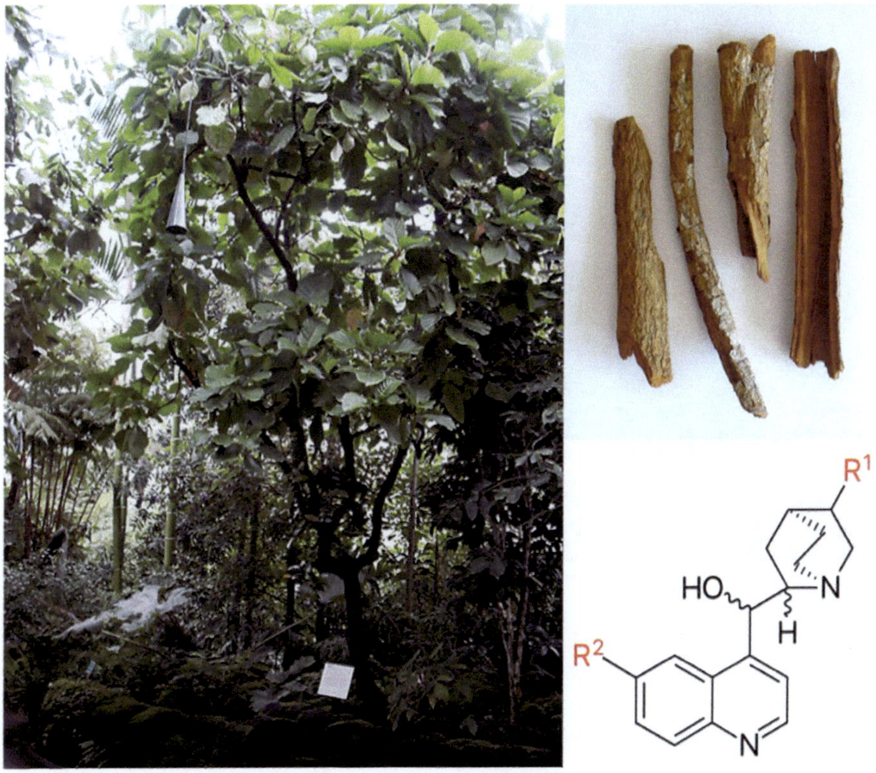

Fig. 14.7

Historically important medicinal tree genus, Cinchona, indigenous to Andean forests in South America. Cinchona tree bark produces the alkaloid, quinine, used for centuries to treat malaria.

Source: Vaccinationist and H. Zell under CC BY-SA 3.0

Ecological diversity. So far we have reviewed metrics of diversity through numbers and types of species, genus, or chemical constituents. But it is useful to also recognize *ecological* diversity, that is, the total number of different ecosystems across in the entire planet (**Fig. 14.8**). Recall the section on desertification and the Sahara Desert with once fertile, arable land supporting vegetative crops and its plentiful lakes, streams, prairies, and even forests. The entire ecosystem disappeared. With the loss of an ecosystem there is a concurrent loss of interactions between species and biological productivity that the ecosystem provided.

In this chapter we reviewed a North American disappearing ecosystem: the prairie ecosystem. Only two centuries ago the range extended from the tallgrass prairies west of the Appalachian Mountains to the Midwest's mixed and shortgrass

(a)

(b)

Fig. 14.8

Two extremely different ecosystems in (a) coral reefs and (b) prairies.

Source: Modification of Jim Maragos & Jim Minnerath under CC BY 4.0

prairie and western to the steppe prairie grasslands. This massive prairie ecosystem was bordered on the north by the Canadian boreal forests and to the south by Mexico's Chihuahua Desert. The horizon-spanning prairie, capable of functioning as a world-class carbon sequestration system, has been mostly converted into commercial crops, animal pastures, or encroached and defragmented by suburban sprawl. A number of species still exist in the altered ecosystems in smaller populations, like bison and prairie dogs. However, diverse and dynamic grassland ecosystems now exist in small, isolated regions. As a consequence rich topsoil is disappearing or must be maintained at greater expense and effort, often using polluting chemical fertilizers.

Fortunately the US federal government owns and protects the remaining prairies in North and South Dakota, Wyoming, and Colorado. But legislators, uneducated in the importance of protected national grasslands, are lobbied to open grasslands to mineral extraction, grazing, and hunting activities, each of which serves to diminish the value of any conservation efforts. It is worthwhile mentioning that the vast prairie grassland soils, replenished by one of the world's largest freshwater systems (Ogallala Aquifer), were responsible in many ways for the US's growth into one of the most economically stable economies during its first two centuries.

Species Distribution Patterns

Biodiversity is expressed in a mixed-distribution pattern across the planet. For example, endemics species have highly restricted distributions and are more vulnerable to extinction.

In North America Lake Huron is home to approximately 79 different species of fish. Most of these species can be found in other lakes in the North American climate, as they are not endemic to Lake Huron. There is quite a different distribution pattern in Africa's Lake Victoria, where 500+ cichlids are unique and endemic. The distribution pattern of endemic Lake Victoria cichlids is spread across the lake's many unique niches. Differences between Lake Huron and Lake Victoria habitats that factor into distinct cichlid distribution and diversity patterns include climate variability and ecosystem age (**Table 14.1**).

Table 14.1. Lake age and latitude are correlated with biodiversity in Lake Huron vs. Lake Victoria. Lake Huron is much younger. and outside the equatorial biodiversity hot spots, whereas Lake Victoria is older and near the equator.

	Lake Climate	Age	Latitude
Lake Victoria	Tropical	15,000 yrs.	1°S
Lake Huron	Temperate	7,000 yrs.	44°N

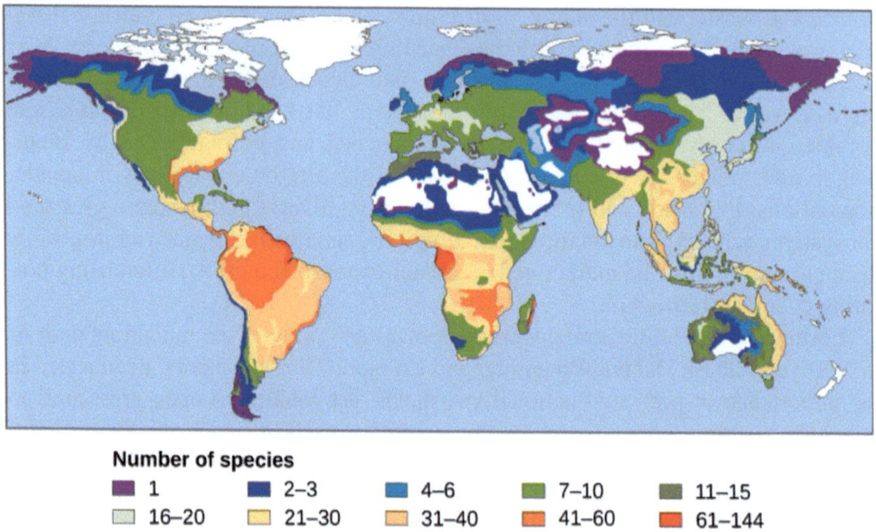

Fig. 14.9

Trend toward higher amphibian biodiversity (orange/reds) is evident at lower latitudes.

Source: OpenStax Biology, under CC By: Source: 4.0

Long before the discipline of ecology was established naturalists described that in most taxonomic group species biodiversity increases as latitude declines (equatorial), as shown in the orange and red regions in **Fig. 14.9**. Biodiversity by latitude explains the differences in species' richness between Lake Huron in North America and Lake Victoria in Africa.

Biogeographers consider there are four primary factors that delineate biodiversity patterns on the planet:

- latitude
- temperature
- stability
- age

The first three factors (temperature, latitude, and stability) are logical conclusions, given what we know about climate in the equatorial biomes with its greater access to energy by direct solar radiation. Tropical regions are generally more stable habitats for species to flourish compared to the fluctuating day-length seasonality and climate extremes of temperate regions.

The fourth factor is based on the hypotheses that greater diversity occurs with increased ecosystem age. According to geologic time temperate regions were most recently devoid of life. This accounts for the last glaciation retreat in the Northern

Hemisphere during the Younger Dryas just 11,700 years ago. The Younger Dryas was a colder and drier climatic period in higher latitudes. The lower latitude tropical regions experienced lower temperatures but are believed to have been teeming with life.

The hypothesis correlating evolutionary age with greater diversity states that advanced age lends more time for habitat specialization to occur. The complexity of tropical ecosystems and their multitude of ecological niches promote speciation events. Greater heterogeneity in ecological niches provides more selective pressures and opportunities for evolution or coevolution to occur. With this concept in mind we can compare climate type to biodiversity between Lake Victoria and Lake Huron. Higher biodiversity occurs in tropical Lake Victoria whereas lower biodiversity occurs in temperate Lake Huron. This observation supports the age = diversity hypothesis. This theory is also illustrated in the relatively small Peruvian Fitzcarraldo region of the Amazon River Basin (recall Chapter 12, Section 12.1) where the region holds the record for the highest number of fish species on Earth. It is reported that the fish ancestor species in the Fitzcarraldo predate the Late Miocene-Pliocene (*c.* 4 Ma).

Regardless of which mechanism may have contributed the most to speciation events, by every metric of biodiversity the tropics are the most profuse, species-rich locations on Earth.

Cataloging Biodiversity

Despite considerable effort cataloging the entirety of species that inhabit the planet, our data base remains incomplete and limited. The number of taxonomically identified eukaryote species (multicellular organisms) is currently around 1.5 million species. This is estimated to be only 20% of the total number of eukaryote species on the planet. Biologists have only scratched the surface of cataloging the many millions of prokaryote species (unicellular). The task is even more daunting because oftentimes conventional methods are not suitable to culture and identify microscopic species. Molecular methods are required to document microscopic species, but many regions are uninhabitable and prove difficult to study. For example, deep sea vents, space, and hot springs shelter a rich diversity of microorganisms. Given that Earth is losing species at an alarming pace scientists are racing against the clock, as was the case under which ichthyologists attempted to catalog Lake Victoria cichlid species. **Table 14.2** presents recent estimates of biodiversity in different taxonomic groups.

There are several international initiatives to catalog extant species. Ironically, the monumental task becomes increasingly diminished as extinction rates continue to rise. The field of taxonomy can be complex and lengthy, as biologists must assess a species' unique morphological, behavioral, and molecular characteristics. Once cataloged it is easier for biologists to locate a species of interest to learn about its ecological niche and species interactions. The unique characteristic of each species

Table 14.2 Estimates of the numbers of described and predicted species by taxonomic group

	Mora et al 2011[1]		Chapman 2009[2]		Groombridge & Jenkins 2002[3]	
	Described	**Predicted**	**Described**	**Predicted**	**Described**	**Predicted**
Animalia	1,124,516	9,920,000	1,424,153	6,836,330	1.225,500	10,820,000
Chromista	17,892	34,900	25,044	200,500	—	—
Fungi	44,368	616,320	98,998	1,500,000	72,000	1,500,000
Plantae	224,244	314,600	310,129	390,800	270,000	320,000
Protozoa	16,236	72,800	28,871	1,000,000	80,000	600,000
Prokaryotes	—	—	10,307	1,000,000	10,175	—
Total	1,438,769	10,960,000	1,897,502	10,897,630	1,657,675	13,240,000

defines its potential value to humans or other species on which humans depend. Cataloging also aids in prioritizing candidates for conservation efforts.

The fields of molecular genetics and bioinformatics make cataloging all known species more accessible and cost effective. **DNA barcoding** is one molecular genetic method. Since mitochondrial DNA (mtDNA) is present in all eukaryote cell organelles and known to experience rapid evolution it makes a good candidate for fingerprinting to identify species. Plants possess an additional organelle, the chloroplast, which can also be barcoded. New methodologies in rapid mass sequencing make molecular biology approaches relatively inexpensive and quick. Databases store the "fingerprints" in a format available to global researchers. DNA barcoding is the same technique used to catalog museum specimens. Databases provide accessible information for known species and a framework for adding new species. But given the present extinction rates many species may have been lost before they could be identified.

There is also the need to develop a system to classify the extent by which species are threatened with extinction. The nonprofit International Union for Conservation of Nature (IUCN) assesses the health of 75,000 species, including the 20,000 at risk.

The IUCN manages the **Red List Index**, which is an online listing of endangered species categorized taxonomically according to threat type.

The threat levels are defined as follows:

- vulnerable (VU)
- endangered (EN)
- critically endangered (CR)
- extinct in wild (EW)
- threatened in wild (TW)

The **Living Planet Index (LPI)** is maintained by organizations that measure the world's biological diversity and threat level, based on population trends in vertebrate species from terrestrial, freshwater, and marine habitats. The LPI has been adopted by the **Convention of Biological Diversity (CBD)**, a multilateral treaty, as its indicator of progress toward the 2011–2020 target to "take effective and urgent action to halt the loss of biodiversity." Currently the LPI group monitors the health of 4271 species and 22,779 populations.

The LPI in partnership with the **WWF** has provided global indicators of biodiversity changes occurring since 2008. The most recent report (2018) states overall species decline was 60% between 1970 and 2014 (**Fig. 14.10**). Species indices are subdivided into five major biogeographic realms: Nearctic, Palearctic, Neotropical, Afrotropical, and Indo-Pacific. Subcategories make it easier to compare how biodiversity is changing in different parts of the world. For example, the Neotropical realm (South and Middle America) experienced the most dramatic decline, with an average of 89% biodiversity loss.

According to LPI the biggest drivers of biodiversity decline (habitat loss, degradation, and overexploitation) accounted for more than two-thirds of all threats to populations in each taxonomic group (**Fig. 14.11**). In addition to these primary anthropogenic activities, introduction of invasive species, pollution, and climate change were sources of pressure.

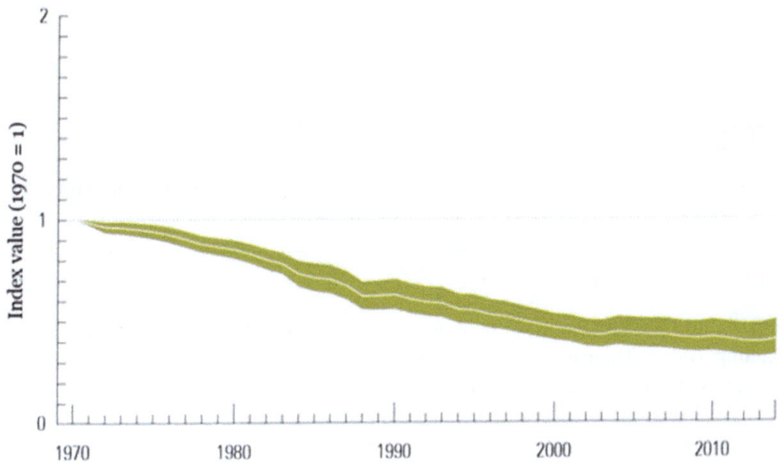

Fig. 14.10

LPI indices indicate a 60% decline (1970 and 2014) in global biodiversity.

Source: WWF/ZSL (2018) under CC BY 4.0

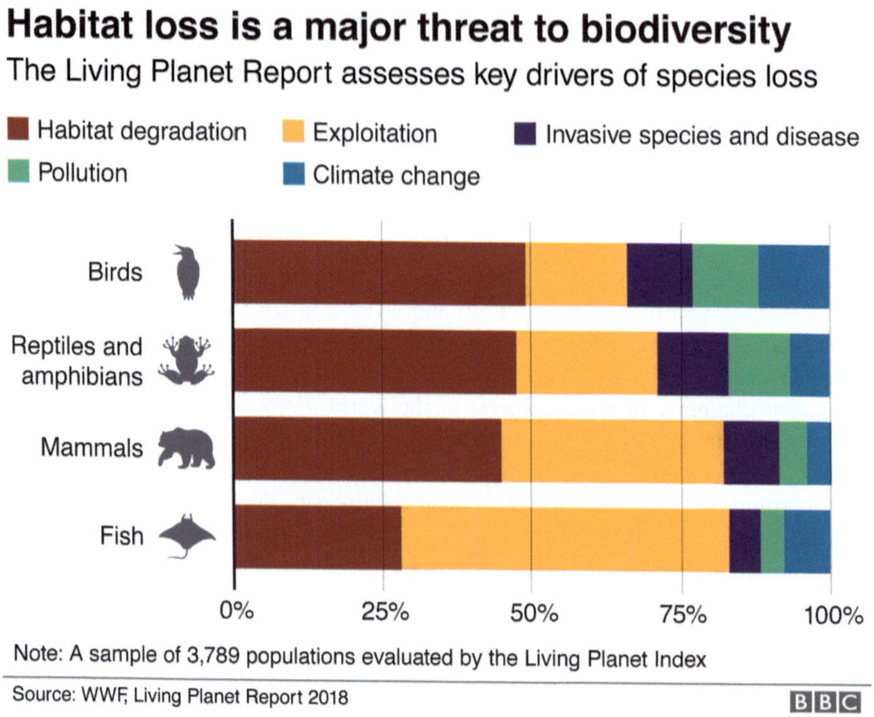

Fig. 14.11

Relative frequency of major threats by taxonomic populations.

Source: WWF/ZSL (2018) under CC BY 4.0

Glossary

Chemical diversity **Genetic diversity** **Ecological diversity**

Citations

1. IPBES, In: J.M. Fromentin, M.R. Emery, J. Donaldson, M.C. Danner, A. Hallosserie, D. Kieling, (Eds.), Thematic Assessment Report on the Sustainable Use of Wild Species of the Intergovernmental Science-Policy Platform on Biodiversity and Ecosystem Services, IPBES secretariat, Bonn, Germany, 2022. https://doi.org/10.5281/zenodo.6448567
2. A.W. King, D.J. Hayes, D.N. Huntzinger, T.O. West, W.M. Post, North American carbon dioxide sources and sinks: magnitude, attribution, and uncertainty, Front Ecol Environ 10 (2012) 512–519. https://doi.org/10.1890/120066
3. L.R. Pertierra, Ecosystem services in Antarctica: global assessment of the current state, future challenges and managing opportunities, Ecosyst Serv (2021) 49. https://doi.org/10.1016/j.ecoser.2021.101299
4. L.C. Braat, R. de Groot, The ecosystem services agenda: bridging the worlds of natural science and economics, conservation and development, and public and private policy, Ecosyst Serv 1 (1) (2012) 4–15. https://doi.org/10.1016/j.ecoser.2012.07.011

5. C.M. Vidal, et al., Age of the oldest known Homo sapiens from eastern Africa, Nature 601 (2022) 579–583. https://doi.org/10.1038/s41586-021-04275-8
6. P.A. Aloo, J. Njiru, J.S. Balirwa, C.S. Nyamweya, Impacts of Nile Perch, Lates niloticus, introduction on the ecology, economy and conservation of Lake Victoria, East Africa, Lakes Reserv Res Manage 22 (2017) 320–333. https://doi.org/10.1111/lre.12192
7. T.J. Lark, Protecting our prairies: research and policy actions for conserving America's grasslands, Land Use Policy (2020) 97. https://doi.org/10.1016/j.landusepol.2020.104727
8. Minnesota Board of Water and Soil Resources. Carbon sequestration in grasslands. https://bwsr.state.mn.us/carbon-sequestration-grasslands Accessed 19 Mar 2024.
9. A.R. Harborne, Honduras: Caribbean Coast, Mar Pollut Bul 942 (12) (2001) 1221–1235.
10. J.B.C. Jackson, M.K. Donovan, K.L. Cramer, V.V. Lam, (Eds.), Status and trends of Caribbean coral reefs: 1970–2012, Global Coral Reef Monitoring Network, IUCN, Gland, Switzerland, 2014.
11. A.R. Pilnick, et al., A novel system for intensive Diadema antillarum propagation as a step towards population enhancement, Sci Rep 11 (2021) 11244. https://doi.org/10.1038/s41598-021-90564-1
12. J.L. Heeney, et al., Prevalence and implications of feline coronavirus infections of captive and free-ranging cheetahs (*Acinonyx jubatus*), J Virol 64 (5) (1990) 1964–1972. https://doi.org/10.1128/JVI.64.5.1964-1972.1990
13. M.A. Birk, Temperature-dependent RNA editing in octopus extensively recodes the neural proteome, Cell (2023). https://doi.org/10.1016/j.cell.2023.05.004
14. N. Nasim, I.S. Sandeep, S. Mohanty, Plant-derived natural products for drug discovery: current approaches and prospects, Nucleus (Calcutta) 65 (3) (2022) 399–411. https://doi.org/10.1007/s13237-022-00405-3
15. U. Anand, N. Jacobo-Herrera, A. Altemimi, N. Lakhssassi, A comprehensive review on medicinal plants as antimicrobial therapeutics: potential avenues of biocompatible drug discovery, Metabolites 9 (11) (2019) 258. https://doi.org/10.3390/metabo9110258
16. H. Konuma, In: H. Ehara, Y. Toyoda, D. Johnson, (Eds.), Status and outlook of global food security and the role of underutilized food resources: Sago Palm, Sago Palm. Springer, Singapore, 2018. https://doi.org/10.1007/978-981-10-5269-9_1
17. C.K. Khoury, A.D. Bjorkman, H. Dempewolf, J. Ramirez-Villegas, L. Guarino, A. Jarvis, L.H. Rieseberg, P.C. Struik, Increasing homogeneity in global food supplies and the implications for food security, Proc Natl Acad Sci 111 (11) (2014) 4001–4006. https://doi.org/10.1073/pnas.1313490111
18. M. Hopkin, Biodiversity: frozen futures, Nature 452 (2008) 404–405. https://doi.org/10.1038/452404a
19. S. Theodoridis, et al., Evolutionary history and past climate change shape the distribution of genetic diversity in terrestrial mammals, Nat Commun 11 (2020) 2557. https://doi.org/10.1038/s41467-020-16449-5
20. J.S. Pigati, K.B. Springer, Hydroclimate response of spring ecosystems to a two-stage Younger Dryas event in western North America, Sci Rep 12 (2022) 7323. https://doi.org/10.1038/s41598-022-11377-4
21. J. Elith, J.R. Leathwick, Species distribution models: ecological explanation and prediction across space and time, Annu Rev Ecol Evol Syst 46 (2009) 677–697.
22. S. Shokralla, J.F. Gibson, H. Nikbakht, D.H. Janzen, W. Hallwachs, M. Hajibabaei, Next-generation DNA barcoding: using next-generation sequencing to enhance and accelerate DNA barcode capture from single specimens, Mol Ecol Resour 14 (5) (2014) 892–901. https://doi.org/10.1111/1755-0998.12236

CHAPTER

Drivers of Biodiversity Loss

15

Chapter Outline

15.1 Biodiversity Loss in the Anthropocene

After reading this section students should be able to:

- Explain the term Anthropocene with respect to human influence on biogeography.
- Discuss several of the ways human action causes loss of biodiversity.
- List the major drivers to biodiversity loss.

The Anthropocene

The name Anthropocene is informally used in scientific context and is proposed to cover the latest geological epoch, dominated by significant human impact on Earth's geology and ecosystems. The timeline is still under scientific debate, awaiting final approval by the International Commission on Stratigraphy (ICS) geologists [1]. But the epoch is more or less settled to have begun with the detonation of the first atomic bomb. It is characterized as an age of irreversible human impacts on the planet and anthropogenic climate change (**Fig. 15.1**). The rapid species loss occurring globally through human action is of grave concern to conservationist scientists. And environmentalists have been sounding the alarm that we have now entered an era of rapidly accelerating species extinction.

Ecology for the 21st Century. DOI: https://doi.org/10.1016/B978-0-443-24031-7.00015-5

Fig. 15.1

Resource overexploitation. A school of 400-ton Pacific jack mackerel encircled by a Chilean purse seiner includes by-catch nontarget species like cod and juveniles. Use of industrial purse seines puts pressure on local fish stocks.

Source: C. Ortiz Rojas, Public Domain

Empirically Indigenous elders have known something was changing. Some elders shared with ecologists how fish once swam into their clothing when as children they bathed in lakes and coastlines. Today dense schools of fish at the water's edge are mostly nonexistent, and young Indigenous children consider fish stories told by their elders to be just myths or exaggerations. That's because huge wildlife populations have mostly retreated from places with a human presence. But this recurring story of lost abundant wildlife is repeated in many regions of the world, differing only in the species and Indigenous tribes. My family's ancestry, along the coastal waters of the state of Maine, spoke of lobster being so plentiful in the late 1800s they were used as bait by fishermen and clams being quickly harvested by the bucket load from intertidal mud flats using weir lines.

According to the history of life on Earth extinction is a natural process with an average occurrence rate of 1:1,000,000 species extinctions per year. The estimation is derived from the fossil record. Of the approximately 8 million species on Earth today, estimates of species under threat for extinction range from 15,000 to 20,000. This is 1000 to 10,000 times faster than the natural rate. In the fossil record of Earth there are five known periods of mass extinction, each with high rates of species and biomass loss (50%–90% loss). The current rate of species loss is comparable to extinction rates occurring during those periods of mass extinction [2]. The single major difference between previous mass extinctions and the contemporary mass extinction is the role of human activity.

Macroevolution is the evolution of a large group of species, for example, whole taxonomic groups rather than an individual species. Trends and major evolutionary transformations are easier to study by phylogenic groups, like the evolution of mammals or flowering plants. A macroevolutionary approach is useful to analyze causes of mass extinction because it describes evolution occurring above the species level. If only one species was evolving, this approach wouldn't be feasible or practical. For example, in the Lake Victoria cichlid diversity, instead of focusing on a single cichlid species, using the macroevolutionary approach, analysis is done on the combined phylogeny of all 500+ of its Neotropical cichlids.

In past mass extinctions various catastrophic events annihilated species. The known causative agents were deadly asteroids, comets, or volcanism. It was in 1796 that Georges Cuvier made the groundbreaking suggestion that life was not static and that all life could disappear (i.e., species extinction). Recall Cuvier opposed the proponents of evolution (Chapter 6). Of the previous five mass extinctions, the single most severe occurred at end-Permian times (250 Mya, "The Great Dying") when an estimated 95% of marine species and 70% of land vertebrates went extinct. Notably each of the previous major mass extinctions altered the course of history of life on Earth. We are living now in a time of accelerated extinction rates. The predictions of total species lost within the last century range from an estimated 10% to 50% of all species. In the 21st century human activities have had a major impact on the sustainability of ecosystems. Threats to biodiversity are called anthropogenic because they are influenced by human activities that directly impact biodiversity.

Drivers of Biodiversity Loss

Some of the most severe threats to biodiversity are caused by habitat loss and climate change and its interrelated effects on ocean acidification and marine oxygen depletion. An in-depth, comprehensive analysis by the **Convention on Biological Diversity** identifies the direct and indirect drivers of biodiversity as:

- Direct: invasive exotic species, climate change, nutrient loading, pollution, and overexploitation
- Indirect: demographic, economic, socio-political, cultural, religious, scientific, and technological

The indirect drivers of biodiversity loss are multifaceted with economics playing a major role in per capita consumption, demographics influencing populations, and science and technology having the greatest influence on resource intensity. They in turn lead to increasing demands for food and energy, which are the primary direct

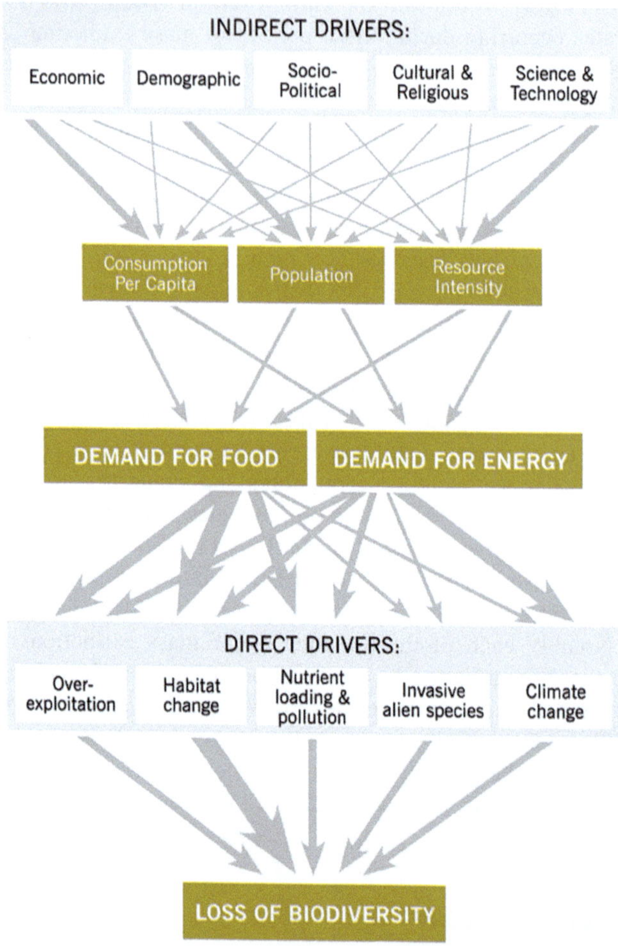

Schematic representation of the links between biodiversity loss, the direct and indirect drivers of change, and the demand for food and energy. The width of the arrows gives a broad and approximate illustration of the importance of the economic sectors in driving biodiversity loss.

Fig. 15.2

Demand for food and energy are dominant triggers driving biodiversity loss. Combined factors have unintended consequence of species becoming threatened or disappearing altogether.

Source: Global Biodiversity Outlook 2 under CC By 2.5

triggers of biodiversity loss (**Fig. 15.2**). When several factors are combined species often become threatened or disappear.

In the diagram habitat change (i.e., habitat loss) has the widest arrow and indicates its importance as a driver of species extinction. Fragmentation of habitats results in reduction of resident populations, making its living members more vulnerable to other factors (recall the bumble bee decline with habitat loss of its preferred pollen sources). No matter which factors are attributed as weakening a population, once made vulnerable a smaller external perturbation (e.g., disease) could entirely wipe out the remaining individuals.

Anthropogenic climate change, currently absorbed by Earth's ecosystems, is predicted to become far more significant in the coming years and decades. The latest IPPC reports project the tipping point for recovery is less than two decades away. Global climate change is related to, and a consequence of, humanity's need for energy and its heavy dependence on fossil fuels to meet those needs. The overwhelming evidence is clear: the burning of fossil fuels causes dramatic increases in atmospheric CO_2 to levels never before seen in human history. Climate scientists see no further debate on the role of fossil fuel burning and predict that the addition of rising atmospheric carbon will significantly impact biodiversity in the coming century.

Nutrient Loading and Pollution

Pollution takes many forms and can target air, water, or soil to pose a serious threat to ecological biodiversity. Effects of pollution on biodiversity are immense. Here are just a few well-known examples:

- Burning of fossil fuels (oil, coal, and gas) causes particle pollutants in the atmosphere or fall to the ground as acid rain. Acid rain (sulfuric and nitric acid) then causes acidification of lakes, streams, and sensitive forest soils, which alters the pH, species survival, and decline in forest growth with tree damage, particularly at higher elevations.
- Chemical pollutants from pesticides and herbicides leach into soils and groundwater, causing depletion of fresh groundwater and developmental disorders in aquatic species.
- Oil pollution negatively affects development of marine organisms [3].
- Nutrient pollution from fertilizers leaching into waterways (phosphorus and nitrogen) causes overgrowth of toxic algae, outbreaks of fish diseases, dead zones, and eutrophication.
- Terrestrial and aquatic plants absorb pollutants and pass them up the food chain to consumer animals and humans (biomagnification).

Chemical pollutants—insecticides, herbicides, plastics. The heavy use of pesticides like neonicotinoids has led to a growing concern for the health of the world's pollinators, including domesticated and wild bees. The ecological damage to biodiversity from systemic pesticides, described in more than 1000 peer-reviewed studies, has shown negative impacts on terrestrial and aquatic invertebrates, amphibians, and birds, as well as their ecosystem dynamics and important ecosystem services [4]. In 2016 the IPBES estimated the annual value of crops directly affected by pollinators to be 235 to 577 billion USD, with greater than 40% of invertebrate pollinators threatened and facing extinction. The long-life use of pesticides and their potential for spray-drift show unintentional residual effects in nontargeted species.

Eutrophication. Increased loading of nutrients (runoff from agricultural fertilization and septic system affluent) and fossil fuel pollution causes a cascade of undesirable effects on estuaries and coastal waters [5]. The result is algal blooms, lowered pH, depletion of aquatic oxygen, or hypoxia, and eutrophication with its endpoint being aquatic dead zones (**Fig. 15.3**).

The process of **eutrophication** is a major threat to the global aquatic food supply. The US NOAA estimates that 65% of US estuaries and coastal waters are moderately to severely degraded by nutrient runoff and pollution.

Excessive nutrient runoff into waterways causes overgrowth of phytoplankton and algae, leading to harmful algal blooms. The phytoplankton rapidly consume resources and die off, falling to the ocean floor to be digested by heterotrophic

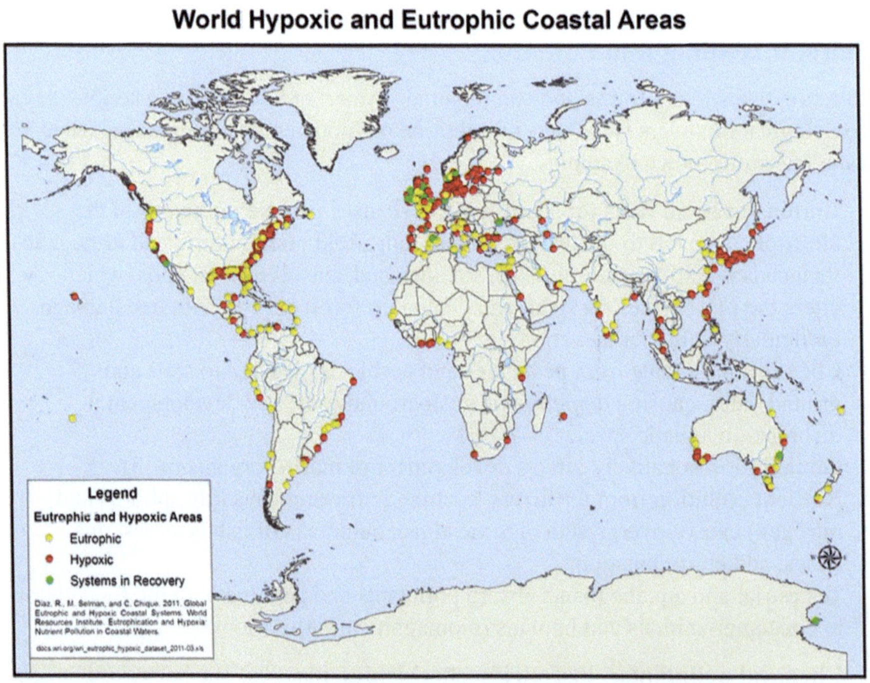

Fig. 15.3

Dead zones (*red*) and threatened areas with excess nutrients (*yellow*). Green dots represent areas once designated dead zones that are being restored.

Source: NOAA, Public Domain

microorganisms. Because the overgrowth of phytoplankton led to accumulation of biomass the growth at the second trophic level also consumes excessive oxygen and releases excessive metabolic carbon. The combined accelerated processes of photosynthesis and cellular respiration create an imbalance, depleting oxygen availability, which leads to hypoxic zones. The excessive biomass in the lower levels of the food chain (phytoplankton and heterotrophic microorganisms) also impedes sunlight entry required to continue photosynthesis. This process constitutes the conditions for eutrophication. Without photosynthesis oxygen levels reach a dangerous, unsustainable level for marine life. And because warm waters cannot hold as high a concentration of oxygen as cool waters the process is exacerbated in the tropical zones with climate change and rising temperatures.

Most aquatic life cannot survive under low oxygen conditions, leading to massive fish kills (**Fig. 15.4**). Mobile fish and other creatures often swim away, abandoning what is developing into an inhabitable region designated as a **dead zone**. Mobile species may not have escaped the crises. This is because during migration they risk becoming vulnerable to new predators or face a different set of environmental stresses.

The ecosystem service impact of excessive chemical pollutants is that the world's marine food supply declines with increasing regions of dead zones. Furthermore, wherever dead zones exist no photosynthesis can occur. Recall that

Fig. 15.4

Oxygen depletion is the most common cause of fish kills.

Source: USFWS, Public Domain

Fig. 15.5

Satellite images of Lake Tai, Gulf of Mexico, and the Adriatic and Baltic Seas with visible teal blue shading wherever dead zones exist. Extension into bodies of water from coastlines indicates size of the dead zone.

Source: Image by author

approximately 50% of Earth's oxygen comes from aquatic primary producers. In the process of photosynthesis carbon is consumed (i.e., sequestered). This means that a rise in dead zones contributes to rising atmospheric carbon and the climate crises (**Fig. 15.5**). The deoxygenation of waterways is unsustainable and may result in ecosystem collapses [6].

Offshore drilling. Offshore drilling results in oil spills and another form of chemical pollutants entering waterways. Oil drillers on the North Sea are mandated to report any spillage. In a summer of 2013 Technology Offshore publication there were more than 60 recorded chemical spills into the North Sea in just over a 30-day period. The chemical pollutants released include:

- methanol
- calcium bromide brine
- triethylene glycol

This one monthly report is indicative of the scale of chemical pollutants entering waterways across the 65,000 oil drilling sites worldwide. According to Dr. Simon Boxall, University of Southampton National Oceanography Centre, even leaks on a small scale, if they are occurring chronically, impact the health of nearby ecosystems. Ecological studies in regions surrounding offshore oil drilling confirm they experience poor growth of wildlife and minimal biodiversity. Dr. Boxall explains, "If you compare, for example, an estuary close to an oil and gas terminal that is, say, a mile away, even with no evidence of a spill, you will find a completely different biodiversity that is very, very limited."

According to the UN's Agenda for Sustainable Development, the oceans are in a state of emergency from increasing eutrophication, acidification, ocean warming, and plastic pollution. Among the proposed policies to address oxygen depletion in the oceans and coastal regions are:

- Nutrient pollution and climate change—improve septic systems and sanitation to keep pollution out of the water. Cut fossil fuel emissions to reduce their impact on climate change and dangerous air pollutants.
- Protect vulnerable marine life—some hypoxic conditions will be unavoidable but protection efforts to prevent further stress to at-risk fisheries can be accomplished by extending marine protected areas and no-catch zones in locations animals use to escape dead zones.
- Improve tracking at sites of low oxygen worldwide—scientists have modeled future oceanic oxygen loss but have little data on locations where low-oxygen zones will occur. Enhanced monitoring could help pinpoint high-risk locations to implement the most effective responses.

Overexploitation

There is no getting around that human population size and resource exploitation present a core threat to biodiversity on the planet. The human population requires continual resources for its survival and growth. Unfortunately nonrenewable resources are being depleted from the environment for the benefit of mankind, or at least to benefit a subset of humans. **Unsustainable resources** include fossil fuels and freshwater aquifers. Though both can be replenished, it takes millions of years so they are classified as unsustainable for our use. **Sustainable resources** include solar, wind, geothermal, and hydropower, which are still underutilized toward solving humanity's energy needs. Humans also deplete their food resources without a concerted effort toward sustainable agricultural, harvesting, and hunting practices. These practices are known as **overexploitation**.

Overharvesting, or **overfishing**, is a particularly serious threat to aquatic species. Even managed commercial fisheries have collapsed after the introduction of high-yield fishing practices. The Western Atlantic Cod Fishery, in operation and productive for more than 400 years, collapsed in the 1980s after permitting the use of factory-made trawlers. Unsustainable overharvesting and fishery collapse have both an economic and a political component that challenges common resource treaties. In open, unregulated waters there is little incentive for a fisherman to exercise restraint in harvesting, a situation known as "the tragedy of the commons" [7,8].

The whaling industry has thrived for 700 to 800 years. It is thought to have started in the 12th century when the Basque began hunting Northern Right whales, but Norwegians were also hunting whales in that period. Whale hunting by the late 19th century entered an era of steam-powered ships and explosive harpoons. By the end of the 20th century population sizes of these large mammals were decimated. It is estimated that 3 million whales were killed in the 20th century alone. It is estimated also that only 10% of all large fish—including tuna, swordfish, marlin, cod, halibut, skates, and flounder—are left in our oceans [9].

Purse seine fishing is a nonselective deep sea fishing method that encircles fish with a net and strings to close around its prey (**Fig. 15.1**). While seine fishing techniques date back to the stone ages the modern-day use of motorized vessels and gear to locate fish have drastically changed catch sizes. And similar to what happened to whale populations with the adoption of steam ships and explosive harpoons, fish populations have declined drastically from commercial fishing practices. According to the 2013 Encyclopedia of Biodiversity an estimated catch of 2 million tons in 1800 increased to ten-fold in 1950. By 2006 catches were reaching 144 million tons. Through the nonselective practices of pelagic trawls, by-catch occurs in some of the most endangered marine species (dolphins, whales, sharks, sea turtles, etc.). Once captured, the dead carcasses of by-catch species are thrown back into the sea.

Habitat Loss

It could be said that it is in the nature of the human species to exploit their ecological niche. Humans design and implement innovative technologies that modify habitats to suit their needs. Often to accomplish a goal ecosystems will be reengineered or removed, in which case many species become locally extinct. The exception would be the few species that thrive in human-built environments (e.g., rats, pigeons, cockroaches, etc.). Human activities incur not only biodiversity loss but an exchange of natural ecosystem services for whatever is gained after altering the environment (farms, cities, resource extraction, etc.). Modern forms of warfare weigh heavily on all ecosystems wherever they are waged [10].

Habitat destruction has accelerated in the latter half of the 20th century, partly due to human population growth but just as significantly due to increased technologies and demand to modify habitats. The island Sumatra, located in western Indonesia, was once known for its exceptional biodiversity as a tropical hot spot. Its many endemic species include the critically endangered orangutan (*Pongo abelii*), a subspecies of the Asian elephant (*Elephas maximum sumatranus*), and the majestic Sumatran tiger (*Panthera tigri*) shown in **Fig. 15.6**. Today half of Sumatra's rainforests have been demolished and the rate of habitat loss in the neighboring island of Borneo, home to another endemic orangutan, is just as alarming. Here the slightly larger **arboreal** orangutan, *Pongo pyamaeus*, is listed as endangered by the IUCN.

Fig. 15.6

Endemics: (a) Orangutan species, *Pongo pygmaeus* exist only in the rainforests of Borneo Island. (b) Sumatran tiger *Panthera tigris* is critically endangered. (c) Proximity of two western Indonesian islands: Sumatra and Borneo. (d) Rainforest removal for oil palm plantations in Borneo's Sabah Province.

Source: Modification of T. Bachner; Dick Mudde; Dr. Lian Pin Koh under CC BY 4.0

In reality the swinging great apes are simply the more visible of the many thousands of species expected to disappear with the loss of the Indonesian island rainforests. The reason for clearing forest timber from the islands of Sumatra and Borneo is the expansion of commercial palm oil plantations [11]. The planting of palm trees is an example of an indirect driver of habitat loss for human consumption (e.g., *economic*) (**Fig. 15.2**).

Palm oil is found in almost 50% of packaged products in grocery stores, including foods and cosmetics. In Europe palm oil is burned as biodiesel. Over a 5-year period (2000–2005) the estimated total loss of global forest cover is 3.1%. According to the World Resource Institute, between 2021 and 2022 the tropics lost another 4.1 million hectares, *the equivalent of 11 football/soccer fields of forest per minute*.

In 2018 the eastern cougar (puma), a top predator that once roamed from Quebec and Manitoba to South Carolina, was officially declared extinct in eastern North America by the US Fish and Wildlife Service. This cougar subspecies (*Puma concolor couguar*) was genetically related to mountain lions that still roam parts of the eastern United States and the Florida panther of the Everglades (**Fig. 15.7**). Cougars are large carnivores that hold an important place in the food web of eastern forests and meadows (not unlike the gray wolves of the Northwest). Without the cougar deer overpopulation leads to overgrazing of shrubs and trees and destruction of the natural

Fig. 15.7

Eastern cougar.

Source: USFWS, Public Domain

habitat of small animals and birds. Any increase in deer populations also means an increase in the spread of tick-borne diseases. Restoration of the eastern puma is analogous to the goals for restoration of the gray wolf in the Yellowstone National Park. In eastern Canada the Ontario Puma Foundation estimates there are 550 pumas still in existence. This is largely attributed to their protected status and conservation efforts to achieve a sustainable puma population in the province of Ontario.

Recently Penn State researchers analyzed sequenced mtDNA taken from preserved samples of eastern cougars and compared the data to mountain lion National Center for Biotechnology Information (NCBI) sequences [12]. Results showed that the eastern cougar is no more similar to each other than subspecies in the western United States and Florida, which strengthened the debate that all North American mountain lions are of a single subspecies, *Puma concolor*. This is promising news for reintroduction of the cougar into the eastern United States.

The mission of the Cougar Rewilding Foundation of West Virginia is to facilitate recolonization of the cougar in suitable eastern wild habitats from Western prairie stock. The foundation hopes to reintroduce the cougar to the Adirondacks, Great Smokey Mountain National Park, and Monongahela and George Washington National Forests on the Virginia/West Virginia border. The Center for Biological Diversity supported these efforts to quickly restore the cougar, a keystone species that plays a critical role in controlling deer herds.

Habitat loss is a major driver of species extinction. Anthropogenic encroachments on habitat range impact a wide range of species and their environmental niches. Recall the plight of the bumble bee (Chapter 9) and both generalist and specialist species (e.g., snow leopard and northern spotted owl) (Chapter 11).

A serendipitous event surrounding the years 2019–2020 offers a convincing argument to aid public awareness with respect to the dire consequences of climate change in the 21st century. While humanity dealt with a global pandemic crisis the natural world responded in an unexpected way. During this period, known as the **Anthropause**, human "lock down" led to increased wildlife populations with animals on the move and frequenting new locales [13,14]. Birds began using habitats in different ways than when humans were present. If you frequented parks, forests, or fields you may have noticed the increased sounds as animal populations increased with a sharper vibrancy than felt in the surrounding landscapes.

15.2 Preserving Biodiversity

After reading this section students should be able to:

- Identify major sources of environmental pollution.
- Explain how resource extraction impacts biodiversity.
- List environmental agencies that monitor and recommend strategies to mitigate biodiversity loss.
- Discuss the landmark achievement of personhood rights to natural resources.

Fig. 15.8

Nation of Change in 2019 stated, "Hundreds of thousands of whales, dolphins, seals, and turtles are killed by ocean plastic pollution every year, including single-use plastics and abandoned plastic gear from the fishing industry."

Source: Facebook, Public Domain

Plastics are forever. Oceans have become the default national dumping grounds for plastic waste all across the globe. This is particularly problematic in east Asia where the largest human populations exist. As waters become more polluted with nondegradable plastic debris aquatic animals can't avoid ingesting the synthetic wastes (**Fig. 15.8**). Whales have been found washed up on shores with pounds and pounds of plastic in their stomachs, clogging their digestive pathways [15]. New legislation has been introduced by activists and environmentally aware politicians to ban single-use plastics, which would tremendously curb its threat to biodiversity.

Legislative Action to Preserve Wildlife

Since 1975 the first treaty to provide protections to international wildlife trade across borders was established by the **Convention on International Trade in Endangered Species of Wild Fauna and Flora (CITES)** treaty. While CITES does not replace national laws it does certify any trade as legal, sustainable, and traceable in accordance with articles of agreement. Wildlife trade is estimated to be worth billions of dollars internationally. Participation in CITES buffers against the illegal trade of endangered species like tigers and elephants.

Preserving the planet's wildlife depends on achieving a greater understanding of biodiversity, an informed public, and a drastic change in human attitudes and behaviors. Increasing quantities of evidence have been observed, monitored, and reported on by international agencies and foundations. The data demonstrates unequivocally that the drivers of biodiversity loss negatively impact the availability and delivery of ecosystem services. We know that alterations in ecosystems and biodiversity have unavoidable consequences.

Fortunately the human response is growing toward a concept of securing our future generations by saving the planet's biodiversity and ecosystems. Internationally legislation and treaties continue to be enacted to protect species. Many nations have created national laws that protect endangered species and have dedicated funding for the establishment of wildlife preserves to protect species and entire ecosystems. The purpose of restoration projects is to return ecosystems to their predisturbance state. Preservation strategies include practices like captive breeding of rare and endangered species in controlled, monitored environments. There are a number of cryobank facilities constructed to preserve genetic data on endangered species by freezing their genome.

Regulatory agencies within the United States established to protect endangered species and to keep hunting and fishing in sustainable numbers have been in operation since 1973. These services are regulated under the **Endangered Species Act (ESA)**. ESA includes the **US Fish & Wildlife Service (USFWS)**, which provides much of the statistical data used to describe the status of ecosystems and biodiversity in this course. The ESA and its facsimile in many other countries provide a framework for recovery mechanisms once a species has been targeted for conservation management.

Legislative Action on Climate Change

Climate change can only be effectively managed on a global scale. A number of agencies are directly involved in these efforts, including the United Nations, Natural Resources Defense Council (NRDC), Greenpeace, Earthjustice, and the Environmental Defense Fund (EDF). An important international treaty by the **United Nations Framework Convention on Climate Change** is committed to addressing greenhouse gases. Unfortunately the powerful fossil fuel lobby and other economic factors have mitigated its potential reach by omission of two of the world's largest contributors to carbon emissions: the United States and China. It remains critical that governments agree on timelines and benchmarks as the drivers of climate change are arguably the single most costly venture the world's population faces.

Individual efforts have significant value in spirit and by its momentum. This includes lifestyle choices that reduce your personal carbon footprint through applying conservation practices in personal consumption and waste production. Conservation can be implemented in small ways, like recycling and organic farming, or it can take a more drastic committed approach, like off-grid living.

In the off-grid living video Rob's lifestyle seems quite extreme. And while most of us wouldn't alter our lives so completely his simple life purpose is nonetheless inspiring. More modest lifestyle changes include supporting legislature that protects our wildlife and waterways or avoiding the purchase of palm oil products to discourage more rain forest destruction.

Legislative Action to Preserve Marine Life

The **Oceans and the Law of the Seas** is the only legal framework created to protect the oceans from activity at the regional level (United Nations Convention on the Law of the Oceans and Seas, **UNCLOS**). Since its groundbreaking work in 1982 UNCLOS extended international law to include the shared water resources of the planet. Rights of sovereignty and usage issues were resolved to include freedom of navigation, 12-mile territorial sea boundaries, and 200-mile off-shore EEZs. The framework even provides a mechanism for resolution of the inevitable conflicts.

In 1986 a moratorium on whale hunting was adopted by 88 country members of the International Whaling Commission. The Humane Society International estimates the ban resulted in preventing tens of thousands of whale deaths. The Humane Society is involved because of the suffering during the painful process of killing whales using explosive harpoons. Indigenous whale hunts continue in parts of the world, including the United States, using traditional subsistence practices that are not subject to the moratorium. Commercial whale hunting has remained legal in three countries (Norway, Japan, and Iceland) [16]. Of the three only Japan is actively developing more deadly methods to kill whales. Norway and Iceland are moving away from the trade altogether. Considering the size of whale biomass it's surprising that whale meat is not the reason they are killed. The main products are whale sperm oil (used for soaps, candles, and cosmetics), whale oil (extracted from blubber for lamps and explosives), and a few expensive products like ambergris, which is an intestinal waxy substance popular in perfumes.

In 1989 and 1991 the UN General Assembly also called for a moratorium to halt large-scale pelagic drift net fishing vessels on the high seas. And in most countries purse seine fishing is only permitted in restricted areas and on 1 or 2 days a week.

Nonetheless all these efforts are still insufficient as overfishing remains a key threat to marine wildlife, particularly because not all countries monitor or enforce the international moratoriums. Conservationists advocate the end of government-subsidized industrialized fishing enterprises altogether. The preferred alternative is to restore local, traditional fisherman rights to save our marine life and to fight hunger in small villages. This is an important message in Sir David Attenborough's video on how to save the ocean's wildlife.

Solutions to overfishing according to Earth.org include:

- Ban fishing subsidies, which encourage fuel-intensive fishing and destructive fishing practices.

- Encourage rights-based fishing, which entitles persons/communities to fish in a particular place at a particular time.
- Enforce regulations on fishing nets, requiring the top of nets to be lowered 2 meters, reducing by-catches up to 98%.
- Protect essential predator species, including shark (by-catch) and tuna, which are threatened keystone species.
- Increase Marine Protected Areas (MPA), which increases protected ocean areas from 7% to 30%.
- Implement traceability standards, which provide transparency and compliance standards to document legal sourcing.
- Impose a ban on fishing in international waters, where the high seas do not belong to any nation.

Personhood Rights for Natural Resources

Personhood Rights. In 2017 the New Zealand Parliament bestowed personhood rights on a waterway, the Whanganui River, called *Te Awa Tupua* by the Indigenous Māori Tribe (**Fig. 15.9**). The government passed legislation that gave the Whanganui River a legal identity with all corresponding rights, duties, and liabilities of a legal

Fig. 15.9

Whanganui River on the North Island of New Zealand.

Source: James Shook,CC BY 5.0

person. The battle for protection, through Māori ownership of the 90-mile river in New Zealand, actually began in the 1870s and remains one of the longest legal battles in New Zealand's history [17]. The unique ancestral relationship of the river to the Māori people is difficult to fully appreciate but was eventually fully acknowledged by the government.

According to a Māori elder, "The river and the land and its people are inseparable. And so, if one is affected the other is affected also. The river is the heartbeat, the pulse of our people. If the river dies, we die as a people. *Ka mate te Awa, ka mate tatou te Iwi*." The tribe recognizes the river as "kin" and an "indivisible and living whole." In the settlement by Parliament restoration of the river was awarded $30 million and another $80 million in damages to the Māori.

This monumental award was necessary because since the 19th century the government had permitted a steamer service on the river for the purpose of mineral extraction along the riverbed. The activity had the crown's approval but failed to seek consultation with the Māori people. The venture ultimately led to erosion of the river's ecology and destruction of its fisheries—and, just as important to the Māori, the river's cultural and spiritual value.

The Whanganui River is the first river to be granted legal personhood; however, Indigenous peoples around the globe have been actively engaged in establishing protective rights for the land and waters that are their livelihood. New legislation granting natural resource rights have been enacted in Bangladesh, Ecuador, and in Indigenous communities around the world. For example, in India the Ganges River recently received personhood status. In Ecuador its constitution has granted nature the "right to integral respect for its existence and for the maintenance and regeneration of its life cycles, structure, functions, and evolutionary processes," which gives all individuals the right to protect the environment. And in Bangladesh, known as "the land of rivers," all of its rivers were given personhood rights.

With the growing success of the rights of nature some efforts are being made to counteract the ability to protect natural resources from exploitation. These efforts to block progress come from special interests that seek to benefit from resource extraction and other forms of irresponsible stewardship for profit. For example, in February of 2022 the Idaho State House of Representatives in the United States was lobbied successfully and introduced a bill specifically designed to prevent animals and natural resources from being granted personhood status. The purpose was to bar environmental groups from legal standing. In contrast in Ohio the voters approved legal rights for Lake Erie, to act as its legal guardians to ensure that it will "exist, flourish, and naturally evolve." This latest act was done to offset toxic algal blooms from agricultural runoff pollution in 2014 that was destroying the lake's ecology and had cut off the drinking water supply for half a million people.

In the 21st century the movement to protect natural resources and ecological sanctuaries will increasingly need voices and actions by diligent stewards that are

Fig. 15.10

Dr. O. Carter, examining a log jam at Oneonta Falls, Colombia River Gorge, Oregon, United States.

Source: Image by author

intent to restore the balance of all things. The fight will not be won easily but the odds are the highest they've ever been (**Fig. 15.10**).

> *I hope this class has been informative and inspired you to pursue a cohesive existence with other species on Earth. After all, we are all in this together. As Edward O. Wilson, a famous scientist at the heart of ecological research, said in* The Diversity of Life, *"No comfort should be drawn from the spurious belief that because extinction is a natural process, man is merely another Darwinian agent."*

Glossary

Arboreal **Indigenous elders**

Citations

1. A. Witze, Geologists reject the Anthropocene as Earth's new epoch after 15 years of debate, Nature 627 (2024) 249–250. https://doi.org/10.1038/d41586-024-00675-8 \

2. G. Ceballos, et al., Accelerated modern human–induced species losses: entering the sixth mass extinction, Sci Adv 1 (2015). e1400253(2015). https://doi.org/10.1126/sciadv.1400253
3. A.J. Mearns, A.M. Morrison, C. Arthur, N. Rutherford, M. Bissell, M.A. Rempel-Hester, Effects of pollution on marine organisms, Water Environ Res 92 (2020) 1510–1532. https://doi.org/10.1002/wer.1400
4. M. Chagnon, D. Kreutzweiser, E.A. Mitchell, C.A. Morrissey, D.A. Noome, J.P. Van der Sluijs, Risks of large-scale use of systemic insecticides to ecosystem functioning and services, Environ Sci Pollut Res Int 22 (1) (2015) 119–134. https://doi.org/10.1007/s11356-014-3277-x
5. W.A. Wurtsbaugh, H.W. Paerl, W.K. Dodds, Nutrients, eutrophication and harmful algal blooms along the freshwater to marine continuum, WIREs Water 6 (2019) e1373. https://doi.org/10.1002/wat2.1373
6. D. Breitburg, et al., Declining oxygen in the global ocean and coastal waters, Science 359, eaam7240. (2018). https://doi.org/10.1126/science.aam7240
7. T. Long, S. Widjaja, H. Wirajuda, et al., Approaches to combatting illegal, unreported and unregulated fishing, Nat Food 1 (2020) 389–391. https://doi.org/10.1038/s43016-020-0121-y
8. P. Sullivan. Fishermen, fish, and the tragedy of the commons. Global Futures. Georgetown blog (2016). https://globalfutures.georgetown.edu/responses/fishermen-fish-and-the-tragedy-of-the-commons.
9. SeaWeb. "Cover Study of Nature Provides Startling New Evidence That Only 10% of All Large Fish Are Left in Global Ocean." *ScienceDaily*. <www.sciencedaily.com/releases/2003/05/030515075848.htm>
10. M.J. Lawrence, H.L.J. Stemberger, A.J. Zolderdo, D.P. Struthers, S.J. Cooke, The effects of modern war and military activities on biodiversity and the environment, Environ Rev 23 (4) (2015) 443–460. https://doi.org/10.1139/er-2015-0039
11. E. Meijaard, et al., The environmental impacts of palm oil in context, Nat Plants 6 (2020) 1418–1426. https://doi.org/10.1038/s41477-020-00813-w
12. M.N. Evanitsky, et al., Mitochondrial Genomes of The Regionally Extinct Nittany Lion (*Puma Concolor* From Pennsylvania). bioRxiv 214510. https://doi.org/10.1101/214510
13. C. Rutz, et al., COVID-19 lockdown allows researchers to quantify the effects of human activity on wildlife, Nat Ecol Evol 4 (9) (2020) 1156–1159. https://doi.org/10.1038/s41559-020-1237-z
14. "Welcome to the Anthropause." *Nature*. 4(9). September 2020. Retrieved 20 Mar 2024.
15. A. Borunda, This young whale died with 88 pounds of plastic in its stomach, National Geographic, Environment, 2019. https://www.nationalgeographic.com/environment/article/whale-dies-88-pounds-plastic-philippines
16. Whale hunting still exists in 3 countries, but it is declining, The Sentient: Climate (11 Sept 2023). https://sentientmedia.org/whale-hunting/ Accessed 20 Mar 2024.
17. M. Kramm, When a river becomes a person, J Human Dev Capab 21 (4) (2020) 307–319. https://doi.org/10.1080/19452829.2020.1801610

Index

Page numbers followed by *f* indicate figures and *t* indicate tables.

B

C

D

E

F

H

I

J

K

L

Q

R

S

T